Lecture Notes in Physics

New Series m: Monographs

The Editorial Policy for Monographs

The series Lecture Notes in Physics reports new developments in physical research and teaching - quickly, informally, and at a high level. The type of material considered for publication in the New Series m includes monographs presenting original research or new angles in a classical field. The timeliness of a manuscript is more important than its form, which may be preliminary or tentative. Manuscripts should be reasonably self-contained. They will often present not only results of the author(s) but also related work by other people and will provide sufficient motivation, examples, and applications.
The manuscripts or a detailed description thereof should be submitted either to one of the series editors or to the managing editor. The proposal is then carefully refereed. A final decision concerning publication can often only be made on the basis of the complete manuscript, but otherwise the editors will try to make a preliminary decision as definite as they can on the basis of the available information.
Manuscripts should be no less than 100 and preferably no more than 400 pages in length. Final manuscripts should preferably be in English, or possibly in French or German. They should include a table of contents and an informative introduction accessible also to readers not particularly familiar with the topic treated. Authors are free to use the material in other publications. However, if extensive use is made elsewhere, the publisher should be informed. Authors receive jointly 50 complimentary copies of their book. They are entitled to purchase further copies of their book at a reduced rate. As a rule no reprints of individual contributions can be supplied. No royalty is paid on Lecture Notes in Physics volumes. Commitment to publish is made by letter of interest rather than by signing a formal contract. Springer-Verlag secures the copyright for each volume.

The Production Process

The books are hardbound, and quality paper appropriate to the needs of the author(s) is used. Publication time is about ten weeks. More than twenty years of experience guarantee authors the best possible service. To reach the goal of rapid publication at a low price the technique of photographic reproduction from a camera-ready manuscript was chosen. This process shifts the main responsibility for the technical quality considerably from the publisher to the author. We therefore urge all authors to observe very carefully our guidelines for the preparation of camera-ready manuscripts, which we will supply on request. This applies especially to the quality of figures and halftones submitted for publication. Figures should be submitted as originals or glossy prints, as very often Xerox copies are not suitable for reproduction. In addition, it might be useful to look at some of the volumes already published or, especially if some atypical text is planned, to write to the Physics Editorial Department of Springer-Verlag direct. This avoids mistakes and time-consuming correspondence during the production period.
As a special service, we offer free of charge LaTeX and TeX macro packages to format the text according to Springer-Verlag's quality requirements. We strongly recommend authors to make use of this offer, as the result will be a book of considerably improved technical quality. The typescript will be reduced in size (75% of the original). Therefore, for example, any writing within figures should not be smaller than 2.5 mm.
Manuscripts not meeting the technical standard of the series will have to be returned for improvement.
For further information please contact Springer-Verlag, Physics Editorial Department V, Tiergartenstrasse 17, D-69121 Heidelberg, FRG.

E. Abdalla M.C. B. Abdalla D. Dalmazi A. Zadra

2D-Gravity in Non-Critical Strings

Discrete and Continuum Approaches

Springer-Verlag
Berlin Heidelberg New York
London Paris Tokyo
Hong Kong Barcelona
Budapest

Authors

E. Abdalla
Instituto de Física, Universidade de São Paulo, CP 20516
São Paulo, Brazil

M. C. B. Abdalla
IFT-UNESP, Rua Pamplona 145, CEP 01405
São Paulo, Brazil

D. Dalmazi
UNESP, Campus de Guaratinguetá, CP 205
Guaratinguetá, S. P., Brazil

A. Zadra
Instituto de Física, Universidade de São Paulo, CP 20516
São Paulo, Brazil

ISBN 3-540-57805-6 Springer-Verlag Berlin Heidelberg New York
ISBN 0-387-57805-6 Springer-Verlag New York Berlin Heidelberg

CIP data applied for.

Printed in Germany

SPIN: 10080426 55/3140-543210 - Printed on acid-free paper

Foreword

The aim of this book is to bring together and compare two different approaches to the problem of the quantization of two-dimensional gravity and supergravity: we begin with the Liouville approach, which is formulated in the continuum two-dimensional space-time and makes extensive use of conformal field theory techniques; the second approach is based on a discretization of two-dimensional random surfaces and described in terms of matrix models, where loop equations and large N methods play an important role.

Over the last six years we have witnessed the interacting development of these different formulations. As if one model were a laboratory for its counterpart, one could test its ideas, conjectures, approximations and main results. From such experiments we learned that each method has its own virtues: for instance, matrix models are excellent for higher genus surfaces and indeed they provide a non-perturbative definition of quantum gravity; on the other hand, the $N = 2$ supergravity theory misses a discrete formulation and, in this case, the continuum approach is more advanced. We hope this healthy "competition" between methods will continue improving our understanding of the quantum gravity problem. Here lies the importance and usefulness of powerful alternative techniques.

To represent the various applications of the study of two-dimensional gravity, we have selected the String Theories. We have learned that, in *critical* string theories (which describe bosonic strings propagating in a 26-dimensional space-time, or superstrings in 10 dimensions), the two-dimensional world-sheet metric field decouples. On the other hand, if one needs to build up a *non-critical* string (which eventually propagates in a "physical" dimension), one will have to face the problem of quantizing the world-sheet gravity. We dedicate Chapter 1 to an overview on the origins of quantum gravity in strings and superstrings, and the following chapters make use of standard concepts of string theory. In this way we unify the language of the whole text; nevertheless we remind the reader that many results have an application to other physical theories, such as the statistical mechanics of random surfaces, Ising systems, physical membranes, large N chromodynamics, higher dimensional quantum gravity, and more.

There are quite a number of reviews and some books in this area. We would like to mention the reviews by d'Hoker[98] and Seiberg[21] on Liouville theory, and those by Alvarez-Gaumé[121], Bilal[125], Gross[136], Lozano and Mañes[226] about matrix model techniques and applications in string theory. Among the more recent reviews we recommend those by David[123], Ginsparg and Moore[227], and di Francesco, Ginsparg and Zinn-Justin[228]. We must mention, however, that these works are mostly concerned with the purely bosonic theories, while this monograph is especially devoted to the extension of results to the supersymmetric case, including Neveu-Schwarz and Ramond sectors, an area hardly covered by most authors.

It is worth saying that a reader not sufficiently familiar with string theory techniques may find the text rather arid, in which case we suggest that he begins with Appendix B on conformal field theories and references therein.

We have come to realize that the problems involved here are very complex and profound. Indeed, Liouville theory and its relation to the uniformization problem of Riemann surfaces is more than one century old. Nevertheless, the very idea of efficiently using the Liouville theory in order to study string theories away from the critical point is very rewarding. The extension of these methods to the supersymmetric case certainly deserves our attention. Whether these ideas will prove useful in order to understand the large N limit of realistic non-abelian gauge theories is not yet proved, especially in the supersymmetric case, but we hope that one step towards the elucidation of these issues might be achieved in the comparison between the different, often complementary, approaches studied here.

Special thanks are due to Prof. D. Dey, who kindly reviewed the manuscript, to Dr. J.C. Brunelli for the encouragement to transform a long review into a Lecture Notes Monograph, and to Dr. A. Saa for his help with computer-related problems. We also acknowledge FAPESP, CNPq and FUNDUNESP for financial support, and the Brazilian Physical Society for the use of software for printing the figures. Two of us (E.A. and M.C.B.A.) are grateful for the hospitality of the International Centre for Theoretical Physics, Trieste and CERN, Geneva, where part of this book was written.

Contents

1. Introduction

The issue of string theory has already a long and well known history, which has been intensively discussed in several reviews and textbooks[1,2]. Just to mention one step in the middle of such development, we quote the heterotic string theory[3] describing several phenomena which one expects to understand[4].

Nevertheless, there are still some loopholes in string theory: we shall mention the fact that one eventually needs a four dimensional theory rather than a string at critical dimension, while the compactification schemes in general spoil the predictability of the theory[5]. Moreover, an important aspect of string theory concerns the description of strong interactions[6], and in that case the expected "stringy" behaviour should be present at low energy levels.

Therefore, it is essential to scratch into the realm of non-critical string theory in order to find the relevant description it gives for strong interactions. Indeed, using strings as a phenomenological model for the Wilson loop, one can find the static potential for quark interaction[7], which gives us some confidence in the effectiveness of the approach.

String theory was born as a consequence of the development of dual theory, whose aim was to present an alternative to quantum field theory, specially not relying on perturbation theory, which fails in the case of strong interactions. The story of those developments has been covered by several authors, and our aim here is not to follow the thread until it reaches today's developments, but an attempt to pave a way from string theory to several physical processes involving the statistical mechanics of random surfaces, as physical membranes, three-dimensional Ising systems and large N chromodynamics, among others.

In the fifties, dynamical calculation involving quantum field theory were restricted to perturbation theory, rendering computations in the framework of strong interactions unreliable. Moreover, higher-order results involving weak interactions were, at least at that time, unnecessary, in such a way that the successes scored by quantum field theory were restricted to Quantum Electrodynamics (a quantum theory of gravity was completely unknown). This led ideas related to quantum field theory to stagnation and discredit in the late fifties.

That was the main motivation for S-Matrix theory, which in fact played a rather dominant role in subsequent developments. The informations required to formulate the theory were very little: it was based on kinematical principles, analyticity and on the bootstrap idea. However, this quality turned out to be a weakness, since there was no dynamical principle involved. The most important concept, arising from analyticity in the complex angular momentum plane, was the "duality", which implies relations between given scattering amplitudes in different channels. In particular, it implies a very rich spectrum, realized by the famous Veneziano formula, remarkable by its simplicity, leading to the development of dual models.

It is not difficult to understand, from the Regge behaviour, that the theory in question describes a string moving in space-time. Indeed, for a relativistic string, computation of energy and angular momentum leads to Regge behaviour. Moreover, scattering of plane waves can be easily seen to lead to the Veneziano formula.

In spite of these theoretical achievements, the theory failed to describe strong interactions in some important physical aspects, and the first is the high energy behaviour of the amplitudes predicted to fall off exponentially. The alternative to the description of strong interactions is indeed Quantum Cromodynamics, whose perturbative expansion provides a very good description of strongly interacting particles at very high energies, as an outcome of the study of the Callan-Symanzik and renormalization group equations. The perturbative expansion for strong interactions is legitimate for the high energy regime, since these theories (and essentially only these) share the property of asymptotic freedom.

On the other hand, string models have been studied in the framework of grand unification, especially after the interpretation of the spin two field as the graviton, the GSO projection implementing space-time supersymmetry of strings with spin, and later the discovery of the anomaly cancellation for specific gauge groups associated to the string interactions. This led to a tremendous effort in the eighties towards an understanding of (critical) string theory, i.e., string theory at the critical dimension as a theory unifying all interactions, including gravity.

However, once again some problems arose. First there was a difficulty of general type, concerning string theory as a physical theory. It is very hard at present to perform experiments to test the theory, due to the extremely high energy level involved. On the other hand, from the conceptual point of view, there are problems in building a Kaluza-Klein theory, in order to obtain results in the physical space-time dimension, starting from the critical dimension (which is 10 for superstrings).

Moreover, as stated previously, the idea of strings underlies several important physical processes. There is a general belief that string theories have in general more than one phase, as indicated by the gravitational dressing to be computed at the end of the Chapter 1, namely, the bosonic string has two critical points at dimensions d=1 and d=25, while the supersymmetric strings displays these points at d=1 and d=9.

Therefore it is of utter importance to consider string theory as a function of space-time dimension. This is only possible considering the Liouville mode. Indeed, naive reparametrization invariance holds only at the critical dimension. In general this symmetry is anomalous. In the past, anomalous gauge theories, have simply been discarded on grounds of self-consistency. However, one can also make sense of them, if one includes the Wess-Zumino term associated with the anomaly. For the gravitational interaction, this amounts to including the Liouville interaction as a consequence of the computation of the determinant of the Laplace operator.

The classical dynamics of strings is entirely based on geometry. The Nambu-Goto action is proportional to the area swept by the string as it evolves in space-time, that is

$$S = \lambda \int \mathrm{d}^2\xi \sqrt{(\dot{X}\cdot X')^2 - \dot{X}^2 X'^2} = \lambda \int \mathrm{d}^2\xi (\det \gamma_{ab})^{1/2} \quad , \tag{1.1}$$

where $X^\mu(\xi_0, \xi_1)$ describes the position of the string and $\gamma_{ab} = \partial_a X^\mu \partial_b X_\mu$ is the so-called induced metric.

As it turns out (see ref. [9]), this action describes a field X^μ which obeys the equation $\partial_a[(\det\gamma)^{1/2}\gamma^{ab}\partial_b X^\mu] = 0$, the minimum area equation, which is equivalent to the two-dimensional Klein-Gordon equation, supplemented by a number of constraints, the so-called Virasoro conditions obtained imposing the metric to be diagonal. It is not difficult to see that this same set of conditions is obtained from a free field action for X^μ, but integrating with a two-dimensional gravitational field $g_{\alpha\beta}$, which is the Polyakov[8] string action

$$S = \lambda \int \mathrm{d}^2\xi \sqrt{|g|} g^{\alpha\beta} \partial_\alpha X^\mu \partial_\beta X_\mu \quad . \tag{1.2}$$

Since two-dimensional gravity has classically zero degrees of freedom one can choose the metric to be flat, in which case X^μ obeys the Klein-Gordon equation. The equation of motion for metric field $g_{\alpha\beta}$ turns out to be a constraint. If we plug it back into the action we re-obtain the Nambu-Goto action. Otherwise, in the conformal gauge one gets directly the Virasoro constraints.

Reparametrization invariance of the Polyakov action is very important. Physically, it obviously means that the string does not depend on how one parametrizes the theory. In fact, it is related to the fact that one is introducing gravity in two dimensions. While considering the partition function one has to integrate over the gravitational field as well. However, due to reparametrization invariance, we have the usual problem of picking up the infinite gauge volume, and we have to consider the partition function

$$Z = \int \frac{\mathcal{D}X^\mu \mathcal{D}g}{\text{diff. volume}} \mathrm{e}^{-S} \quad . \tag{1.3}$$

This partition function can be computed as a sum over geometries. Indeed, for a fixed topology of the surface, the metric field can be chosen to be locally flat, by choosing appropriately the diffeomorphism transformation. However, once the topology is fixed, by a smooth diffeomorphism transformation it can not be changed. Given the topology, the non-equivalent gravitational metrics are described by a small number of parameters, the so called Teichmüller parameters. We shall not enter into the technical details of this complex computation, but once one fixes the topology, the gravity field is locally flat, one has to integrate only over the conformal factor σ, the ghosts, and the topological parameters (which we call P), and the partition function becomes

$$Z = \sum_{\text{topologies}} \int \mathcal{D}\sigma\, \mathcal{D}X^\mu\, \mathrm{d}P\, \mathcal{D}\mathrm{gh}\; \mathrm{e}^{-S_{\text{string}} - S_{\text{ghost}} - S_{\text{gauge-fixing}}} \quad . \tag{1.4}$$

At the critical dimension, the conformal part decouples, and we are left with a sum over topologies. However, for non-critical strings this is not true, and we are forced to consider the contribution of the conformal part, leading to the Liouville action, as we shall see.

At this point we are contemplated with two possibilities, either we consider the continuous theory, together with the Liouville mode, or one discretizes the theory taking advantage of its interpretation as a theory of random surfaces. These are the two avenues that we shall follow, and compare results.

However, the opened avenues are rather large, leading to important concepts, as theories of vertex operators, Riemann surfaces, and number theory.

Let us study the Liouville mode integration in detail. Indeed, when a gauge symmetry is anomalous, it turns out that the pure gauge degrees of freedom do not decouple. This is the case of the Liouville mode, which couples to the trace of the energy momentum tensor. When one takes into account the pure gauge degree of freedom, the field space gets larger, and one can formulate the theory in a gauge invariant language, see ref. [11], especially Chapters 13 and 14, and references therein.

We consider the string moving freely in space-time and write the partition function in terms of the Polyakov action (1.2) as

$$Z = \int \mathcal{D}X\mathcal{D}g_{\alpha\beta}\mathrm{e}^{-\int \mathrm{d}^2\xi \, \frac{1}{2}\sqrt{|g|}g^{\alpha\beta}\partial_\alpha X^\mu \partial_\beta X_\mu} \quad . \tag{1.5}$$

The integral over the metric $g_{\alpha\beta}$ implies, in particular, a sum over all topologies of the two-dimensional surface described by the string $X_\mu(\xi)$. Classically, the gravity equation of motion leads to the well known Virasoro constraints[9]. However, a full quantum theory requires performing the integration over the metric. First, one has to fix the gauge. If one proceeds naively, counting the degrees of freedom, one finds that the metric has none. In fact, choosing the conformal gauge $g_{\alpha\beta} = \eta_{\alpha\beta}\mathrm{e}^{2\sigma}$ leads to a metric independent action, and one is free from gravity. Quantum mechanically this is a wrong procedure due to the presence of the conformal anomaly, see [10] and Chapter 15 of ref. [11], and references therein.

In general, we have to deal with the determinant of the Laplace operator arising from the integration of (1.5) over the matter fields. This is not a simple task. The main point is that we have to use conformal invariance as a guideline. It is a gauge invariance, and ghosts are unavoidable. Some general remarks are in order. In general we work in the Euclidian space, and the Laplace operator is given by

$$\Delta = \frac{1}{\sqrt{|g|}}\partial_\mu \left(\sqrt{|g|}g^{\mu\nu}\partial_\nu\right) \quad , \tag{1.6}$$

where we kept the upper and lower indices to remind about the use of the inverse metric, while comparing to a fiducial metric $\hat{g}$, as given by the expressions

$$g_{\mu\nu} = \hat{g}_{\mu\nu}\mathrm{e}^{2\sigma} \quad , \quad g^{\mu\nu} = \hat{g}^{\mu\nu}\mathrm{e}^{-2\sigma} \quad . \tag{1.7}$$

The quantum definition of the model requires writing down the most general Lagrangian with all possible counterterms[1)]. There are two of them: the cosmological term

$$\mathcal{S}_{cosm} = \mu \int \mathrm{d}^2z\sqrt{|g|} \quad , \tag{1.8a}$$

and the Einstein action

$$\mathcal{S}_E = \frac{\gamma}{4}\int \mathrm{d}^2z\sqrt{|g|}R \quad . \tag{1.8b}$$

However, the latter is a topological invariant, proportional to the Euler characteristic. Eventually we have to perform a summation over all topologies, implying a sum over the possible Euler characteristics, and in fact γ must be an integer.

---

1) For an open string the situation is more complex, see ref. [10].

However, for our purposes it is enough to restrain to the case of a fixed Euler characteristic, and we may abandon that term. The cosmological term will now be chosen as zero, but will be reintroduced later in the discussion of Liouville theory.

We have to consider the measure for the metric, that is[10]

$$Dg_{\alpha\beta} = 2g_{\alpha\beta}\mathrm{d}\tau + \mathrm{d}h_{\alpha\beta} \quad , \tag{1.9}$$

where the first term represents a conformal reparametrization, while $\mathrm{d}h_{\alpha\beta}$ is traceless

$$g^{\alpha\beta}\mathrm{d}h_{\alpha\beta} = 0 \quad , \tag{1.10}$$

in such a way that we can find a measure for the metric in terms of the expressions

$$\begin{aligned} ||\mathrm{d}g||^2 &= \int \mathrm{d}^2 z\sqrt{|g|}(G^{\alpha\beta\gamma\delta} + ug^{\alpha\beta}g^{\gamma\delta})\mathrm{d}g_{\alpha\beta}\mathrm{d}g_{\gamma\delta} \quad , \\ G_{\alpha\beta}{}^{\gamma\delta} &= \frac{1}{2}(\delta^\gamma_\alpha\delta^\delta_\beta + \delta^\delta_\alpha\delta^\gamma_\beta - g_{\alpha\beta}g^{\gamma\delta}) \quad , \end{aligned} \tag{1.11}$$

where u is an arbitrary parameter. Therefore we can independently integrate over the traceless part $h_{\alpha\beta}$ and the trace τ that is,

$$\mathcal{D}g = \mathcal{D}h\,\mathcal{D}\tau \quad . \tag{1.12}$$

The Fadeev-Popov procedure must be followed now, and we have to fix the gauge. It is a known fact that every metric in two-dimensional space-time is conformally equivalent to a constant curvature metric[2)]. Therefore, using (1.7) we find for a variation in σ

$$\mathrm{d}g_{ab} = 2g_{ab}\mathrm{d}\sigma \quad , \tag{1.13}$$

while a diffeomorphism transformation with infinitesimal parameters $\mathrm{d}V_\alpha$ implies

$$\mathrm{d}g_{\alpha\beta} = \mathrm{d}h_{\alpha\beta} = 2G_{\alpha\beta}{}^{\gamma\delta}D_\gamma \mathrm{d}V_\delta \equiv (P_1\mathrm{d}V)_{\alpha\beta} \quad . \tag{1.14}$$

The change of variables from $\{\mathcal{D}\tau\,,\mathcal{D}h\}$ to $\{\mathcal{D}\sigma\,,\mathcal{D}V\}$ is given by

$$\mathcal{D}\tau\,\mathcal{D}h = \mathcal{D}\sigma\,\mathcal{D}V \det\begin{bmatrix} 1 & \cdots \\ 0 & P_1 \end{bmatrix} \quad , \tag{1.15}$$

that is, we have to compute the Jacobian

$$\det P_1 = \left[\det P_1^\dagger P_1\right]^{1/2} \quad ; \tag{1.16}$$

in this way we have separated the diffeomorphism volume $\Omega_{diff} = \int \mathcal{D}V$, i.e., the integration over the gauge parameters V_α. The operator P_1 is only onc out of a set of operators P_n acting on tensors in two dimensions, which transform as

$$P_n \to P_n' = \left(\frac{\partial z'}{\partial z}\right)^n P_n \quad , \tag{1.17}$$

where n may be a positive or negative integer[3)] .

---

2) Indeed, if (1.7) holds, $\hat{R} = \mathrm{e}^{2\sigma}[R + 2\Delta\sigma]$, and we have to solve $\Delta\sigma - \hat{R}\mathrm{e}^{-2\sigma} = -R$, which is a second order differential equation, and has solutions. For a more detailed discussion see ref. [10]. Notice however that this is a local statement.

3) In fact, this is a very general type of tensor, since we can choose the metric locally to be such that $g_{zz} = g_{\bar{z}\bar{z}} = 0$, $g_{z\bar{z}} = g_{\bar{z}z} = \frac{1}{2}\mathrm{e}^{2\sigma}$; therefore, there is no difference between upper z and lower $\bar{z}$ indices.

Expression (1.16) is nothing but the Fadeev-Popov determinant. In light-cone coordinates it can be written as

$$\det D_+ \det D_- = \int \mathcal{D}c^+ \mathcal{D}c^- \mathcal{D}b_{++} \mathcal{D}b_{--} e^{-\frac{1}{\pi}\int d^2z (c^+ D_- b_{++} + c^- D_+ b_{--})} \quad , \tag{1.18}$$

or also covariantly[2],

$$\int \mathcal{D}c^\mu \mathcal{D}b_{\mu\nu} e^{-\frac{1}{2\pi}\int d^2z \sqrt{|g|} g^{\alpha\beta} c^\gamma D_\alpha b_{\beta\gamma}} \quad , \tag{1.19}$$

where c^μ and $b_{\mu\nu}$ are the ghost fields[2]. The computation of the effective action for the gravity field boils down to the computation of the determinants of differential operators.

The determinant of the Laplace operator is not difficult to compute, and the fact that we have d (space-time dimensions) bosonic fields X_μ, means that the determinant is raised to the power d. The question is to find the contribution of the ghosts. However we do not enter into details of the computation. If we are not careful enough we might conclude from (1.18) that the contribution should be that of two fermions, or one extra boson, with wrong sign. Nevertheless, we have to take into account the fact that the power of the determinant measures the lack of conformal invariance – full local invariance means that we may erase the $g_{\alpha\beta}$ field. In such a case we are measuring the central charge of the Virasoro generators[2]. For the ghosts we have

$$T_{++} = \frac{1}{2} {:} c^+ \partial_+ b_{++} {:} + {:} \partial_+ c^+ b_{++} {:} \quad , \tag{1.20}$$

and a simple computation leads to the operator product expansion (OPE)

$$T_{++}(z) T_{++}(w) = \frac{-13}{(z-w)^4} + \frac{2}{(z-w)^2} T + \frac{1}{(z-w)} \partial T \quad , \tag{1.21}$$

which implies that the ghost system contributes with –26. Thus, in the bosonic case, the determinant is raised to the power $d - 26$.

The last piece of information concerns the computation of the determinant itself. We look at

$$\ln \det(\Delta) = -\int_\epsilon^\infty \frac{dt}{t} \operatorname{tr} \exp(t\Delta) = -\int_\epsilon^\infty \frac{dt}{t} G_\Delta(0,0;t) \quad , \tag{1.22}$$

where we introduced the heat-kernel expression in the right hand side, with a cut-off ϵ, and the diffusion evolution operator is defined by[10,11]

$$G_\Delta(z, z'; t) = \langle z | e^{t\Delta} | z' \rangle \quad , \tag{1.23}$$

with the initial condition

$$\lim_{t \to 0} G(z, z'; t) = \delta(z - z') \quad . \tag{1.24}$$

The expansion of G in a Laurent series in t is well known. In fact, if $\Delta = \partial^2 - \Psi$, the computation is performed using the second term Ψ as a perturbation, that is

$$\Psi = (\mathrm{e}^{-2\sigma} - 1)\partial^2 \quad . \tag{1.25}$$

For the diffusion operator we find the following expression

$$G(0,0;t) = \frac{1}{4\pi t} - \frac{1}{12\pi}\partial^2\sigma + \cdots \quad , \tag{1.26}$$

where we have used the free field result

$$G(z,z';t) = \frac{1}{4\pi t}\mathrm{e}^{-\frac{|z-z'|^2}{4t}} \quad , \tag{1.27}$$

and the local coordinates are such that $\sigma(0) = 0 = \partial\sigma(0)$ holds.

In fact, in order to compute the determinant we need only the first few terms of the perturbative expansion of the diffusion operator. However we need a general coordinate invariant expression for the heat kernel, which is obtained from (1.26) using $R = -2\partial^2\sigma$, leading to

$$G(0,0;t) = \frac{1}{4\pi t} + \frac{1}{24\pi}R + \cdots \quad . \tag{1.28}$$

The determinant is obtained upon variation with respect to its defining parameters, as derived from the expression

$$\delta \ln \det \Delta = \mathrm{tr}\,\mathrm{e}^{\epsilon\Delta}\Delta^{-1}\delta\Delta = -2\mathrm{tr}\,\mathrm{e}^{\epsilon\Delta}\delta\sigma \quad , \tag{1.29}$$

which after integration over σ (see Chapter 4 of ref. [11]) with measure $\int \mathrm{d}^2x\sqrt{|g|}$ yields the determinant. Notice that we introduce a cut-off at $t \to \infty$; a proper discussion of these subtleties can be found in refs. [10, 11, 12]. Using (1.28) and (1.29), as well as (A.2) it is not difficult to compute the expression

$$\begin{aligned} \ln \det\Delta = &-\frac{1}{4\pi\epsilon}\int \mathrm{d}^2z\sqrt{|\hat{g}|}\mathrm{e}^{2\sigma} - \frac{1}{12\pi}\int \mathrm{d}^2z\sqrt{|\hat{g}|}\hat{g}^{\alpha\beta}\partial_\alpha\sigma\partial_\beta\sigma \\ &-\frac{1}{12\pi}\int \mathrm{d}^2z\sqrt{|\hat{g}|}\hat{R}\sigma \quad . \end{aligned} \tag{1.30}$$

The first term is divergent and leads to an arbitrariness in the cosmological constant. We are thus led to the Liouville theory:

$$S_{eff} = \frac{26-D}{24\pi}\int \mathrm{d}^2z\sqrt{|\hat{g}|}\left(\hat{g}^{\alpha\beta}\partial_\alpha\sigma\partial_\beta\sigma + \mu\mathrm{e}^{2\sigma} + \hat{R}\sigma\right) \quad . \tag{1.31}$$

The last term may be dropped for flat space. In fact, the above expression may be interpreted[13] also as a change of variables in the sense of Fujikawa, that is integration over a matter field with metric $\mathrm{e}^{2\sigma}\hat{g}_{\mu\nu}$ is equal to integration with metric $\hat{g}_{\mu\nu}$ if we take into account a Jacobian given by the exponential of the Liouville action, that is

$$\mathcal{D}_{\mathrm{e}^\sigma\hat{g}}X = \mathrm{e}^{-S_L}\mathcal{D}_{\hat{g}}X \quad . \tag{1.32}$$

The Liouville equation is known for more than a century. In classical mathematics it has been used in order to study the uniformization problem by Poincaré[14]. Such problem consists of the conformal map from a Riemann surface into a plane. This reduces to the problem that given a function $F(z,w)=0$ (see ref. [15]) defining a surface, one has to find $\varphi(t)$ and $\psi(t)$, such that $F(\varphi(t),\psi(t))=0$. In other words, given a not necessarily single valued relation $w=f(z)$, one finds $z=\varphi(t)$, $w=\psi(t)$ in terms of the so-called uniformizing parameter z.

In terms of objects which are closer to the present problems in physics - more specifically of string theory - given a Riemann sphere with n missing points, one ought to find a function from the upper half plane into the above surface, such that the inverse is locally single valued and holomorphic. Such sort of problems is a century old and still poses difficulties[16].

Nevertheless, the knowledge of properties of Liouville theory is essential in order to advance towards the solution of that important mathematical problem.

The uniformization problem, in mathematics, has been studied, by means of the use of Liouville theory, by Poincaré[14], Klein[15], Koebe and Picard[17]. The problem consists in proving that "every Riemann surface can be represented as a domain in the plane factored by a fixed point free Fuchsian group - a finite group leaving invariant a certain region in the plane"[15,18,19].

In general, we can consider a Riemann sphere $S^2 \equiv \mathbf{C} \cup \{\infty\}$, with n missing points, that is $\Omega = S^2 - \{p_i, \cdots, p_n\}$. For $n=1$ we have a disk, and for $n=2$ the cylinder. If $n \geq 3$ we have more interesting and difficult cases. Indeed, to be more precise, we consider the mapping $\lambda(\tau)$, which maps the upper half plane $\mathcal{H}$[18,19] into Ω,

$$\lambda : \mathcal{H} \rightarrow \Omega \quad , \quad z = \lambda(\tau) \quad . \tag{1.33}$$

The mapping must have a locally single valued inverse, be holomorphic and it must be possible to continue it analytically along a closed path in Ω, if (and only if) it is homotopic to zero. A closed path $\gamma \in \Omega$ with winding number $+1$ with respect to p_n (and zero with respect to other p_j's, $j \neq n$) lifts to a path in $\mathcal{H}$ which joins τ to $T_n\tau$, where T_n depends only on the homotopy class of γ. This implies[19] that $T_n\tau$ is a Möbius transform of τ. It is possible to choose a set of T_n such that $T_1 \cdots T_n = \mathbf{1}$. This set generates a so-called Fuchsian group. Analyzing the properties of the Schwarzian derivative

$$\{\psi, z\} = \left(\frac{\psi''}{\psi'}\right)' - \frac{1}{2}\left(\frac{\psi''}{\psi'}\right)^2 \quad , \tag{1.34}$$

and examining the behaviour of the transformation function near each singularity p_k[4)], one proves that[19] if $z=\lambda(\tau)$ is a conformal universal covering map of Ω, then the Schwarzian derivative $\{\tau, z\}$ is given by the expression

$$\{\tau, z\} = \frac{1}{2}\sum_{k=1}^{n}\frac{1}{(z-p_k)^2} + \sum_{k=1}^{n}\frac{m_k}{z-p_k} \quad , \tag{1.35}$$

---

4) Near the puncture the function $\zeta(z) = \exp\left(2\pi i c_k^{-1}(\tau-\alpha_k)^{-1}\right)$ is single valued, therefore it may be expanded in a Taylor series around the point p_k as $\sim \sum a_i (z-p_k)^i$.

where m_k fulfill the conditions

$$\sum_{k=1}^{n} m_k = 0 \quad ,$$
$$\sum_{k=1}^{n}(2m_k p_k + 1) = 0 \quad , \tag{1.36}$$
$$\sum_{k=1}^{n} p_k(m_k p_k^2 + 1) = 0 \quad .$$

Notice that if $n = 3$ the system is determined and the problem solved. In the proof of this theorem, one uses the transformation property of the Schwarzian derivative under coordinates redefinition

$$\{\psi, \zeta\}\mathrm{d}\zeta^2 = \{\psi, z\}\mathrm{d}z^2 + \{z, \zeta\}\mathrm{d}\zeta^2 \quad , \tag{1.37}$$

(see refs. [18,19]) and the behaviour $\{\tau, z\} \sim z^{-4}$ indicating that it may be the energy-momentum tensor of a classical field theory. Thus we can interprete the second order poles in (1.35) as sources of curvature. The whole expression is the partial fraction decomposition of the energy momentum tensor.

The relation of the above problem with the Liouville equation comes next, if we consider an ingredient arising from the theory of differential equations, and later a diffeomorphism invariant metric in the plane, showing that the latter satisfies a certain differential equation, such that the function (1.35) acts as a"potential" and a Fuchsian group is the corresponding monodromy.

One notices first that the function ψ is the quotient η_1/η_2 of linearly independent solutions of the following differential equation

$$\eta'' + \frac{1}{2}\{\psi, z\}\eta = 0 \quad . \tag{1.38}$$

Indeed, one has $\eta_1 = \psi(\psi')^{-1/2}$, $\eta_2 = (\psi')^{-1/2}$, as one readily verifies.

It is trivial to prove that a Möbius transformation

$$S\tau = \frac{a\tau + b}{c\tau + d} \quad , \quad ad - bc = 1 \quad , \tag{1.39}$$

is such that

$$\mathcal{I}m\ (S\tau) = \mathcal{I}m\,\tau \times |S'(\tau)| \quad ; \tag{1.40}$$

consequently, the function

$$\rho(z) = (|\lambda'(z)| \mathcal{I}m\,\tau)^{-1} \quad , \tag{1.41}$$

defined on Ω, is single-valued. It does not depend on the particular choice of the covering map λ, and can be used as a (diagonal) metric

$$\rho(z)|\mathrm{d}z| = \frac{1}{\mathcal{I}m\,\tau}|\mathrm{d}\tau| \quad , \tag{1.42}$$

in which the Euclidean semicircles orthogonal to the real line are geodesics[19]. As an immediate consequence $\phi = 2\ln\rho$ satisfies the Liouville equation

$$\Delta\phi = 2\mathrm{e}^{\phi} \quad . \tag{1.43}$$

Indeed,

$$\begin{aligned}\partial\bar{\partial}\ln\rho &= \partial\bar{\partial}\left[-\ln|\lambda'| - \ln\mathcal{I}m\,\tau\right] \\ &= \partial\left\{\frac{\mathrm{d}\tau}{\mathrm{d}z}\left(-\frac{\lambda''}{2\lambda'} - \frac{1}{2i\mathcal{I}m\,\tau}\right)\right\} \\ &= -\frac{\mathrm{d}\tau}{\mathrm{d}z}\frac{\mathrm{d}\bar{\tau}}{\mathrm{d}\bar{z}}\frac{\partial}{\partial\bar{\tau}}\frac{1}{\tau-\bar{\tau}} = -\frac{1}{|\lambda'|^2}\frac{1}{(\tau-\bar{\tau})^2} \\ &= \frac{1}{4}\rho^2 \quad .\end{aligned} \tag{1.44}$$

The solution of the above differential equation depends on boundary conditions which must be given since Ω has singularities at $\{p_1 \cdots p_n\}$. Near a puncture the function $\zeta(z) = \mathrm{e}^{2\pi i c^{-1}(\tau-\alpha)^{-1}}$ is regular and may be expanded in a power series around $z \sim p$. This allows one to obtain[19]

$$\left|\frac{\mathrm{d}z}{\mathrm{d}\tau}\right| \sim |z-p| \ln R|z-p| \quad , \tag{1.45}$$

with $R = |a_1|$; enforcing that ρ obeys (1.44) we have the behaviour

$$\rho(z) = -\frac{1}{|z-p|\ln R|z-p|} + \mathcal{O}(|z-p|^0) \quad . \tag{1.46}$$

The function $\rho^{-1}(z)$ satisfies a Fuchsian differential equation[19]

$$\left(\frac{\mathrm{d}^2}{\mathrm{d}z^2} + \frac{1}{4}\sum_k \frac{1}{(z-p_k)^2} + \frac{1}{2}\sum_k \frac{m_k}{z-p_k}\right)\rho^{-1} = 0 \quad , \tag{1.47}$$

with the boundary condition (1.46).

This problem has a unique solution. In fact, one can study the general solutions of the Liouville differential equation as arising from the action

$$S = \frac{1}{4\pi}\int \mathrm{d}^2z\sqrt{|\hat{g}|}\left\{\frac{1}{2}\partial^{\mu}\phi\partial_{\mu}\phi + \frac{2\mu}{\gamma^2}\mathrm{e}^{\gamma\phi} + \frac{1}{\gamma}\phi\hat{R}(\hat{g})\right\} \quad , \tag{1.48}$$

dropping an overall factor, which is trivial in the classical version, but quantum mechanically amounts to a redefinition of $\hbar$. Here $\hat{g}$ is a given fiducial metric, γ is a coupling constant introduced by redefining the field σ as $\sigma = \frac{\gamma}{2}\phi$. We are thus led to the equation of motion

$$\Delta\phi = \frac{2\mu}{\gamma}e^{\gamma\phi} + \frac{1}{\gamma}\hat{R} \quad . \tag{1.49}$$

In a flat fiducial metric the general solution of (1.49) is

$$e^{\gamma\phi} = \frac{4}{\mu}\frac{A'(z)B'(\overline{z})}{(1-AB)^2} \quad , \tag{1.50}$$

in such a way that Liouville theory is solved in terms of right and left moving fields.

If we consider the Liouville equation of motion (1.49), we see that neglecting the cosmological term, the Liouville field obeys the free Green functions equation, with solution

$$\phi = \frac{a-1}{\gamma}\ln(z\overline{z}) \quad , \tag{1.51}$$

where a parametrizes the strength of the curvature at the origin. We have to check that the dropping of the cosmological term does not affect the above solution near the singularity. That term behaves as

$$e^{\gamma\phi} \sim |z|^{-2(1-a)} \quad , \tag{1.52}$$

which is integrable only for $a > 0$. This means that one can not localize too much curvature at a single point. As a matter of fact, the results below show that the classical solution to the field equations exactly matches the above condition. This result will also be used to characterize the conditions to be fulfilled by the constants in the quantum theory.

To quantize the Liouville theory, we can use the canonical method. The energy momentum is traceless, i.e. $T_{+-} = 0$, therefore implying conformal invariance. We compute the space of states imposing that the other components of the energy momentum tensor are zero on the physical space of states. Later on, the conformal dimension of the primary operator $e^{\beta\varphi}$ will be fixed.

In fact, Eq. (1.47) can be seen as the definition of a null vector in conformal field theory; in particular considering the solution

$$e^{\gamma\phi}|dz|^2 = \frac{4}{\mu}\frac{|dw|^2}{(1-|w|^2)^2} \quad , \tag{1.53}$$

where $w = A(z)$, $\overline{w} = B(\overline{z})$, and taking the quotient $w = \eta_1/\eta_2$ of two solutions of (1.38), one finds

$$e^{-\frac{1}{2}\gamma\phi} = \frac{\sqrt{\mu}}{2} = \left|\frac{|\eta_1|^2 - |\eta_2|^2}{\eta_1\eta_2' - \eta_2\eta_1'}\right| \quad , \tag{1.54}$$

which satisfies

$$\left(\partial_z^2 + \gamma^2 T(z)\right)e^{-\frac{1}{2}\gamma\phi} = 0 \quad , \tag{1.55}$$

showing that $e^{-\frac{1}{2}\gamma\phi}$ is a solution of the null vector equation.

To have a better understanding of Liouville theory, as well as of the set of states and operators, we shall consider the Schrödinger problem associated with the Hamiltonian derived from the Liouville action.

Let us consider the Lagrangian

$$\mathcal{L} = \frac{\sqrt{|\hat{g}|}}{4\pi}\left(\frac{1}{2}\hat{g}^{\mu\nu}\partial_\mu\phi\partial_\nu\phi + \frac{2\mu}{\gamma^2}e^{\gamma\phi} - \frac{Q}{2}\phi\hat{R}\right) \quad , \tag{1.56}$$

where Q is classically related to the coupling constant γ as $Q = -\frac{2}{\gamma}$. This relation will acquire quantum corrections. We have, for the $(++)$ component of the energy momentum tensor[21]

$$T_{++} = \frac{1}{16\pi}(\phi' + 4\pi P)^2 + \frac{Q}{64\pi}(\phi' + 4\pi P)' + \frac{Q}{64\pi}\left(\dot{\phi}' + 4\pi\dot{P}\right) \quad , \tag{1.57}$$

where P is the momentum canonically conjugated to ϕ. Using the equations of motion, we can rewrite (1.57) as

$$8\pi T_{++} \equiv \hat{T}_{++} = \frac{1}{2}(\phi' + 4\pi P)^2 + Q(\phi' + 4\pi P)' - \frac{\mu Q}{4\gamma}e^{\gamma\phi} + \frac{Q^2}{16}\hat{R} \quad . \tag{1.58}$$

We can transform to a more complicated space adding the corresponding Schwarzian derivative of the transformation function. Since we shall later renormalize the relevant quantities due to quantum contributions, we define the Hamiltonian as

$$H = \frac{1}{2}(\phi' + 4\pi P)^2 + Q(\phi' + 4\pi P)' + \frac{\mu}{2\gamma^2}e^{\gamma\phi} + \frac{Q^2}{16}\hat{R} + \frac{Q^2}{8} \quad . \tag{1.59}$$

For flat space $\hat{R} = 0$. In the so-called minisuperspace approximation, where the quantities do not depend on the space variable σ, we find

$$H \equiv \frac{1}{2}p^2 + \frac{\mu}{2\gamma^2}e^{\gamma\phi} + \frac{Q^2}{8} \quad . \tag{1.60}$$

We can now discuss the quantum mechanical problem

$$H\psi = \left(\frac{1}{2}p^2 + \frac{\mu}{2\gamma^2}e^{\gamma\phi} + \frac{Q^2}{8}\right)\psi = \Delta\psi \quad , \tag{1.61}$$

where now $p = -i\frac{\partial}{\partial\phi}$.

Since the potential reflects the wave function to $-\infty$, we have

$$\psi\,(\phi \to -\infty) \approx \sin p\phi \tag{1.62}$$

and $\Delta = \frac{1}{2}p^2 + \frac{1}{8}Q^2$.

As suggested in ref. [20], the spectrum consists of a sum over all momenta p of the Virasoro irreducible representation with conformal dimension $\Delta(p) = \frac{1}{2}p^2 + \frac{1}{8}Q^2$. Nevertheless, as we shall see in Chapter 2, there are operators such as $\mathcal{O} = e^{\beta\phi}$ which have dimension $\Delta = -\frac{1}{2}(\beta + \frac{Q}{2})^2 + \frac{Q^2}{8}$. We are faced with the problem of finding the corresponding wave function $\mathcal{O}(z=0)|0\rangle$. We solve the Schrödinger Eq. (1.61) for the above Δ and find

$$\psi_{\mathcal{O}}(\phi) = e^{(\frac{1}{2}Q+\beta)\phi} \tag{1.63}$$

which diverges as $\phi \to +\infty$, being not normalizable. This is the general case in Liouville field theory as discussed[21] by Seiberg[5)].

Let us now consider the puncture operator, obtained as a solution of Eq. (1.61) with $\Delta = Q^2/8$. There are two solutions: $\psi = 1$ and $\psi = \phi$, corresponding to the operators $e^{\frac{Q}{2}\phi}$ and $\phi e^{\frac{Q}{2}\phi}$ respectively. We shall see in Chapter 2 that the latter is the correct puncture operator leading to the expected scaling behavior.

One can study several solutions of the generalized field equation (including one source of curvature)

$$\Delta\phi - \frac{2\mu}{\gamma} e^{\gamma\phi} + \frac{1-a}{\gamma} 8\pi\delta^{(2)}(z) = 0 \quad , \tag{1.64}$$

and find some solutions of the type (1.50) when $A(z) = z^a$, $B(z) = \bar{z}^a$, as given by

$$e^{\gamma\phi} = \frac{4}{\mu} \frac{a^2}{(z\bar{z})^{1-a}[1-(z\bar{z})^a]^2} \quad , \tag{1.65}$$

which for $a \to 0$ leads us to

$$e^{\gamma\phi} = \frac{1}{\mu}\left(\frac{1}{|z|\ln|z|}\right)^2 \quad , \tag{1.66}$$

which corresponds to (1.46) when $\mu = 1$. Notice that in (1.65) we can interchange a into $-a$. The singularity is related to the behaviour for $z \to 0$. We have seen in (1.52) that Eq. (1.64) is only fulfilled for $a > 0$. For $a \to 0$ we can see that (1.52) is again integrable due to the logarithmic correction. Thus we assume from now on that $a \geq 0$.

Equation (1.65) is a class of solutions, labelled by the parameter a; in the case where a is real, the solution is called elliptic, with monodromy $A \to e^{2\pi i a}A$. On the other hand, (1.66) is called a parabolic solution, with monodromy $A \to A - 2\pi$. A remaining class of solutions is called hyperbolic and corresponds to imaginary values for $a = im$, and reads

$$e^{\gamma\phi} = \frac{1}{\mu} \frac{m^2}{\bar{z}z\left[\sin\frac{m}{2}\ln\bar{z}z\right]^2} \quad , \tag{1.67}$$

with monodromy $A \to e^{-2\pi m}A$.

The elliptic solution corresponds to the cusp of figure 1.1*a*, while the parabolic has the puncture singularity (figure 1.1*b*). The hyperbolic solution looks different (figure 1.1*c*).

[5)] Our conventions can be translated into Seiberg's by replacing $\gamma \to -\gamma$, $\phi \to -\phi$. Moreover the transformation of coordinates from $(z, \bar{z})$ to (x_0, x_1) amounts to a factor of 2.

(a) (b) (c)

Figure 1.1: Elliptic solution (a), parabolic solution (b) and hyperbolic solution (c).

Quantum Liouville theory has been studied by several authors. It has been discussed as a classical integrable model with boundary conditions in ref. [22], the string spectrum has been analysed in ref. [23]. The full quantum operator solution has been studied in paper [24] where conformal invariance and the issue of Virasoro generators has been considered in detail. The model including boundary terms was also subject of research for several authors[25].

We shall be interested in the issue of Liouville theory in the framework of non-critical strings, as a device to compute dressed vertices correlation functions. In fact, in the general case, reparametrization invariance is broken for strings if one is not in the critical dimension, as we have seen in the previous discussion. Breaking of local invariances has in general been thought as a dead end to the theory. However, inclusion of the gauge mode can bring back several good features of the theory, which realizes the gauge symmetry in an enlarged space, as is exemplified in detail in the case of chiral gauge theories in two dimensions[11,26]. The germ of this sort of insight has been provided by the discussion of the Liouville mode in string theory, in the way we have done at the beginning of this Chapter, namely as the outcome of the determinant of the matter plus ghost Laplacian.

The fact that Liouville theory is conformally invariant and gives a contribution to the Virasoro operator is known since a long time[27].

At the classical level, one may write the Liouville solution[22] in the form (1.50). Negative integer powers of it may be written using the expansion

$$e^{-j\gamma\phi} = \left(\frac{\mu}{4}\right)^j \left(\frac{1}{\sqrt{\partial A}}\frac{1}{\sqrt{\overline{\partial} B}} - \frac{A}{\sqrt{\partial A}}\frac{B}{\sqrt{\overline{\partial} B}}\right)^{2j} = \left(\frac{\mu}{4}\right)^j \sum_{m=-j}^{j} \psi^j_m(z)\psi^{jm}(\overline{z}) \quad (1.68)$$

where ψ^j_m transforms as the $(2j+1)$-dimensional spin-j representation of $SL(2,R)$ (see refs. [22, 23, 28]).

The previous prolegomena warrant the importance of the Liouville theory in the study of two-dimensional random surfaces, with applications to quantum gravity. Several technical questions can be overcome with this technique: as we shall see, it is possible to obtain exactly the correlation function of the dressed vertex operator using the well known Coulomb gas method[29,30]. Moreover, the physical insight gains light. Indeed, one is able to treat[31], at least partially, the Wheeler-de Witt equation[32,33], allowing a better understanding of time in this simplified model, a concept which suffers its own ill definiteness in general relativity[34]. Moreover, it will be possible to compare different approaches to the same problem, namely Liouville continuous approach at one hand, and matrix models (discrete approach) on the other hand. This will be the central question in this work.

The equation of motion

$$\Delta\phi - \frac{2\mu}{\gamma}e^{\gamma\phi} + \sum \beta_i 8\pi\delta^{(2)}(z_i) = 0 \tag{1.69}$$

yields, when integrated in the whole space, upon using $R = -\gamma e^{-\gamma\phi}\Delta\phi$ and the Gauss-Bonnet theorem, the relation

$$\gamma\sum \beta_i + 2h - 2 - \frac{\mu}{8\pi}A = 0 \quad , \tag{1.70}$$

where h is the number of handles of the Riemann surface, and A its area, ensuring that a classical solution exists if and only if

$$\gamma\sum \beta_i + 2h - 2 > 0 \quad . \tag{1.71}$$

In fact, an equation of motion as (1.69) may be obtained from the study of correlation functions of the type

$$\left\langle \prod_i e^{\beta_i\phi(z_i)} \right\rangle = \int \mathcal{D}\phi e^{-S[\phi]} \prod_i e^{\beta_i\phi(z_i)} \quad . \tag{1.72}$$

Therefore, in the quantum theory it is important to analyze the operators $e^{\beta\phi(z)}$; they are responsible for the local curvature strength β. This means that they produce elliptic solutions such as (1.65), with $a = 1-\gamma\beta$. From the discussion following (1.65) we see that $\beta\gamma < 1$, and for $\gamma < 0$ (this is the case in the quantum theory, according to our conventions) $\beta > 1/\gamma$. Classically this is equivalent to

$$\beta \geq -\frac{Q}{2} \quad , \tag{1.73}$$

which will be the correct inequality even taking into account the quantum corrections.

Canonical quantization of the theory has been performed by Braaten, Curtright and Thorn[20], and one can find the anomalous dimension of the above operators

$$\Delta(e^{\beta\phi}) = -\frac{1}{2}\beta(\beta + Q) = -\frac{1}{2}(\beta + \frac{Q}{2})^2 + \frac{c-1}{24}, \tag{1.74}$$

where $c = 1 + 3Q^2$ is the central charge of the Virasoro operator.

This result mirrors the fact that Liouville theory is not, and cannot be treated as a free theory. In fact, Liouville theory cannot be treated in perturbation theory due to the lack of a normalizable ground state at finite values of the field (for $\gamma\phi \to -\infty$ all derivatives drop to zero, and the theory is trivial[35]). In fact, a look at the full action of string theory and the Liouville field, shows that the latter can be thought as a target space coordinate, and the full action is that of a string theory in a non-trivial background[36].

In such a case, we have to move towards quantum theory and sum over the geometry. By looking at the Liouville interaction with curvature, and recalling that in string theory the sum over geometries corresponds to a string perturbation theory in terms of the genus, that is

$$Z \sim \sum Z_h g_{st}^{2(h-1)} \quad , \tag{1.75}$$

we see, recalling (1.56) that the string coupling constant must be related to the Liouville field by

$$g_{st} = g_0 e^{-\frac{Q}{2}\phi} \quad . \tag{1.76}$$

This will be important while defining the relation between the tachyon vertex and the corresponding wave function, as for example in the discussion concerning (1.63).

We thus definitely move to the discussion of quantum gravity in two dimensions. Aiming at this whole set of problems, we shall first consider the general problem of quantum gravity in two dimensions in the operator approach, in the light cone gauge, namely computing the correlation functions of the gravitational fields, and analysing the consequences of the interaction with matter fields. In this approach, one is not able to compute general correlation functions of dressed operators, but the relation of matter central charge and gravitational dressing is rather clear. Later, we consider the more difficult case of gravity in the conformal gauge. However, consequences are far reaching, permitting the aforementioned calculation of correlation functions of dressed vertex operators, and a thorough comparison with matrix models results.

The effective action for two-dimensional gravity is given by

$$S_g = \frac{c}{96\pi}\int d^2x \sqrt{|g|} R \frac{1}{\Delta} R \quad . \tag{1.77}$$

In the conformal gauge, this is the kinetic term of the Liouville action. In the light-cone gauge (see Appendix A), with the help of an auxiliary field φ, this action can be written in a local form[38],

$$S = \int d^2x\, \mathcal{L} = \int d^2x (\partial_+\varphi\partial_-\varphi - g_{++}(\partial_-\varphi)^2 - \alpha\partial_- g_{++}\partial_-\varphi) \quad , \tag{1.78}$$

where $\alpha^2 = c/12\pi$. Notice that from now to Eq. (1.152) we will be working in Minkowski space, which is more suitable in the light-cone gauge.

The Lagrangian defined above is explicitly covariant, and under an infinitesimal coordinate transformation $x^\mu \to x^\mu + \epsilon^\mu(x)$, the fields φ and $g^{\mu\nu}$ transform as

$$\delta\varphi = \epsilon^\mu \partial_\mu \varphi \quad , \tag{1.79a}$$

$$\delta g^{\mu\nu} = \epsilon^\alpha \partial_\alpha g^{\mu\nu} - g^{\alpha\mu}\partial_\alpha \epsilon^\nu - g^{\alpha\nu}\partial_\alpha\epsilon^\mu \quad . \tag{1.79b}$$

We summarize the whole set of symmetry transformations by the following

$$\delta\varphi = \varepsilon + \epsilon^\mu\partial_\mu\varphi + \frac{1}{2}\alpha\Lambda \quad , \tag{1.80a}$$

$$\delta g^{\mu\nu} = \epsilon^\alpha\partial_\alpha g^{\mu\nu} - g^{\mu\alpha}\partial_\alpha\epsilon^\nu - g^{\alpha\nu}\partial_\alpha\epsilon^\mu - \Lambda g^{\mu\nu} \quad , \tag{1.80b}$$

where $\square\varepsilon = 0$ and the Weyl transformation parameter Λ satisfies

$$R\Lambda = \nabla_\alpha \xi^\alpha \quad . \tag{1.80c}$$

In the light cone gauge we have

$$\delta\varphi = \epsilon^+\partial_+\varphi + \epsilon^-\partial_-\varphi - \frac{1}{2}\alpha(\partial_+\epsilon^+ + \partial_-\epsilon^-) \tag{1.81a}$$

$$\delta g_{++} = \partial_+\epsilon^- + \epsilon^-\overleftrightarrow{\partial}_- g_{++} + \partial_+(\epsilon^+ g_{++}) \equiv \nabla_+\epsilon^- + \partial_+(\epsilon^+ g_{++}) \quad , \tag{1.81b}$$

where we disregarded the ε symmetry, which is trivial. Above, in order to maintain the light cone gauge, we have imposed

$$\delta g^{+-} = -2(\partial_-\epsilon^- + \partial_+\epsilon^+ + \Lambda) = 0 \quad , \tag{1.82}$$

which fixes Λ, and

$$\delta g^{++} = -2g^{+-}\partial_-\epsilon^+ = -4\partial_-\epsilon^+ = 0 \quad , \tag{1.83}$$

constraining $\epsilon^+ = \epsilon^+(x^+)$. Using now (1.80$c$), with an appropriate choice of ξ^μ, we have

$$\partial_-^3\epsilon^- = 0 \quad . \tag{1.84}$$

The use of the symmetries discussed so far allows one to obtain the Green functions of the quantum theory. The symmetry (1.81b) implies a useful Ward identity: the effective gravity action under that transformation changes as[37]

$$\delta S = -4\int T_{--}\delta g_{++}\mathrm{d}^2x = 4\int \mathrm{d}^2x\epsilon^-\nabla_+ T_{--} \quad . \tag{1.85}$$

Using now

$$-\frac{\alpha^2}{8}\partial_-^3 g_{++} = (\partial_+ - g_{++}\partial_- - 2\partial_- g_{++})T_{--} = \nabla_+ T_{--} \quad , \tag{1.86}$$

we are left with

$$\delta S = -\frac{\alpha^2}{2}\int \mathrm{d}^2x\epsilon^-\partial_-^3 g_{++} \quad . \tag{1.87}$$

According to formulae (A.2) and (A.5), the scalar curvature R is proportional to $\partial_-^2 g_{++}$ in the light cone gauge. The resulting equation obeyed by the energy momentum component T_{--}, namely $\nabla_+ T_{--} \sim \partial_- R$, is a manifestation of the conformal anomaly.

For the n-point correlator of g_{++}, the infinitesimal variation (1.81b) implies

$$\begin{aligned}\int \mathcal{D}g\{\sum_i \langle g_{++}(x_1)\cdots\nabla_+\epsilon^-(x_i)\cdots g_{++}(x_n)\mathrm{e}^{iS(g)}\rangle \\ +\langle g_{++}(x_1)\cdots g_{++}(x_n)(-\frac{i\alpha^2}{2})\int \mathrm{d}^2x\partial_-^3 g_{++}\epsilon^-\mathrm{e}^{iS(g)}\rangle\} = 0 \quad .\end{aligned} \tag{1.88}$$

Equation (1.88) should hold independently of ϵ^-, thus the following Ward identity must be satisfied[6]:

$$\frac{i\alpha^2}{2}\partial_-^3\langle g_{++}(x)g_{++}(x_1)\cdots g_{++}(x_n)\rangle =$$

$$= \sum_i \Big\{ \langle g_{++}(x_1)\cdots \not{g}_{++}(x_i)\cdots g_{++}(x_n)\rangle \partial_{+_i}\delta(x-x_i)$$

$$+\left[\delta(x-x_i)\partial_{-_i} - \partial_{-_i}\delta(x-x_i)\right]\langle g_{++}(x_1)\cdots g_{++}(x_n)\rangle \Big\} \quad , \tag{1.89}$$

where the dash means that the field has been deleted.

In order to obtain short distance expansions, we must get rid of the delta functions. We use

$$\partial_-\frac{1}{x^+} = 2i\pi\delta^{(2)}(x) \quad , \tag{1.90a}$$

or, as a consequence of this

$$\frac{1}{4\pi i}\partial_-^3\frac{(x^-)^2}{x^+} = \delta^{(2)}(x) \quad , \tag{1.90b}$$

which is a more useful identity, because of the operator ∂_-^3 which also appears on the r.h.s. of Eq. (1.89); after rescaling $g_{++} \to \frac{1}{2\pi\alpha^2}g_{++}$, we find

$$\langle g_{++}(x)g_{++}(x_1)\cdots g_{++}(x_n)\rangle =$$

$$= -2\pi\alpha^2\sum_{j=1}^{n}\frac{(x^- - x_j^-)^2}{(x^+ - x_j^+)^2}\langle g_{++}(x_1)\cdots \not{g}_{++}(x_j)\cdots g_{++}(x_n)\rangle$$

$$-\sum_{j=1}^{n}\Big[\frac{(x^- - x_j^-)^2}{x^+ - x_j^+}\partial_-^j + 2\frac{x^- - x_j^-}{x^+ - x_j^+}\Big]\langle g_{++}(x_1)\cdots g_{++}(x_n)\rangle \quad , \tag{1.91}$$

where the slash in the jth field in the first term of (1.91) means that it has been deleted. This implies the following operator product expansion (OPE):

$$g_{++}(x)g_{++}(y) = -\frac{c}{6}\Big(\frac{x^- - y^-}{x^+ - y^+}\Big)^2 - \frac{(x^- - y^-)^2}{x^+ - y^+}\partial_-^y g_{++}(y) - 2\frac{x^- - y^-}{x^+ - y^+}g_{++}(y) \quad , \tag{1.92}$$

where $2\pi\alpha^2 \equiv c/6$. In the case of a matter field which transforms as

$$\delta\chi = \epsilon^-\partial_-\chi + \lambda\partial_-\epsilon^-\chi \quad , \tag{1.93}$$

we have

$$g_{++}(x)\chi(y) = -\frac{(x^- - y^-)^2}{x^+ - y^+}\partial_-\chi(y) + 2\lambda\frac{x^- - y^-}{x^+ - y^+}\chi(y) \quad . \tag{1.94}$$

[6] Here we use $\partial_{\pm_i} \equiv \frac{\partial}{\partial x_i^{\pm}}$.

Let us recall the action (1.77); the corresponding equations of motion imply that the gravitational field obeys

$$\partial_-^3 g_{++} = 0 \quad . \tag{1.95}$$

This equation can be seen as the on-shell anomaly cancellation. On the other hand, it suggests the expansion

$$g_{++} = J^-(x^+) - 2x^- J^0(x^+) + (x^-)^2 J^+(x^+) \quad . \tag{1.96}$$

Substituting the expansion above into (1.94), we find the following OPE

$$J^a(x)\chi(y) = \frac{l^a(y)}{x^+ - y^+}\chi(y) \quad , \tag{1.97}$$

where

$$l^+(x) = -\partial_- \tag{1.98a}$$

$$l^0(x) = -x^-\partial_- - \lambda \tag{1.98b}$$

$$l^-(x) = -(x^-)^2\partial_- - 2\lambda x^- \quad , \tag{1.98c}$$

are generators of the $SL(2,R)$ algebra. We can also calculate product expansions among the J^a operators, using Eq. (1.92). The resulting $SL(2,R)$ Kac-Moody algebra,

$$J^a(x^+)J^b(y^+) = -\frac{c}{12}\eta^{ab}\frac{1}{(x^+ - y^+)^2} - 2\frac{\varepsilon^{abc}\eta_{cd}}{x^+ - y^+}J^d(y) \quad , \tag{1.99}$$

was originally found by Polyakov[37]. This algebra can be expressed in terms of the mode expansion,

$$J^a(x) = \sum J_n^a e^{inx} \tag{1.100a}$$

from which we have

$$[J_n^a, J_m^b] = f^{abc}J_{n+m}^c + n\kappa\eta^{ab}\delta_{n,-m} \quad , \quad \kappa = c/6 \quad , \tag{1.100b}$$

which is the loop algebra corresponding to the quantized version of[38]

$$\{J^a(x), J^b(y)\} = -2\epsilon^{abc}\eta_{cd}J^d(x)\delta(x-y) + \frac{1}{2}\alpha^2\eta^{ab}\delta'(x-y) \quad , \tag{1.101}$$

obtained from the canonical quantization of the theory; an appropriate choice of η_{ab} and $\varepsilon^{-0+} = 1$ defines the $SL(2,R)$ structure constants (see Appendix A).

The appearance of the $SL(2,R)$ symmetry is quite unexpected. It is not explicit in the defining action (1.78), but has arisen later in a natural way as a consequence of the constraint (1.84), or alternatively of the gravity equation of motion (1.95). The relation between this symmetry and the conformal symmetry is not clear either. Although one can in general define operators such as the currents in (1.96), they are not conserved[39] in arbitrary gauges. On the other hand, the Liouville theory - or gravity in the conformal gauge can be related to the Wess-Zumino-Witten $SL(2,R)$ invariant gauge theory[40]. Choosing appropriately the gauge one can have a relation from that model with either the light cone gauge gravity or with Liouville theory. Moreover, this construction has been proved to be even more general, and the $SL(n,R)$ WZW theory could be related to the Toda theory by means of the so-called Hamiltonian reduction[41]. That construction provides a link between the light-cone formalism and Liouville theory through the $SL(2,R)$ symmetric WZW model. In fact, we will re-obtain several results of the light cone gauge in the Liouville theory, which, as previously discussed, is more appropriate to describe the gravitational dressing of vertex operators.

We consider now a general matter field interacting with gravity. In 2-dimensional space-time, it is sufficiently general to consider a set of Majorana fermions, interacting with gravity via the usual coupling of the zweibein $\hat{e}^{a\mu}$, with the Lagrangian[38]

$$i\overline{\chi}\hat{e}^{a\mu}\gamma_a\partial_\mu\chi = i\psi_-(\partial_+ - g_{++}\partial_-)\psi_- + i\psi_+(\partial_- - g_{--}\partial_+)\psi_+ \quad , \tag{1.102a}$$

where

$$\psi_+ \equiv (2\hat{e}^{+-})^{1/2}\chi_+ \tag{1.102b}$$

$$\psi_- \equiv (2\hat{e}_{-+})^{1/2}\chi_- \tag{1.102c}$$

$$g_{--} = -\frac{\hat{e}^{++}}{\hat{e}^{+-}} \tag{1.102d}$$

$$g_{++} = -\frac{\hat{e}^{--}}{\hat{e}^{-+}} \quad . \tag{1.102e}$$

We can use reparametrization invariance to get rid of g_{--} by fixing it to be zero. In the gauge $g_{--} = 0$, $g_{+-} = 1/2$, the effective action obtained upon integration of the matter fields takes the form

$$S_{eff}(g_{++}) = \frac{-iN}{2}\ln\det(\partial_+ - g_{++}\partial_-) \quad , \tag{1.103}$$

where N is the number of fermions. To implement this gauge, we need a principle of gauge invariance. The total energy momentum tensor is zero or else, the equation of motion of $g_{\mu\nu}$ must hold. Therefore at $g_{--} = 0$, $g_{+-} = 1/2$, we impose

$$\left.\frac{\delta Z}{\delta g_{--}}\right|_{g_{--}=0} = \left.\frac{\delta Z}{\delta g_{+-}}\right|_{g_{+-}=1/2} = 0 \quad . \tag{1.104}$$

We need a set of ghost fields, which we introduce as follows. Reparametrization invariance

$$\delta g_{--} = \nabla_-\epsilon_- \tag{1.105a}$$

$$\delta g_{+-} = \frac{1}{2}(\nabla_+\epsilon_- + \nabla_-\epsilon_+) \quad , \tag{1.105b}$$

requires a ghost pair (-1,2) (that is, ghosts with Lorentz spin -1 and +2, respectively) for g_{--}, with Lagrangian

$$\mathcal{L}_{gh}^{(1)} = \eta_{++}\nabla_-\epsilon_- \quad , \tag{1.106}$$

and a ghost (1,0) for g_{+-}, with Lagrangian

$$\mathcal{L}_{gh}^{(2)} = \zeta(\nabla_+\epsilon_- + \nabla_-\epsilon_+) \quad . \tag{1.107}$$

Going back to (1.104) we have for the first equation

$$T_{++}^{tot}(g, \text{ghosts}) = 0 \tag{1.108}$$

and for the second,

$$\theta(g, \text{ghosts}) = T_{+-}^{tot} = 0 \quad . \tag{1.109}$$

where $T_{\pm}^{tot}$ is the total energy momentum tensor of matter, gravity and ghosts. We have thus

$$T_M + T_{gr} + T_{gh}^{(1)} + T_{gh}^{(2)} = 0 \quad . \tag{1.110}$$

As a consequence, also the sum of the central charges vanishes,

$$c + c_{gr} + c_{gh}^{(1)} + c_{gh}^{(2)} = 0 \quad . \tag{1.111}$$

The above central charges are ($c = \#\text{of bosons} + \frac{1}{2}\#\text{of fermions}$),

$$c_{gh}^{(1)} = -26 \quad (\text{reparametrization ghost}) \tag{1.112}$$

and

$$c_{gh}^{(2)} = -2 \quad (\text{dilatation ghost}) \quad . \tag{1.113}$$

For a ghost field of "spin" $(1-j, j)$, the central charge is computed to be

$$c_j = \pm\frac{1 - 3(1-2j)^2}{2} = \mp[1 + 6j(j-1)] \quad , \tag{1.114}$$

where the lower value is valid for anticommuting ghosts. In order to complete the computations, we need c_{gr}, which depends on the explicit form of the energy momentum tensor of pure gravity, T_{gr}. It can be calculated by canonical methods[39], and one can show that it has a piece of the Sugawara form and an improvement, linear in the current

$$T_{++} = T_S + \partial_+ J^0 \quad , \tag{1.115}$$

where

$$T_S = \frac{-1}{\kappa + 2}\eta_{ab} \colon J^a J^b \colon \tag{1.116}$$

is the Sugawara term and κ is the renormalized coupling constant. The constant $\frac{1}{\kappa+2}$ arises from the Kac-Moody algebra obeyed by $J^a{}'s$, and the Sugawara construction. The Sugawara term contributes to the central charge with the quantity

$$c_S = \frac{3\kappa}{\kappa + 2} \quad , \tag{1.117}$$

as usual. Here we remember that $c = \frac{\kappa \dim G}{c_V + \kappa}$ with $\dim SL(2,R) = 3$, $c_V|_{G=SL(2,R)} = 2$. For the derivative contribution we have

$$J^{0\prime}(x) J^{0\prime}(y) = \frac{\kappa\eta^{00} 2\cdot 3}{2(x-y)^4} = \frac{(-6\kappa)/2}{(x-y)^4} \quad . \tag{1.118}$$

Thus we obtain the gravity contribution to the total central charge as

$$c_{gr} = \frac{3\kappa}{\kappa+2} - 6\kappa \quad . \tag{1.119}$$

Equation (1.111), together with (1.112, 113) and (1.119) implies:

$$c - 28 + \frac{3\kappa}{\kappa+2} - 6\kappa = 0 \quad , \tag{1.120}$$

which means

$$\kappa = -2 + \frac{1}{12}(c - 13 \pm \sqrt{(c-1)(c-25)}) \quad . \tag{1.121a}$$

The plus sign reproduces the semi-classical result for $c \to -\infty$[43]. Thus

$$\kappa + 2 = \frac{1}{12}(c - 13 + \sqrt{(c-1)(c-25)}) \quad . \tag{1.121b}$$

Relation (1.121) makes a profound link between the central charge in flat space, and the renormalized constant κ, which is related to the gravitational dressing of the theory, and is one of the most striking relations in the framework of non-critical strings and two-dimensional gravity. Indeed, from now on we start to obtain the critical behavior of the dressed theory in terms of the critical behavior in flat space. This correspondence turns out to be independent of the method used, since it reappears in the discussion of the Liouville theory later on[13].

One of the characteristics of this relation is the special role played by the flat central charge values $c = 1$ and 25. Between these critical values the relation seems to cease due to the appearance of an imaginary parameter in the action. This signalizes a phase transition, such that for $c \leq 1$ or $c \geq 25$ we have a string theory related to grand unification, while for $1 < c < 25$ we have a strongly coupled phase, probably important for the issue of large N chromodynamics.

We can define "eigenstates" of the operators by the conformal equations

$$g_{++}(x^-, x^+)\phi^{(j)}(y^-, y^+) = \left(2j\frac{x^- - y^-}{x^+ - y^+} - \frac{(x^- - y^-)^2}{x^+ - y^+}\partial_-\right)\phi^{(j)}(y^-, y^+) \quad . \tag{1.122}$$

From (1.116), we have the Virasoro operator

$$L_0^{gr} = \frac{\eta_{ab}}{\kappa+2}\sum :J^a_{-n}J^b_n: \quad . \tag{1.123}$$

The physical states are defined by

$$J_n|Phys\rangle = 0 \quad , \quad n > 0 \tag{1.124a}$$

and

$$L_n|Phys\rangle = 0 \quad , \quad n > 0 \quad . \tag{1.124b}$$

Therefore we act with L_0 on a physical state obtaining the physical conformal dimension Δ which has two pieces, due to the decomposition (1.110), supposing that the physical state is a product of gravity and matter, in the zero ghost sector, namely Δ_0 arising from the matter piece, and $\frac{j(j+1)}{\kappa+2}$, where $j = -\Delta$ from the gravity contribution. Therefore we have

$$\Delta_0 = -\frac{j(j+1)}{k+2} - j = \Delta + \frac{\Delta(1-\Delta)}{k+2} \quad , \tag{1.125}$$

where $k = -\kappa - 4$, thus completing the relation between the free matter theory defined by Δ_0 and c, and the gravity interacting theory defined by Δ and κ.

We further examine the relation between the conformal dimension of the matter fields with and without gravity, by solving Eq. (1.125)

$$\Delta = -\frac{k+1}{2} \pm \frac{1}{2}\sqrt{(k+1)^2 + 4(k+2)\Delta_0} \quad . \tag{1.126}$$

We use Kac's formula[44,45], with

$$c = -\frac{6(k+\frac{1}{2})(k+\frac{3}{2})}{k+2} \quad , \tag{1.127}$$

as follows from Eq. (1.121*b*), with $k = -\kappa - 4$ and

$$\Delta_0 = \frac{c-1}{24} + \frac{1}{4}(n\alpha_+ + m\alpha_-)^2 \quad , \tag{1.128}$$

obtaining for Δ the result

$$\Delta = -\frac{k+1}{2} \pm [n(k+2) - m] \quad , \tag{1.129}$$

which acquires discrete values.

Correlation functions may be computed in closed form, following ref. [46] where the correlators for the Kac-Moody symmetry $SL(2,R)$ have been computed.

Up to now the only issue that has been addressed to is the one of induced 2-D gravity, where conformal invariance was used to obtain information about the operator products and hence about the Green functions of the theory. As an example, 2-D gravity could be described in terms of a SL(2,R) Kac-Moody algebra, and the critical behavior of a given theory in the presence of gravity, Eq. (1.126), could be related to the behavior in flat space.

From now on we generalize this construction[47], by including a graviton/ - gravitino pair, described in the light cone gauge as[48]

$$g^{\mu\nu} = \begin{pmatrix} 0 & 2 \\ 2 & -4g_{++} \end{pmatrix} \quad , \quad \Psi_- = 0 \qquad \Psi_+ = \begin{pmatrix} \psi_+ \\ 0 \end{pmatrix} \quad , \tag{1.130}$$

with the following notation

$$\gamma_+ = 2i\begin{pmatrix} 0 & 0 \\ 1 & 0 \end{pmatrix}, \qquad \gamma_- = -2i\begin{pmatrix} 0 & 1 \\ 0 & 0 \end{pmatrix} \quad . \tag{1.131}$$

The root of the discussion is the issue of the diffeomorphism anomaly, which must vanish for consistency reasons.

We suppose that the theory contains two main symmetries. The first one is a gauge symmetry associated with diffeomorphic transformations surviving the gauge fixing in (1.130). It is given by the transformations

$$\delta\Psi_+ = \epsilon_+\partial_-\Psi_+ - \frac{1}{2}\partial_-\epsilon_+\Psi_+ \tag{1.132a}$$

$$\delta g_{++} = \partial_+\epsilon_+ + \epsilon_+\partial_- g_{++} - \partial_-\epsilon_+ g_{++} \ , \tag{1.132b}$$

where ϵ_+ is a parameter labelling the diffeomorphisms. The second one is local supersymmetry, which after fixing the gauge as in (1.130), reads

$$\delta\Psi_+ = -\partial_+\epsilon - \frac{1}{2}\epsilon\partial_- g_{++} + \partial_-\epsilon g_{++} \tag{1.133a}$$

$$\delta g_{++} = -i\overline{\Psi}_+\gamma_+\epsilon \quad , \tag{1.133b}$$

with $\gamma_-\epsilon = 0 \quad \left(\epsilon = \begin{pmatrix}\varepsilon\\ 0\end{pmatrix}\right)$.

The action including matter fields may be defined as

$$S = \int \mathrm{d}^2x \left\{ \partial_+\phi\partial_-\phi + \frac{i}{2}\overline{\chi}\gamma_-\partial_+\chi + 2\overline{\chi}\Psi_+\partial_-\phi - g_{++}\left((\partial_-\phi)^2 - \frac{i}{2}\overline{\chi}\gamma_-\partial_-\chi\right)\right\} \quad , \tag{1.134a}$$

with the symmetries (1.132) and (1.133) supplemented by

$$\delta\phi = \epsilon_+\partial_-\phi - i\varepsilon\chi \tag{1.134b}$$

$$\delta\chi = \epsilon_+\partial_-\chi + \frac{1}{2}\partial_-\epsilon_+\chi + \varepsilon\partial_-\phi \tag{1.134c}$$

and $\chi = \begin{pmatrix}0\\ \chi\end{pmatrix}$.

The action (1.134a) may be obtained truncating two dimensional supergravity with a matter multiplet[48,49].

Let us turn to the issues of the operator product expansion (OPE) and quantum solution. There is a well known anomaly in the quantum theory of two-dimensional supergravity, generalizing the reparametrization anomaly. It is given by the expression[42]:

$$\delta S = -\frac{c}{24\pi}\int \mathrm{d}^2x\epsilon_+\partial_-^3 g_{++} + \frac{ic}{6\pi}\int \mathrm{d}^2x\bar{\epsilon}\gamma_+\partial_-^2\Psi_+ \quad . \tag{1.135}$$

where δS is the anomalous response of (1.134a) to the variations (1.132). Notice that we can eliminate the anomaly by requiring

$$g_{++} = J^-(x^+) - 2x^-J^0(x^+) + (x^-)^2J^+(x^+) \tag{1.136}$$

$$\psi_+ = \psi^{-1/2}(x^+) + \psi^{1/2}(x^+)x^- \quad . \tag{1.137}$$

We will come back to this later on.

The anomaly may be used to derive the short distance expansion of the fields. Starting from the correlator

$$\langle g_{++}(x_1)g_{++}(x_2)\cdots g_{++}(x_n)\psi_+(y_1)\cdots\psi_+(y_m)\rangle =$$
$$\int \mathcal{D}g_{++}\mathcal{D}\psi_+ g_{++}(x_1)g_{++}(x_2)\cdots g_{++}(x_n)\psi_+(y_1)\cdots\psi_+(y_m)e^{iS_{eff}(g,\psi)} \quad (1.138)$$

we make variations of g_{++} and ψ_+ under (1.132) and (1.133); as a consequence, we have a variation of the action given by the anomaly, Eq. (1.135). Thus, we obtain the Ward identities

$$\frac{ic}{24\pi}\frac{\partial^3}{\partial x^{-3}}\langle g_{++}(x)\underline{X}\rangle = \sum_{i=1}^{n}\frac{\partial}{\partial x^+}\delta(x-x_i)\left\langle\underline{X}_{\hat{x}_i}\right\rangle$$
$$+\sum_{i=1}^{n}\left(\frac{\partial}{\partial x^-}\delta(x-x_i)+\delta(x-x_i)\frac{\partial}{\partial x_i^-}\right)\langle\underline{X}\rangle$$
$$+\sum_{j=1}^{n}\left(\frac{1}{2}\frac{\partial}{\partial x^-}\delta(x-y_j)+\delta(x-y_j)\frac{\partial}{\partial x_i^-}\right)\langle\underline{X}\rangle \quad ,(1.139)$$

and

$$\frac{c}{6\pi}\frac{\partial^2}{\partial x^{-2}}\left\langle\gamma_+\psi_+(x)\underline{X}\right\rangle =$$
$$-2i\sum_{i=1}^{n}\delta(x-x_i)\left\langle\gamma_+\psi_+(x_i)\underline{X}_{\hat{x}_i}\right\rangle + i\sum_{j=1}^{m}(-1)^j\partial_+\delta(x-y_j)\left\langle\underline{X}_{\hat{y}_i}\right\rangle$$
$$-i\sum_{j=1}^{m}(-1)^{j+1}\left\{\frac{1}{2}\delta(x-y_j)\frac{\partial}{\partial y_j^-}+\frac{\partial}{\partial z^-}\delta(z-y_j)\right\}\left\langle\underline{X}_{\hat{y}_i}g_{++}(y_j)\right\rangle \quad ,(1.140)$$

where $\underline{X} = g_{++}(x_1)g_{++}(x_2)\cdots g_{++}(x_n)\psi_+(y_1)\cdots\psi_+(y_m)$, and $\underline{X}_{\hat{x}_i}$ is $\underline{X}$ with the field corresponding to x_i being deleted.

Using the manipulations already used in (1.89) and (1.92), we obtain the operator product expansions

$$\frac{c}{6}g_{++}(x)g_{++}(y) = -\left(\frac{x^--y^-}{x^+-y^+}\right)^2 - \left\{2\left(\frac{x^--y^-}{x^+-y^+}\right) + \frac{(x^--y^-)^2}{x^+-y^+}\frac{\partial}{\partial y^-}\right\}g_{++} \quad (1.141a)$$
$$\frac{c}{6}g_{++}(x)\psi_+(y) = -\left\{\frac{x^--y^-}{x^+-y^+}+\frac{(x^--y^-)^2}{x^+-y^+}\frac{\partial}{\partial y^-}\right\}\psi_+ \quad , \quad (1.141b)$$
$$\frac{c}{6}\psi_+(x)\gamma_+\psi_+(y) = \frac{1}{2}\frac{x^--y^-}{(x^+-y^+)^2} - \frac{1}{2}\left\{\frac{1}{x^+-y^+}+\frac{1}{2}\frac{x^--y^-}{x^+-y^+}\frac{\partial}{\partial y^-}\right\}g_{++} \quad (1.141c)$$

The above expressions imply a super Kac-Moody algebra generalizing the Neveu -Schwarz algebra of superstring theory; we define the expansions

$$J^a(x^+) = \sum_{n=-\infty}^{\infty} J_n^a e^{inx^+} \quad , \quad (1.142)$$

$$\psi_+^r(x^+) = \sum_{n=-\infty}^{\infty} \psi_n^r e^{inx^+} \quad . \quad (1.143)$$

In terms of the components J_n^a and ψ_n^r, the short distance expansions turn into the (anti) commutator relations (after a suitable rescaling)

$$[J_n^a, J_m^b] = f^{abc} J_{n+m}^c + \kappa n \eta^{ab} \delta_{n,-m} \quad , \tag{1.144a}$$

$$[J_n^a, \psi_m^r] = h^{ars} \psi_{n+m}^s \quad , \tag{1.144b}$$

$$\{\psi_n^r, \psi_m^s\} = h^{rsc} J_{n+m}^c + \kappa n \eta^{rs} \delta_{n,-m} \quad , \tag{1.144c}$$

where $f^{+-0} = 2$, (f antisymmetric), $\eta^{00} = -\frac{1}{2}$, $\eta^{(1/2)(-1/2)} = -4$, $\eta^{+-} = 2$, $h^{+,-1/2,1/2} = -1$, $h^{-,1/2,-1/2} = 1$, $h^{0,\pm 1/2,\pm 1/2} = \pm 1/2$, $h^{1/2,1/2,+} = -2$, $h^{1/2,-1/2,0} = -2$ and $h^{-1/2,-1/2,-} = -2$.

The finite algebra is realized by the following set of operators

$$l^+ = -\partial_- \tag{1.145a}$$

$$l^0 = -y^- \partial_- - \frac{1}{2}\theta\partial_\theta + j \tag{1.145b}$$

$$l^- = -(y^-)^2 \partial_- - y^- \theta\partial_\theta + 2jy^- \tag{1.145c}$$

$$\mathcal{P}^{1/2} = \partial_\theta + \theta\partial_- \tag{1.145d}$$

$$\mathcal{P}^{-1/2} = y^- \partial_\theta + y^- \theta\partial_- - 2j\theta \tag{1.145e}$$

with the Casimir c_g given by

$$c_g = \eta_{ab} \mathcal{J}^a \mathcal{J}^b + \eta_{rs} \mathcal{P}^r \mathcal{P}^s = j(j + \frac{1}{2}) \quad . \tag{1.146}$$

The energy momentum tensor generating the general coordinate transformations may be readily obtained. It has the Sugawara form, plus a linear term in J^0,

$$T(x) = \frac{1}{\kappa + \frac{3}{2}} \{ : J^a(x^+) J^b(x^+) : \eta_{ab} + : \psi^r(x^+) \psi^s(x^+) : \eta_{rs} \} + J^{0'}(x^+) \quad . \tag{1.147}$$

The coefficient $\frac{1}{\kappa + \frac{3}{2}}$ has been obtained by requiring a super Kac-Moody Virasoro algebra

$$[L_n, J_m^a] = m J_{n+m}^a \quad , \tag{1.148}$$

where we have, for the Sugawara piece of the energy momentum tensor, the expression

$$T_{Sug}(x^+) = \sum_{n=-\infty}^{\infty} L_{-n} e^{inx^+} \quad . \tag{1.149}$$

We may now consider the interaction of matter fields and super gravity. The full Virasoro algebra can be obtained by computing the expectation value of the product of the energy momentum tensors at nearby points. After a lengthy calculation one obtains the following result for the central charge:

$$c_{Sug} = \frac{\kappa}{\kappa + \frac{3}{2}} \quad . \tag{1.150}$$

In order to balance the central charges, we have to consider the following set of fields

i) matter fields, with central charge c;

ii) diffeomorphism ghosts, with central charge -15;

iii) superconformal ghosts, with central charge $-\frac{5}{2}$;

iv) the Sugawara piece of the supergravity multiplet contributing $\frac{\kappa}{\kappa+\frac{3}{2}}$;

v) a linear term contributing -6κ to the central charge.

With the above balance, we have (recalling that in the supersymmetric case, we use $c = \frac{3}{2}\hat{c}$, where $\hat{c} = d$)

$$\kappa + \frac{3}{2} = \frac{\hat{c} - 5 \pm \sqrt{(\hat{c}-1)(\hat{c}-9)}}{8} \quad , \tag{1.151}$$

(recall the bosonic result (1.121)).

With the realization (1.145), the Casimir $c_g = \tau^a \tau^a$, where τ^a are the generators of the symmetry group, and $j = -\Delta$, the anomalous dimensions of the free theory Δ_0 has a relation with the gravitationally interacting theory given by

$$\Delta_0 = \frac{\Delta(\Delta - \frac{1}{2})}{\kappa + \frac{3}{2}} + \Delta \quad . \tag{1.152}$$

Notice that $\Delta = \frac{1}{2}$ is a fixed point (in the same way as $\Delta = 1$ was a fixed point in the bosonic theory). Conclusions similar to the bosonic case may be drawn. A new series of critical indices emerges, and may be compared with the usual results.

The previous discussion of quantum gravity gives us hope that the full problem may be understood by the methods of conformal field theory. Nevertheless, the quantization in the light-cone gauge poses several restrictions. The first concerns the role played by the cosmological constant, which has been discussed in the framework of this gauge in [38,39], but for the computation of correlation functions, it plays a rather unclear role. Moreover, the plethora of information on Liouville theory, both at classical as well as in the quantum level, together with its role in the geometry of Riemann surfaces, practically force us to face a comparison of the results obtained in the light cone gauge with further discussion in the conformal gauge, and compels us to re-obtain all relevant results directly in the Liouville theory. Therefore, we go back to the Liouville theory, and try first to re-obtain the interesting results of the last few sections in the conformal gauge, following the very interesting approach of David, Distler and Kawai (DDK)[13] in the purely bosonic case, as well as Distler, Hlousek and Kawai (DHK)[50] in the supersymmetric case. Moreover, their approach naturally generalizes to higher-genus Riemann surfaces, while the light cone approach is characteristic for the 2 dimensional Minkowski space.

In string theory, or when analysing random surfaces, we are faced with the problem of two-dimensional quantum gravity in an arbitrary genus surface. The sum over all the topologies is an essential ingredient in string theories; since the sum over genera is badly behaved, one expects that each term is not a good representative of the full theory, and that non-perturbative effects play a major role. Moreover, as we have already seen, Eq. (1.121) is not real for $1 < c < 25$, and one expects a phase transition at $c = 1$. Therefore, it is possible that string theory has a phase transition at this point, where the string *degenerates to a branched polymer or is in a crumpled phase*[51]. Therefore at this point one is led to the study of random surfaces, and related critical phenomena[52].

The propagation of strings along random surfaces is a rather general problem in physics. It describes low energy QCD[53] in the large N limit. In fact we may argue that the large N limit is related to two-dimensional gravity (Liouville theory)[54]. Some statistical models have been studied in this context, such as the Ising model on a random lattice[55,56], and the $Q = 0, 1$ Potts models[57]; polymers on a random surface[58], the $O(N)$ model[59], the Yang Lee[60] edge singularity[61], as well as D=1 strings[62] including nonperturbative effects[63] and relations to the KdV hierarchy[64]; W gravity and three-dimensional Ising model[65], non-critical strings[66], and topological field theory[67].

The starting point is the consideration of 2-D gravity with a cosmological term

$$S = \frac{1}{2}\int d^2\xi\sqrt{g}g^{ab}\partial_a X^\mu \partial_b X_\mu + \mu\int d^2\xi\sqrt{g} \quad . \tag{1.153}$$

We have omitted the modulus in the square root of g because from now on we will be dealing with Euclidian space only.

For a genus-h surface, we have the Gauss-Bonnet identity

$$\frac{1}{8\pi}\int d^2\xi\sqrt{g}R = 1 - h \quad , \tag{1.154}$$

such that the Einstein gravity term in two dimensions is trivial.

In order to take into account Weyl and reparametrization transformations, we need to introduce ghosts, (the previously discussed b, c systems), and integrate over them. However, in general, the conformal mode only cancels if we tune the central charge of the matter field X_μ, which is D, to cancel against the ghost contribution, -26, as is well known. Otherwise, we are forced to consider the Liouville mode, whose cation is given in Eq. (1.56). If we compute the energy momentum tensor, we find

$$T_{zz} = -\frac{1}{2}[(\partial\phi)^2 + Q\partial^2\phi] \quad , \tag{1.155}$$

from where we have the central charge $c_\phi = 1 + 3Q^2$; since $c_{tot} = c_\phi + c - 26 = 0$, we have

$$Q = \sqrt{\frac{25 - c}{3}} \quad . \tag{1.156}$$

Critical exponents of the theory may be computed as follows. Consider the partition functions as a function of the area[13] A,

$$Z(A) = \int \mathcal{D}\phi\mathcal{D}X e^{-S}\delta\left(\int e^{\gamma\phi}\sqrt{\hat{g}}d^2\xi - A\right) \sim KA^{\gamma_h - 3} \quad , \tag{1.157}$$

defining the string susceptibility[68] γ_h. Consider now the shift

$$\phi \to \phi + \frac{\rho}{\gamma} \quad , \tag{1.158}$$

with ρ a constant[13]. We have a corresponding shift in the action given by

$$S \to S - Q(1-h)\frac{\rho}{\gamma} \quad ; \tag{1.159}$$

the δ-function in (1.157) transforms as

$$\delta\left(\int e^{\gamma\phi}\sqrt{\hat{g}}d^2\xi - A\right) \to e^{-\rho}\delta\left(\int e^{\gamma\phi}\sqrt{\hat{g}}d^2\xi - e^{-\rho}A\right) \quad . \tag{1.160}$$

Therefore

$$Z(A) = e^{[\frac{Q(1-h)}{\gamma}-1]\rho} Z(e^{-\rho}A) \quad , \tag{1.161}$$

from which we obtain

$$Z(A) = KA^{(1-h)\frac{Q}{\gamma}-1} \quad . \tag{1.162}$$

For the susceptibility γ_h as a function of the genus h we find the expression[7)]

$$\gamma_h = \frac{(1-h)}{12}[c - 25 - \sqrt{(25-c)(1-c)}] + 2 \quad ; \tag{1.163}$$

compare with (1.121-126) for genus $h = 0$ we find also[13]

$$\gamma_h - 2(1-h)(\gamma_0 - 2) \quad . \tag{1.164}$$

We have agreement with the semiclassical result[69]

$$\gamma_h = (1-h)\frac{(c-19)}{6} + 2 \quad . \tag{1.165}$$

The above result, obtained by standard quantum field methods in the light-cone gauge, as well as by simple manipulations of the partition function in the conformal gauge, will be generalized for the expression of correlation functions involving vertex operators dressed by gravity in the continuum formulation of non-critical strings. As already quickly reviewed in the supersymmetric light-cone gauge, a supersymmetric version of the continuum result will also be obtained, some conclusions will be drawn as well in the case of $N = 2$ supersymmetry. The results nicely match those obtained from the discrete matrix model approach, in the bosonic as well as in the $N = 1$ supersymmetric case. The hope is that one can pass the $c = 1$ barrier (resp. $c = 3/2$ if one includes supersymmetry) in order to understand the role of strings in the other phase, which presumably describes strong interactions. This will be briefly discussed in Chapter 8. In the $N = 2$ supersymmetric case the barrier is smooth since both transition points coalesce, and the computations do not show any special feature beyond this point in the continuum formulation.

---

7) In fact this expression is only correct if there are no negative metric states, and the behaviour is dominated by the above simple scaling of the area. Otherwise a correction arises, and we have $\gamma_{h=0} = \frac{c-1-24\Delta-\sqrt{(1-c)(25-c-24\Delta)}}{12(1-\Delta)}$ where Δ assumes the minimal possible value.

2. Correlation Functions in the Bosonic Theory (Continuum Approach for Spherical Topology)

2.1. Introduction and General Overview

Two-dimensional gravity describes important phenomena[11,43], such as random surfaces and string theories away from criticality. Up to recently, the most efficient means of extracting results from the theory of two-dimensional gravity was the matrix model approach[56,70], to be discussed later. In this theory, one obtains a series expansion in the genus with determined coefficients. However, that theory suffers some drawbacks, such as difficulties in the interpretation of real solutions of the Painlevé equation[70] in the light of the "Schwinger-Dyson" equations[71], or the lack of a super matrix model formulation, due to the well known difficulty of defining lattice fermions, which is necessary to deal with two-dimensional supergravity. Therefore, alternative computations and their detailed comparison is mandatory in order that the correct points be settled. Thus, we use the zero mode integration technique for Liouville[72–78] and super Liouville[79–82] theory, in order to be able to compute correlators of dressed vertex operators in the continuum, as well as in the matrix model/KdV[63,71,83] approach, or also with a super eigenvalue model[84], to deal with the discrete theory in subsequent chapters.

In the previous Chapter we have reviewed several pieces of works dealing with the quantization of gravity; in the light-cone gauge it was possible to compute all relevant correlation functions exactly, in the purely bosonic, as well as in the supersymmetric case. However, it is not always clear what is the role of the cosmological constant in this gauge (see discussion in [39]). In the conformal gauge, many results were rederived in [13].

In this Chapter we start with the purely bosonic case, which is well established. Later, we shall develop the $N = 1$ supersymmetric case and compare with the super eigenvalue model of ref. [84].

Observing formulae (1.121) and (1.151) we find two critical values for c, the lowest ones being $c = 1$ for $N = 0$, and $c = \frac{3}{2}$ for $N = 1$. Below such points, a formulation of the theory in terms of vertex operators dressed by the Liouville (resp. later super Liouville) fields make sense. In the present Chapter we shall deal with the formulation of the bosonic case. The results are very rewarding. Even before comparing to the matrix models results, namely staying in the framework of the theory in the continuum formulation we may analyse, as an example, the spectrum of intermediate states. In the former case, namely the bosonic theory, we reobtain the discrete states from Klebanov and Polyakov[85], and Witten[86]. In the latter case, that is $N = 1$ supersymmetry to be analysed in Chapter 5, a similar structure arises, and one foresees again the appearance of discrete states. In the $N = 2$ supersymmetric theory to be analysed in Chapter 7 we have a completely new picture[87], in a sense reminiscent of the fact that in the critical case most of the correlators vanish[88,89]. In this case, the two critical points coincide at $\hat{c} = 1$, and the spectrum is not of the rich type above.

The extremely rich structure of the results thus obtained permits us to draw several new conclusions, and provides a clean method of computation which can be used to check the matrix model results as well as to extend them. The highly non-trivial comparison provides a valuable guide towards an understanding of the non-critical dynamics of strings and superstrings.

2.1.1. Model Description

In this section we are interested in the calculation of correlation functions in two-dimensional bosonic theories conformally coupled to gravity, defined on the sphere, i.e. at genus-zero Riemann surfaces. Such calculations have been performed[70,83,90] in a discrete formulation via matrix models, in cases where the central charge of the matter sector c is less than unit ($c < 1$). In the discrete formulation the calculation is exact (in the sense that it provides a genus expansions to all orders), leading to extremely elegant and enlighting results[70] (for the $c = 1$ case[91], see Sect. 3.7.).

It would be highly desirable to reproduce those results in the continuum approach. The light cone gauge is not very promising for this type of calculus due to the lack of an Euclidean formulation of this gauge, the inherent non-locality, and the difficulty in dealing with the cosmological constant. Thus we remain with the conformal gauge, which in the DDK[13] formulation was useful to compute critical exponents. However, it is important to stress that in the DDK formulation no relevant information about the interaction of Liouville theory, represented by the cosmological term, has been used. The critical exponents have been obtained only taking into account the translational invariance of the measure $[\mathcal{D}_{\hat{g}}\phi]$. To obtain more precise informations, as for example correlation functions, it would be necessary to treat the cosmological constant with more care. However, as we have seen, in spite of a lot of efforts, the quantization of Liouville theory was a difficult task that, for many years, resisted an adequate solution (see however [22-24], [27,28]). In other words, one did not know any non-perturbative way of calculating correlation functions, taking into account the cosmological constant, until Goulian and Li[73] developped a technique, which allowed them to calculate 0- 2- and 3-point functions for minimal unitary models ($c < 1$) conformally coupled to two dimensional gravity. Later, we shall see that the results are in complete agreement with the ones obtained via matrix models.

The main ingredient in this technique is the integration over the zero mode in the Liouville field (ϕ_0). The importance of this zero mode integration had already been noticed previously by Gupta, Trivedi and Wise[72]. They have used it to obtain non-perturbative partition functions for fixed area ($\mathcal{Z}(A)$) in conformal theories coupled to two-dimensional gravity, only for certain values of the central charge.

The aforementioned results are obtained upon the use of the DDK formalism, where the total partition function on the sphere is

$$\mathcal{Z} = \frac{1}{V_{SL(2,\mathbb{C})}} \int \mathcal{D}_{\hat{g}}\phi \mathcal{D}_{\hat{g}} X \, \mathrm{e}^{-(S_L(\phi,\hat{g}) + S_M(X,\hat{g}))} \quad , \tag{2.1}$$

where the Liouville action

$$S_L = \frac{1}{8\pi} \int \mathrm{d}^2 w \sqrt{\hat{g}} [\hat{g}^{ab} \partial_a \phi \partial_b \phi - Q \hat{R} \phi + 8\pi \bar{\mu} \mathrm{e}^{\alpha\phi}] \tag{2.2}$$

was already introduced in the previous Chapter, and μ and Q from Eq. (1.56) have been renormalized respectively into $\bar{\mu}$ and Q. The latter is given as a function of c as

$$Q = 2\sqrt{2 + \alpha_0^2} \quad , \tag{2.3}$$

where we define $c = 1 - 12\alpha_0^2$ and substitute in (1.156). The matter action, with central charge $c \leq 1$ is represented by S_M, and $V_{SL(2,\mathbb{C})}$ denotes the volume of the residual $SL(2,\mathbb{C})$ symmetry in the conformal gauge. To fix the quantum value of α, which replaces the classical value γ, we impose that the Liouville term constitutes a marginal deformation of the conformal theory, in order not to spoil conformal invariance, and therefore must have conformal dimension $(\Delta, \overline{\Delta}) = (1,1)$. Using the methods of [44] (for a simple introduction see appendix B), we impose

$$\Delta\left(\mathrm{e}^{\alpha\phi}\right) = -\frac{1}{2}\alpha(\alpha + Q) = 1 \quad , \tag{2.4}$$

and find as a result

$$\alpha_\pm = -\frac{Q}{2} \pm |\alpha_0| \quad . \tag{2.5}$$

With these definitions we are in a position to defining the zero mode integration technique, and computing the first few critical exponents.

The reason for choosing α_+ is two-fold. The first concerns the classical limit $\alpha_0 \to \infty$, where the exponential term in (2.2) must be finite. The second relies on the discussion following (1.72): a quantum operator $\mathrm{e}^{\alpha\phi}$ should obey the inequality $\alpha \geq -Q/2$.

2.1.2. Zero Mode Integration Technique and First Critical Exponents

To illustrate the zero mode integration technique we shall analyze only the Liouville partition function which can be written as

$$\mathcal{Z}_L = \frac{1}{V_{SL(2,\mathbb{C})}} \int \mathcal{D}_{\hat{g}}\phi \, \mathrm{e}^{-S_L} = \frac{1}{V_{SL(2,\mathbb{C})}} \int_0^\infty \mathrm{d}A \mathrm{e}^{-\bar{\mu}A} \mathcal{Z}_L(A) \quad , \tag{2.6}$$

with

$$\mathcal{Z}_L(A) = \int \mathcal{D}_{\hat{g}}\phi \, \delta\left(\int \mathrm{d}^2 w \sqrt{\hat{g}} \mathrm{e}^{\alpha_+\phi} - A\right) \mathrm{e}^{\frac{1}{8\pi}\int \mathrm{d}^2 w \sqrt{\hat{g}}[\phi\hat{\Delta}\phi + Q\hat{R}\phi]} \quad . \tag{2.7}$$

The idea is to decompose the Liouville field (ϕ) as a sum of the zero modes (ϕ_0) of the Laplacian ($\hat{\Delta}$), which is a constant on the sphere, and orthogonal modes ($\tilde{\phi}$). It is easy to see, after integrating over the zero mode, that we obtain:

$$\mathcal{Z}_L(A) = \frac{A^{\frac{Q}{\alpha_+}-1}}{\alpha_+} \int \mathcal{D}_{\hat{g}}\tilde{\phi} \left(\int \mathrm{d}^2 w \sqrt{\hat{g}} \mathrm{e}^{\alpha_+ \tilde{\phi}} \right)^{-\frac{Q}{\alpha_+}} \mathrm{e}^{\frac{1}{8\pi} \int \mathrm{d}^2 w \sqrt{\hat{g}} \tilde{\phi} \hat{\Delta} \tilde{\phi}} \quad , \tag{2.8}$$

where we used the fact that on the sphere $\int \mathrm{d}^2\xi \hat{R}\sqrt{g} = 8\pi$ (compare with Eq. (1.154)). We drop the tilde on ϕ and rewrite the partition function as

$$\mathcal{Z}_L = \frac{1}{V_{SL(2,\mathbb{C})}} \int_0^\infty \mathrm{d}A A^{\frac{Q}{\alpha_+}-1} \mathrm{e}^{-\bar{\mu} A} \left\langle \left(\int \mathrm{d}^2 w \sqrt{\hat{g}} \mathrm{e}^{\alpha_+ \phi} \right)^{-\frac{Q}{\alpha_+}} \right\rangle_0 \quad . \tag{2.9}$$

In the above equation $\langle \quad \rangle_0$ means that the expectation value is calculated considering the cosmological constant to be zero, ($\bar{\mu} = 0$). Notice that $\mathcal{Z}_L$ can still be rewritten as

$$\mathcal{Z}_L = \frac{(\bar{\mu})^{-\frac{Q}{\alpha_+}} \Gamma\left(\frac{Q}{\alpha_+}\right)}{V_{SL(2,\mathbb{C})}} \left\langle \left(\int \mathrm{d}^2 w \sqrt{\hat{g}} \mathrm{e}^{\alpha_+ \phi} \right)^{-\frac{Q}{\alpha_+}} \right\rangle_0 \quad , \tag{2.10}$$

where we used

$$(\bar{\mu})^{-\frac{Q}{\alpha_+}} \Gamma\left(\frac{Q}{\alpha_+}\right) = \int_0^\infty \mathrm{d}A A^{\frac{Q}{\alpha_+}-1} \mathrm{e}^{-\bar{\mu} A} \quad . \tag{2.11}$$

From above we see that the integration over the zero mode ϕ_0 results in "bringing down" the interaction (cosmological term) to the integrand in the path integral, i.e., the calculation of the 0-point (partition) function for the interacting theory ($\mu \neq 0$) is equivalent to the computation of a ($-Q/\alpha_+$)-point function in the free theory. This argument, however, is valid only for integer non-negative values (n) of $-Q/\alpha_+$; thus we assume for the time being $-Q/\alpha_+ = n$. As Q and α_+ are functions of c we are led to some "magic" values given by $c = \frac{6n^2+25(1-n)}{(n-1)}$.

At this point of the computation, the zero mode integration seems to be a useless technique. Indeed, in spite of the fact that we expect a calculable result for c assuming the above values, one has singularities of the type $\Gamma(-n)$. For this reason, in ref. [72] a more conservative attitude has been adopted, and only fixed area correlators have been computed. These are singularity free. We use free propagators ($\bar{\mu} = 0$), and the method explained in ($B.30$) in order to compute the correlators. Thus we obtain

$$\mathcal{Z}_L(A) = \frac{A^{-n-1}}{\alpha_+} \prod_{i=1}^{n} \int \mathrm{d}^2 w_i \prod_{i<j} |w_i - w_j|^{-2\alpha_+^2} \quad . \tag{2.12}$$

Note that the above result already represents some improvement as compared to the relations obtained by DDK[13] and KPZ[42] where they have only the scaling behavior $\mathcal{Z}_L(A) \sim kA^{-n-1}$. Indeed, further advance in this problem was achieved with the work of Goulian and Li[73]. Before going to this point it is interesting to understand the origin and the meaning of the singularities which had already been observed semi classically[21].

If we introduce a cut-off λ, $(\lambda \to 0)$ for small areas, in (2.11) we have:

$$\bar{\mu}^n \Gamma(-n) \sim \int_\lambda^\infty \mathrm{d}A A^{-n-1} \mathrm{e}^{-\bar{\mu}A} \quad . \tag{2.13}$$

Since $\bar{\mu} > 0$ we can calculate the most divergent term of the integral (2.13) and the λ (cut-off) independent term:

$$\bar{\mu}^n \Gamma(-n) \sim \begin{cases} -\ln\lambda - \ln\bar{\mu} \quad , \quad n = 0 \\ \frac{\lambda^{-n}}{n} + (-)^{n+1} \frac{\bar{\mu}^n}{n!} \ln\bar{\mu} \quad , \quad n = 1, 2, \cdots \quad . \end{cases} \tag{2.14}$$

The above expression is well defined for finite λ, and we infer that the surfaces of small areas are the origin of the singularities in $\Gamma(-n)$, which are essentially classic. According to the delta function in (2.7) small areas are equivalent to configurations where $\phi \to \infty$ (remember that $\alpha_+ < 0$) and those are exactly the configurations that minimize the Liouville action and give the major contribution to the path integral. Because of its classical origin, those singularities could be detected in a semiclassical[21] analysis. Another important feature of such singularities is that the cut-off independent term is universal and could for example be obtained via a cut-off ϵ in the central charge:

$$\bar{\mu}^{n-\epsilon} \Gamma(-n+\epsilon) \sim (-)^{n+1} \frac{\bar{\mu}^n}{n!} \frac{1}{\epsilon} + (-)^{n+1} \frac{\bar{\mu}^n}{n!} \ln\bar{\mu} \quad . \tag{2.15}$$

The universality of the finite piece suggests that the implicit singularity in the factor $\mu^n \Gamma(-n)$ be disregarded such that only the finite piece has physical content. As we shall see in the next section with this interpretation we shall be able to deal with the quantum Liouville theory.

2.2. The 3-Point Function

Before describing the zero mode technique in more detail, it is important to say some words about the results of Bershadsky and Klebanov[92] who have calculated the partition function on the torus in the case of a boson compactified into a circle of radius R coupled to two-dimensional gravity via Liouville theory. They have reproduced the R dependence of the partition function, previously obtained via matrix models. On the torus the Einstein action vanishes, therefore $n = 0$ (recall that in general $n = -(1-h)Q/\alpha_+$, see (1.162) and remember that $\gamma = \alpha_+$). Thus we use $\bar{\mu}^{-n}\Gamma(n) \sim 1/\epsilon - \ln\bar{\mu}$ $(\epsilon \to 0)$. The singular term will be disregarded, i.e., $\bar{\mu}^{-n}\Gamma(n) \sim -\ln\bar{\mu}$, as a consequence of the aforementioned non universality (see also [93]).

The starting point is the result obtained in the matrix model approach for the partition function on a torus of radius R coupled to gravity by the authors of ref. [94] and reads

$$Z = -\frac{1}{24} \left[\frac{R}{\sqrt{\alpha}} + \frac{\sqrt{\alpha}}{R} \right] \ln\bar{\mu} \quad . \tag{2.16}$$

In the continuum version, the sum over surfaces is given by the expression

$$Z = \int d^2\tau \int \mathcal{D}X\mathcal{D}\phi\mathcal{D}b\mathcal{D}c\, e^{-S_X - S_\phi - S_{gh}} \quad , \tag{2.17}$$

where X denotes the matter fields, and τ is the moduli parameter which in the case of genus one is the two-dimensional region $\tau = \tau_1 + i\tau_2$, $\tau_2 > 0$, $|\tau| > 1$, $-\frac{1}{2} \leq \tau_1 < \frac{1}{2}$. This result has been obtained again performing the integral over the moduli with a proper normalization in ref. [92].

For fixed area, the sum over surfaces is given by the expression

$$Z(A) = \int d^2\tau \int \mathcal{D}X\mathcal{D}\phi\mathcal{D}b\mathcal{D}c\, \delta\left[\int d^2\sigma \mathcal{O}_{min} e^{\beta_{min}\phi} - A\right] e^{-S_X - S_\phi - S_{gh}} \quad . \tag{2.18}$$

The subscript *min* means the operator with lowest conformal dimension. Moreover in the torus we can choose a flat background metric. Bershadsky and Klebanov used relations obtained among different partition functions by [95], and found for the $c < 1$ (A_{p-1}, A_{q-1}) models the result

$$Z = \frac{(p-1)(q-1)}{24(p+q-1)A} \quad , \tag{2.19}$$

which also agrees with matrix models predictions. This is a clear sign of the correctness of the procedure, and we proceed to calculate correlation functions for unitary minimal models coupled to gravity. Suppose $\mathcal{O}_{rr'}$ is a primary operator in the Kac table[44] with quantum numbers r and r'. In the DDK formalism, used in ref. [92] each $\mathcal{O}_{r_i r'_i}$ operator which appears in the correlation function should be integrated over the two-dimensional surface and dressed by the Liouville field; this means that one is interested in correlation functions such as:

$$\left\langle \prod_{i=1}^{N} \int d^2 z_i \sqrt{\hat{g}} \mathcal{O}_{r_i r'_i} e^{\beta_i \phi(z_i)} \right\rangle_{S_L + S_M} = \prod_{i=1}^{N} \int d^2 z_i \sqrt{\hat{g}} \langle \mathcal{O}_{r_i r'_i} \rangle_{S_M} \langle e^{\beta_i \phi(z_i)} \rangle_{S_L} \tag{2.20}$$

where the matter and gravitational contributions are calculated separately. The dressing β_i is determined imposing that $\mathcal{O}_{r_i r'_i} e^{\beta_i \phi(z_i)}$ be an operator with conformal dimension one. If $h_{ii'}$ is the dimension of $\mathcal{O}_{r_i r'_i}$ we have:

$$\Delta(\mathcal{O}_{r_i r'_i} e^{\beta_i \phi(z_i)}) = h_{ii'} - \frac{\beta_i}{2}(\beta_i + Q) = 1 \quad , \tag{2.21}$$

where we have used the free field representation for the Liouville field in the presence of a constant background charge Q.

Calculating the gravitational contribution to the correlation function (2.20) is the most difficult point, the path integral representation is

$$\left\langle \prod_{i=1}^{N} e^{\beta_i \phi(z_i)} \right\rangle_{S_L} = \int \mathcal{D}_{\hat{g}}\phi\, e^{-\frac{1}{8\pi}\int d^2 w \sqrt{\hat{g}}[\hat{g}^{ab}\partial_a\phi\partial_b\phi - Q\hat{R}\phi + 8\pi\bar{\mu}e^{\alpha+\phi}]} \prod_{i=1}^{N} e^{\beta_i \phi(z_i)} . \tag{2.22}$$

At this point we use the zero mode technique. Notice that it is not necessary to introduce the notion of partition function for fixed areas to perform the zero mode integration; if we write $\phi = \tilde{\phi} + \phi_0$, we perform the integral

$$\int_{-\infty}^{\infty} \mathrm{d}\phi_0 \mathrm{e}^{\phi_0\left(\sum_{i=1}^{N} \beta_i + Q(1-h)\right) - \mathrm{e}^{\alpha_+ \phi_0}\left(\bar{\mu}\int \mathrm{d}^2 w\sqrt{\hat{g}}\mathrm{e}^{\alpha_+ \tilde{\phi}}\right)} = \frac{\Gamma(-s)}{\alpha_+}\left(\bar{\mu}\int \mathrm{d}^2 w\sqrt{\hat{g}}\mathrm{e}^{\alpha_+ \tilde{\phi}}\right)^s \tag{2.23}$$

where we used

$$\int \mathrm{d}x \mathrm{e}^{ax} \mathrm{e}^{-b\mathrm{e}^{\gamma x}} = \frac{1}{\gamma} b^{-\frac{a}{\gamma}} \Gamma\left(\frac{a}{\gamma}\right) \tag{2.24}$$

and

$$s = -\frac{1}{\alpha_+}\left(\sum_{i=1}^{N} \beta_i + Q(1-h)\right) \quad . \tag{2.25}$$

On the sphere ($h = 0$), we have (we rename $\tilde{\phi}$ by ϕ for convenience)

$$\left\langle \prod_{i=1}^{N} \mathrm{e}^{\beta_i \phi(z_i)} \right\rangle_{S_L} = (\bar{\mu})^s \frac{\Gamma(-s)}{\alpha_+} \int \mathcal{D}_{\hat{g}} \phi \mathrm{e}^{-S_L(\mu=0)} \left[\int \mathrm{d}^2 w\sqrt{\hat{g}}\mathrm{e}^{\alpha_+ \phi}\right]^s \prod_{i=1}^{N} \mathrm{e}^{\beta_i \phi(z_i)}, \tag{2.26}$$

with

$$s = -\frac{1}{\alpha_+}\left(\sum_{i=1}^{N} \beta_i + Q\right) \quad . \tag{2.27}$$

Notice that for $N = 0$ we obtain the result (2.10) from last section for the partition function on the sphere. Actually, at this point it is worthwhile noticing that the only dimensionful parameter in the N-point correlator is the cosmological constant $\bar{\mu}$ ($[\bar{\mu}] = [l]^{-2}$) therefore the dimension of the correlator is given by the factor $\bar{\mu}^s$, i.e.

$$\mathcal{A}_N \sim \bar{\mu}^{-\frac{1}{\alpha_+}(Q + \sum_{i=1}^{N} \beta_i)} \quad , \tag{2.28}$$

thus, the scaling of the partition function ($N = 0$ point correlator) in terms of $\bar{\mu}$ is given by:

$$\mathcal{Z} = \mathcal{A}_0 \sim \bar{\mu}^{-Q/\alpha_+} \quad , \tag{2.29}$$

where the factor $-Q/\alpha_+$ is called the gravitational scaling of the partition function $\mathcal{Z}$. In other words we can associate the factor $\bar{\mu}^{-Q/\alpha_+}$ with the measure of the path integral. Therefore for the one point correlator we expect the scaling behavior

$$\langle \mathcal{O}_i \rangle = \langle \mathcal{O}_{r_i r_i'} \mathrm{e}^{\beta_i \phi} \rangle \sim \bar{\mu}^{-\frac{Q}{\alpha_+} - \frac{\beta_i}{\alpha_+}} \quad . \tag{2.30}$$

We conclude that the gravitational scaling of $\mathcal{O}_i$ is $-\beta_i/\alpha_+$. These definitions will be useful later for a comparison with matrix model results. So far what is really important is the way of facing the expression (2.26). We shall hold a positive attitude towards the singularity represented by $\Gamma(-s)$, not restricting to correlation functions for fixed areas, putting aside for a while the existence of the singularity and proceeding with the calculation. Assuming s to be a non-negative integer we use the free propagator in the calculation of the above expectation value continuing afterwards the final correlation functions to arbitrary complex values of s. Therefore we shall use from now on the free field result

$$\underbrace{\phi(z)\phi(w)} = -\ln(z-w) \quad . \tag{2.31}$$

Performing all possible contractions, and fixing the gauge related to the $SL(2,\mathbb{C})$ symmetry, (what enables us to fix $z_1 = 0$, $z_2 = 1$, $z_3 = \infty$) we obtain for the 3-point function ($N = 3$):

$$\langle A_{r_1} A_{r_2} A_{r_3} \rangle = \left\langle \prod_{i=1}^{3} \mathcal{O}_{r_i r_i'} \right\rangle_{S_M} (\bar{\mu})^s \frac{\Gamma(-s)}{\alpha_+} \prod_{i=1}^{s} \int \mathrm{d}^2 w_i |w_i|^{2\alpha} |1-w_i|^{2\beta} \prod_{i<j} |w_i - w_j|^{4\rho} \tag{2.32}$$

where $A_{r_i} = \mathcal{O}_{r_i r_i'} \mathrm{e}^{\beta_i \phi(z_i)}$, $\alpha = -\alpha_+ \beta_1$, $\beta = -\alpha_+ \beta_2$, $\rho = -\alpha_+^2/2$. Fortunately the s interrelated integrals have been calculated by Dotsenko and Fatteev[30] (see Appendix C, Eq. (C.39)), from where one finds the result:

$$\begin{aligned}\langle A_{r_1} A_{r_2} A_{r_3} \rangle = \left\langle \prod_{i=1}^{3} \mathcal{O}_{r_i r_i'} \right\rangle_{S_M} & (\bar{\mu})^s \frac{\Gamma(-s)}{\alpha_+} \Gamma(s+1) [\Delta(1-\rho)]^s \prod_{i=1}^{s} \Delta(i\rho) \\ & \times \prod_{i=0}^{s-1} \Delta(1+\alpha+i\rho)\Delta(1+\beta+i\rho)\Delta(-1-\alpha-\beta-(s-1+i)\rho) ,\end{aligned} \tag{2.33}$$

where $\Delta(x) = \Gamma(x)/\Gamma(1-x)$.

Notice that in the case of the N-point function with $N > 3$ we still have $N-3$ integrals over $\mathrm{d}^2 z_i$ ($i = 4, \cdots, N$) and such integrals together with the s integrals over $\mathrm{d}^2 w_i$ will result into $s + N - 3$ coupled integrals which in principle are very involved. For this reason let us restrict for a moment to the case $N = 3$.

The expression (2.33) is quite general, and is valid for any conformal theory with central charge $c \leq 1$ and any primary operator $\mathcal{O}_{r_i r_i'}$ of conformal weight $h_{ii'}$ such that the dressing β_i does obey (2.21) with s given in (2.27), a positive integer. It is useful to extend it to arbitrary real values of s. Following [73], let us first restrain to unitary minimal models with $c = 1 - 6/q(q+1)$, and diagonal operators from the Kac table ($r_i = r_i'$) with dimension $\Delta(\mathcal{O}_{r_i r_i}) = h_{ii} = \frac{r_i^2 - 1}{4q(q+1)}$. The value of Q reads, in this case $Q = \sqrt{\frac{25-c}{3}} = \frac{2(2q+1)}{\sqrt{2q(q+1)}}$ and $|\alpha_0| \equiv \frac{1}{\sqrt{2q(q+1)}}$.

Now we can compute the dressing from $\frac{r^2-1}{4q(q+1)} - \frac{1}{2}\beta(\beta + Q) = 1$ and we find, assuming $\beta + \frac{Q}{2} > 0$,

$$\beta_i = \frac{r_i - 2q - 1}{\sqrt{2q(q+1)}} \quad . \tag{2.34}$$

Consequently, plugging into (2.33), we have

$$\begin{aligned}\langle A_{r_1} A_{r_2} A_{r_3} \rangle = \left\langle \prod_{i=1}^{3} \mathcal{O}_{r_i r_i'} \right\rangle & \frac{\bar{\mu}^s \Gamma(-s)\Gamma(s+1)}{\alpha_+} [\Delta(1-\rho)]^s \\ \times \prod_{i=1}^{s} \Delta(i\rho) & \prod_{i=1}^{s} \prod_{j=1}^{3} \Delta(r_j + \rho(r_j + i)) \quad .\end{aligned} \tag{2.35}$$

Now using $\alpha_+ = -\sqrt{\frac{2q}{q+1}}$ and the definition of s

$$s = -\frac{1}{\alpha_+}(\sum \beta_i + Q) = \frac{1}{2q}(r_1 + r_2 + r_3 - 2q - 1) \quad , \tag{2.36}$$

one finds for ρ the expressions

$$\rho = -\frac{\alpha_+^2}{2} = -\frac{q}{q+1} = -\frac{(r_1 + r_2 + r_3 - 1)}{2s + 1 + r_1 + r_2 + r_3} \quad . \tag{2.37}$$

The relation above shows that s can be a rational number and therefore the 3-point function has to be continued for such values of s, or in other words we can say that q in equation (2.37) can not yet be taken as an integer (this is only true after the analytic continuation of (2.35) for arbitrary values of s).

The 3-point function can be continued to non-integer values of s using the gamma function identity

$$\Gamma(nx) = (2\pi)^{\frac{1-n}{2}} n^{nx-\frac{1}{2}} \prod_{k=0}^{n-1} \Gamma(x + \frac{k}{n}) \quad , \tag{2.38}$$

where the right hand side holds only for integer n but the left hand side exists for any value of n. In this way, the three point function (2.35) can be extended to any real values of s. To obtain the matter contribution $\langle \prod_{i=1}^{3} \mathcal{O}_{r_i r_i} \rangle$, in an easier way, one restricts to some values of r_i which obey

$$r_1 + r_2 + r_3 = 2q - 1 \quad . \tag{2.39}$$

Besides, to avoid differences in the normalization between the A_{r_i} operators and the measure in the path integral, one computes the following ratio

$$\frac{\langle A_{r_1} A_{r_2} A_{r_3} \rangle^2 \mathcal{Z}}{\langle A_{r_1} A_{r_1} \rangle \langle A_{r_2} A_{r_2} \rangle \langle A_{r_3} A_{r_3} \rangle} = \frac{r_1 r_2 r_3}{(q+1)(2q+1)} \quad . \tag{2.40}$$

This result has been obtained noticing that the zero- one- and two-point functions can be calculated starting out of the 3-point function by integrating the cosmological constant $\bar{\mu}$, since the A_{r_i} operator reduces to the cosmological term $e^{\alpha_+ \phi}$, in the limit $r_i \to 1$, since $\mathcal{O}_{11}$ reduces to the identity and $\beta_i(r_i \to 1) \to \alpha_+$. In this way we have for example:

$$\langle A_1 A_{r_2} A_{r_1} \rangle = \frac{\partial}{\partial \bar{\mu}} \langle A_{r_1} A_{r_2} \rangle \qquad \text{and} \qquad \langle A_1 A_1 \rangle = \frac{\partial^2}{\partial \bar{\mu}^2} \mathcal{Z} \quad . \tag{2.41}$$

We use (2.38) and $\Gamma(x)\Gamma(1-x) = \frac{\pi}{\sin \pi x}$ in order to compute the identity

$$\begin{aligned}
\prod_{i=1}^{s} \frac{\Gamma(r_j + (r_j + i)\rho)}{\Gamma(1 - r_j - (r_j + i)\rho)} &= (-\rho)^{n^2 \rho^{-1} - n(1-\rho^{-1}) - 2r_j n(1+\rho^{-1}) + 2} \\
&\times r_j^2 (1 + \rho^{-1})^2 \Gamma(r_j(1 + \rho^{-1}))^2 \frac{\sin(-\pi r_j \rho^{-1})}{\pi} \\
&\times \prod_{i=1}^{n} \frac{\sin(i\pi\rho)}{\sin(i\pi\rho^{-1})} \prod_{i=1}^{r_j} \frac{\sin(i\pi\rho)}{\sin(i\pi\rho^{-1})} \frac{\sin(\pi(n+1-i)\rho^{-1}}{\sin \pi(n+1-i)\rho} \\
&\times \prod_{i=0}^{n} \frac{\Gamma(1 + n + r_i + (i - r_j)\rho)^2}{\Gamma(1 + r_j + (r_j - i)\rho^{-1})^2} \quad ,
\end{aligned} \tag{2.42}$$

where $n = \sum r_i - 1$. To perform the $\bar{\mu}$ integration we assume that the 2- and the 3-point functions are functions of some power of $\bar{\mu}$, which is expected from the zero mode trick. Thus we obtain from (2.35), (2.41) and (2.42)

$$\langle A_r A_r \rangle = q(1+\rho^{-1})^2 \rho^4 r \frac{\sin\frac{\pi r}{q}}{\sin\frac{\pi r}{q+1}} \times \frac{\Gamma\left(-\frac{r}{q}\right)^2}{\Gamma\left(\frac{r}{q+1}\right)^2} [\mu\Delta(-\rho)]^{r/q} \quad , \tag{2.43}$$

and for the partition function

$$\mathcal{Z} = \frac{\rho^3\Delta(\rho^{-1})\Delta(\rho)}{(\rho-1)(\rho+1)} [\bar{\mu}\Delta(-\rho)]^{1-\rho^{-1}} \quad , \tag{2.44a}$$

$$\mathcal{Z} = \frac{q^2}{(q+1)(2q+1)} q\Delta\left(-\frac{q}{q+1}\right)\Delta\left(\frac{-(q+1)}{q}\right)\left[\mu\Delta\left(\frac{q}{q+1}\right)\right]^{2+\frac{1}{q}} , \tag{2.44b}$$

where in the last formula we have substituted (2.37). In ref. [77] Dotsenko generalized the above results to the non-minimal models with central charge $c = 1 - 12\alpha_0^2$ and operators out of the diagonal of the Kac table such as $\mathcal{O}_{rr'} = e^{i\alpha_{rr'}X}$, where X is the matter free scalar field in the Coulomb gas representation, and $\alpha_{rr'} = \frac{(1-r)}{2}d_- + \frac{(1-r')}{2}d_+$, with $d_\pm = \alpha_0 \pm \sqrt{2+\alpha_0^2}$; the conformal dimension is computed according to the Coulomb gas technique, given by $\Delta(\mathcal{O}_{rr'}) = \frac{1}{2}(\alpha_{rr'}^2 - 2\alpha_0\alpha_{rr'})$.

One can avoid the constraint (2.39) introducing screening charges in the matter sector (see next section as well as Appendix B), obtaining[77]

$$\frac{\langle A_{r_1r_1'}A_{r_2r_2'}A_{r_3r_3'}\rangle^2\mathcal{Z}}{\langle A_{r_1r_1'}\rangle\langle A_{r_2r_2'}\rangle\langle A_{r_3r_3'}\rangle} = \frac{(r_1'+r_1\rho')(r_2'+r_2\rho')(r_3'+r_3\rho')}{(1+\rho')(1-\rho')(-\rho')} \quad . \tag{2.45}$$

We recover the result (2.40) for $r_i = r_i'$ $\alpha_0 = [2q(q+1)]^{-1/2}$ and $\rho' = (q+1)/q$. Another important aspect contained in [77] is that in the Liouville action a cosmological term with α_- instead of α_+ can be included:

$$S_L^{int} = \int d^2w\sqrt{\hat{g}}[\bar{\mu}e^{\alpha_+\phi} + (a(\bar{\mu}))^{\vartheta}e^{\alpha_-\phi}] \quad , \tag{2.46}$$

where ϑ and the functional form of $a(\bar{\mu})$ are not important. Integrating out the zero mode ϕ_0, Dotsenko obtained, besides the usual cosmological term to the power of s, a second term to the power $\bar{s}$. With this artifact and the inclusion of screening charges in the matter sector, s and $\bar{s}$ turn out to be negative integers, and products of integers $P(\bar{s}) = \prod_{i=1}^{\bar{s}} f(i)$ can be continued to negative values of $\bar{s}$ as follows:

$$P(\bar{s}) = \frac{\prod_{i=1}^{\infty} f(i)}{\prod_{i=\bar{s}+1}^{\infty} f(i)} \quad ; \tag{2.47a}$$

$$\text{for} \quad \bar{s} = -n \quad , \quad P(-n) = \frac{\prod_{i=1}^{\infty} f(i)}{\prod_{i=1-n}^{\infty} f(i)} = \prod_{i=0}^{n-1}\frac{1}{f(-i)} \quad . \tag{2.47b}$$

Although such a trick might a priori not be justified, the final result for the ratios of correlation functions does not seem to depend on the details of the continuation in s or $\bar{s}$.

A big step towards the comprehension of the technique is the computation of the N-point function for tachyonic operators (T_{k_i}) for bosonic matter in the Coulomb gas formulation, with central charge $c = 1 - 12\alpha_0^2$, on the sphere, that is[75]

$$\langle T_{k_1} \cdots T_{k_N} \rangle = \left\langle \prod_{j=1}^{N} \int \mathrm{d}^2 z_j \sqrt{\hat{g}} \mathrm{e}^{ik_j X(z_j) + \beta(k_j)\phi(z_j)} \right\rangle_S \quad , \tag{2.48}$$

with the dressed tachyonic operator given by $T_k = \int \mathrm{d}^2 z \sqrt{\hat{g}} \mathrm{e}^{ikX + \beta\phi}$ and the action for the Liouville and matter fields written as

$$S = \frac{1}{8\pi} \int \mathrm{d}^2 w \sqrt{\hat{g}} [\hat{g}^{ab} \partial_a \phi \partial_b \phi - Q\hat{R}\phi + 8\pi\bar{\mu}\mathrm{e}^{\alpha_+\phi} + \hat{g}^{ab}\partial_a X \partial_b X + 2i\alpha_0 \hat{R} X] \, . \tag{2.49}$$

As before we have $Q = 2\sqrt{2 + \alpha_0^2}$; $\alpha_\pm = -\frac{Q}{2} \pm |\alpha_0|$; moreover, $\alpha_+\alpha_- = 2$.

The momentum associated with the dressing $\beta(k)$ is fixed, again constraining the dressed vertex to have unity conformal dimension $\Delta(\mathrm{e}^{ik_j X + \beta_j \phi}) = \frac{1}{2} k_j (k_j - 2\alpha_0) - \frac{1}{2}\beta_j(\beta_j - Q) = 1$. Consequently we have

$$\beta_j = \beta(k_j) = -\frac{Q}{2} + |k_j - \alpha_0| \quad . \tag{2.50}$$

The sign has been chosen in accordance to the discussion in Chapter 1 concerning the possible perturbation one is able to do in Liouville theory (see Eq. (1.73)).

Before performing the calculation itself, we observe that there exist at least two important reasons to study the tachyonic amplitudes. The first is that the Liouville field can be seen from the space-time point of view as an Euclidean time:

$$ikX + \beta\phi = i(p \cdot f) \quad , \quad \text{with} \quad p_\mu = (\beta, k) \quad \text{and} \quad f_\mu = (-i\phi, X) \quad . \tag{2.51}$$

In order that this interpretation holds correctly, we have, from (2.50)

$$\left(\beta + \frac{Q}{2}\right)^2 - (k - \alpha_0)^2 = 0 \quad , \tag{2.52}$$

defining a massless particle for the tachyonic case. Therefore, we should define the energy as $E = \beta + \frac{Q}{2}$, and the momentum as $p = k - \alpha_0$.[1)] The above interpretation is equivalent to saying that the total action S corresponds to a string embedded in a two dimensional Euclidean space with a non-trivial background. We shall see (a posteriori) that the scattering amplitudes do confirm this interpretation. In this case the only degree of freedom of the string is its center of mass, represented by the operator T_k, which actually corresponds to a massless particle in two dimensions.

---

1) The minimal momentum square $p^2 = 0$ appears for $k = \alpha_0 (\neq 0)$ such that the $c < 1$ theory $(\alpha_0 \neq 0)$ works as infrared cut-off for the $c = 1$ case $(\alpha_0 = 0)$.

The second reason to study such operators is that we can always recover the previous results corresponding to $c < 1$ by simply fixing $k_i = \alpha_{r_i r_i'}$, since k can assume any continuous real value in the Coulomb gas representation.

We now pass to the study of tachyons with $k \in \Re$; we are interested in comparing the three point function

$$\langle T_{k_1} T_{k_2} T_{k_3} \rangle = 2\pi\delta(\sum_{i=1}^{3} k_i - 2\alpha_0)\mathcal{A}_3(k_1, k_2, k_3) \tag{2.53}$$

with the previous results (2.40) and (2.45). Here we have (see also (2.33))

$$\mathcal{A}_3 = \left\langle e^{ik_1 X(0)} e^{ik_2 X(1)} e^{ik_3 X(\infty)} \right\rangle_{S_M} \frac{\bar{\mu}^s}{\alpha_+} \Gamma(-s)\Gamma(s+1)[\Delta(1-\rho)]^s$$
$$\times \prod_{i=1}^{s} \Delta(i\rho) \prod_{i=0}^{s-1} \Delta(1+\alpha+i\rho)\Delta(1+\beta+i\rho)\Delta(-1-\alpha-\beta-(s-1+i)\rho)\,. \tag{2.54}$$

The conservation law for the momenta k_i

$$\sum_{i=1}^{3} k_i = 2\alpha_0 \quad , \tag{2.55a}$$

is obtained using the integration over the zero mode X_0 in the matter sector. Moreover from the Liouville zero mode integration we have

$$\sum_{i=1}^{3} \beta_i = -\alpha_+ s - Q \quad . \tag{2.55b}$$

The quantities α, β, ρ and s are defined in (2.32). Notice that in the Coulomb gas approach, the matter contribution $\langle\rangle_{S_M}$ is trivial (a constant) since $X(z)$ is a free field.

Expression (2.54) only makes sense for integer values of s. In ref. [75] the analytical continuation to non-integer values of s has been done in a different way as compared to those performed before[73,77] as we next show. Observe that the dressing $\beta(k)$ (see (2.50)) suggests the choice of the kinematics, which must be consistent with momentum conservation (2.55); we take as an example

$$k_1\,, k_3 \geq \alpha_0 \quad , \quad k_2 \leq \alpha_0 \quad . \tag{2.56}$$

With this kinematics, the definition of s in (2.55b) and the momentum conservation law (2.55a) permit us to eliminate β

$$\beta = \begin{cases} \rho(1-s) & , \quad \alpha_0 > 0 \\ -1-\rho s & , \quad \alpha_0 < 0 \quad . \end{cases} \tag{2.57}$$

Introducing back to the amplitude $\mathcal{A}_3$ we see that for $\alpha_0 > 0$ the factor $\Delta(1+\beta+i\rho)$ for $i = s-1$ is $\Delta(1) \sim \frac{1}{\Gamma(0)} \sim 0$. For this reason in the case $\alpha_0 > 0$ the amplitude vanishes.

For $\alpha_0 \leq 0$ we have, after some cancellations,[2)]

$$\mathcal{A}_3 = [\bar{\mu}\Delta(-\rho)]^s \Delta(-s)\Delta(\rho-\alpha)\Delta(1+\alpha+(s-1)\rho) \quad . \tag{2.58}$$

Notice that the above expression can now be defined for non-integer values of s. To observe the factorization of the amplitude it is necessary to rewrite it using the following kinematic relations

$$\frac{1}{2}(\beta_1^2-k_1^2) = \rho-\alpha \quad ; \quad \frac{1}{2}(\beta_2^2-k_2^2) = 1+\alpha+(s-1)\rho \quad ; \quad \frac{1}{2}(\beta_3^2-k_3^2) = -s \quad . \tag{2.59}$$

Therefore we have

$$\mathcal{A}_3 = [\bar{\mu}\Delta(-\rho)]^s \prod_{j=1}^{3} \Delta\left(\frac{1}{2}(\beta_j^2-k_j^2)\right) \quad . \tag{2.60}$$

Now we would like to comment about some qualitative features of the above result. First of all observe that in the limit of zero momenta ($k_j \to 0$) we have $\beta_j \to \alpha_+$ (and $\frac{1}{2}(\beta_j^2 - k_j^2) \to -\rho$) which means that the tachyonic operator reduces to the cosmological term for vanishing momentum. Therefore for integer and positive s we can interpret the amplitude as a correlation function of $3+s$ tachyons where s tachyons have zero momenta, that is,

$$\mathcal{A}_3 = \bar{\mu}^s \langle (T_0)^s T_{k_1} T_{k_2} T_{k_3} \rangle_{\bar{\mu}=0} = \bar{\mu}^s \prod_{j=1}^{s+3} \Delta(\frac{1}{2}(\beta_j^2-k_j^2)) \quad . \tag{2.61}$$

In this way we can understand the $[\Delta(-\rho)]^s$ factor in the amplitude $\mathcal{A}_3$ as a multiple point correlator with some points having zero momentum. For further discussion we separate the cases $c=1$ and $c<1$. For the latter case the important operators are the dressed matter fields $T_{k_{nm}}$ of the (p,q) minimal models, for which

$$\begin{aligned} k_{nm} &= \frac{n-1}{2}\alpha_+ + \frac{1-m}{2}\alpha_- \quad ; \\ \alpha_0 &= \frac{\alpha_- - \alpha_+}{2}; \quad \alpha_+ = -\sqrt{\frac{2p}{q}}; \quad \alpha_- = -\sqrt{\frac{2q}{p}} \; . \end{aligned} \tag{2.62}$$

The dressing $\beta_{nm} = \beta(k_{nm})$ given by (2.50) turns into

$$\beta_{nm} = \frac{1-n}{2}\alpha_+ + \frac{1+m}{2}\alpha_- \quad . \tag{2.63}$$

Compare, at this point, with (B.105). Therefore we have, for the external legs contributions to the correlators, the expression

$$\Delta\left(\frac{1}{2}(\beta_{nm}^2 - k_{nm}^2)\right) = \Delta\left(m\frac{q}{p}+1-n\right) \quad . \tag{2.64}$$

---

2) We have absorbed a $\frac{1}{\alpha_+}$ factor in the measure of the path integral.

Since q and p are coprimes and m, n integers, it is clear that inside the Kac table, i.e. $1 \leq n \leq q-1$ and $1 \leq m \leq p-1$, the expression (2.64) never vanishes or diverges, and we can freely redefine the vertices as

$$\widetilde{T}_k = \frac{T_k}{\Delta\left(\frac{1}{2}(\beta^2 - k^2)\right)} \quad , \tag{2.65}$$

and the cosmological constant as

$$\mu = \bar{\mu}\Delta(-\rho) \quad , \tag{2.66}$$

to obtain

$$\mathcal{A}_3 = \left\langle \widetilde{T}_{k_1} \widetilde{T}_{k_2} \widetilde{T}_{k_3} \right\rangle = \mu^s \quad , \tag{2.67}$$

a result which holds for $c < 1$.

The case $c = 1$, or equivalently $\alpha_0 = 0$ requires more care in redefining the vertices T_k and the cosmological constant $\bar{\mu}$. Since $\lim_{\alpha_0 \to 0} \Delta(-\rho) \approx \Delta(1+\alpha_0) \approx \alpha_0$, we see from (2.58) that for positive integer s, the 3-point function $\mathcal{A}_3$ vanishes, indicating the decoupling of the usual cosmological term $\bar{\mu}e^{\sqrt{2}\phi}$. However, if we redefine the cosmological constant as in (2.66), we have a finite amplitude $\mathcal{A}_3$, and the cosmological term becomes in the $\alpha_0 \to 0$ limit

$$\frac{\mu}{\Delta(-\rho)} e^{\alpha_+ \phi} \longrightarrow \frac{\mu}{\sqrt{2}\alpha_0} e^{-\sqrt{2}\phi} + \frac{\mu}{\sqrt{2}} \phi e^{-\sqrt{2}\phi} \quad . \tag{2.68}$$

Then, if we drop the divergent (and non-universal) term on the r.h.s. of (2.68) we end up with $\bar{\mu}\phi e^{-\sqrt{2}\phi}$, which is presumably the correct cosmological term for the $c = 1$ case. Compare with discussion after (1.63).

A more elegant way of understanding the cosmological term in the $c = 1$ theory, is by inspection using the space-time interpretation of the theory. One considers the reparametrization invariant action describing the operator $T(x)$,

$$S = \frac{1}{8\pi} \int d^2\xi \left(\sqrt{\hat{g}} \partial^i X^\mu \partial_i X_\mu - iQX_0 R \right) + \int d^2\xi \int d^d k T(k) \oint e^{ik^\mu X_\mu(\xi)} \tag{2.69}$$

with $-k_0(k_0 + Q) + k^2 = 2$.

Defining $T(X) = \int d^d k T(k) e^{ik^i X_i(\xi)}$ we have the space-time equation of motion

$$\left[-\partial^i \partial_i + iQ \frac{\partial}{\partial X_0} + \frac{\partial^2}{\partial X_0^2} - 2 \right] T(X_0, X^i) = 0 \quad , \tag{2.70}$$

whose X_i independent solutions for $d = 1$, $X_0 = i\phi$, $Q = 2\sqrt{2}$ are

$$T(\phi) = e^{-\sqrt{2}\phi} \quad , \quad \text{or} \quad \phi e^{-\sqrt{2}\phi} \quad . \tag{2.71}$$

After redefining μ we have in the $\alpha_0 \to 0$ limit the result

$$\mathcal{A}_3 = (\mu)^{\sum_{i=1}^3 \frac{|k_i|}{\sqrt{2}} - 1} \prod_{j=1}^3 \Delta(1 - \sqrt{2}|k_j|) \quad . \tag{2.72}$$

The most important feature of (2.72) is the presence of poles at $\sqrt{2}|k_j| = 1, 2, 3, \ldots$. These states are called discrete states and are remnants[96] of the massive states of string in higher dimensions. In our case of a two-dimensional embedding space these discrete states do not have particle interpretation since there is no room for transverse excitations and they only appear for specific values of the momentum[97]. This problem is discussed in more detail in section 5 of this Chapter.

2.3. The 3-Point Function with (n,m) Screening Charges

In the last section we have calculated the tachyon 3-point correlation function with momenta k_j satisfying the conservation law $\sum_i k_i = 2\alpha_0$. This conservation law has to be generalized in order to include more general correlation functions. This is achieved when the so-called screening charges are included in the problem, as we shall see below. In fact this procedure turns out to be especially useful in our case, since there are relations obtained among the correlation functions by means of derivatives with respect to the cosmological constant $\bar{\mu}$, which relates the $(n+p)$-point function to the pth derivative with respect to $\bar{\mu}$ of the n-point function; as an example, with the above conservation law, it would be impossible to take $k_i \to 0\,(i = 1,2,3)$ to calculate the partition function $\mathcal{Z}$ starting from the 3-point function $\mathcal{A}_3$ using $\frac{\partial^3 \mathcal{Z}}{\partial \mu^3} = \mathcal{A}_3(0,0,0)$ for $\alpha_0 \neq 0\,(c \neq 1)$. Modifying the conservation law we could extend our calculation to the case of conformal minimal models, where α_0 and k assume special values[29,30]. The modification is made introducing screening charges (s.c.) $V_\pm$ in the matter sector, where we define $V_\pm = \mathrm{e}^{id_\pm X}$ with $d_\pm$ defined in such way that the vertices $V_\pm$ correspond to (1,1) conformal operators, that means, $\frac{1}{2}d(d-2\alpha_0) = 1$, $d_\pm = \alpha_0 \pm \sqrt{\alpha_0^2+2}$. For a general discussion concerning screening charges, see $(B.99)$ and the discussion following it.

The $V_\pm$ operators can be introduced directly in the matter sector of the action, defining

$$S_M^\lambda = \frac{1}{8\pi}\int \mathrm{d}^2 w\sqrt{\hat{g}}[\hat{g}^{ab}\partial_a X\partial_b X + 2i\alpha_0 \hat{R}X + 8\pi\lambda_+ \mathrm{e}^{id_+X} + 8\pi\lambda_- \mathrm{e}^{id_-X}] \quad . \tag{2.73}$$

When we calculate the 3-point function using the above action we expand $\mathrm{e}^{-S_M^\lambda}$ in λ_- and λ_+ and the integration over the matter zero mode leads us to a new conservation law, which is satisfied only[3)] by the term with n V_+ charges and m V_- charges from the above expression.

$$\begin{aligned}\langle T_{k_1}T_{k_2}T_{k_3}\rangle_{S_L+S_M^\lambda} &= \lambda_+^n\lambda_-^m\left\langle T_{k_1}T_{k_2}T_{k_3}\left(\frac{1}{n!}\prod_{i=1}^n\int \mathrm{d}^2t_i\sqrt{\hat{g}}\mathrm{e}^{id_+X(t_i)}\right)\right.\\ &\qquad\left.\times\left(\frac{1}{m!}\prod_{i=1}^m\int \mathrm{d}^2r_i\sqrt{\hat{g}}\mathrm{e}^{id_-X(r_i)}\right)\right\rangle_{S_L+S_M} \\ &= 2\pi\delta(\sum_{i=1}^3 k_i + nd_+ + md_- - 2\alpha_0)\mathcal{A}_3^{nm}(k_1,k_2,k_3) \quad ,\end{aligned} \tag{2.74}$$

leading to the new conservation law

$$\sum_{i=1}^3 k_i + nd_+ + md_- - 2\alpha_0 = 0\ . \tag{2.75}$$

3) One admits that d_- and d_+ are incommensurable, in order that nd_+ be not cancelled by md_-.

Since the gravitational piece has not been modified with respect to the $n = m = 0$ case, we again obtain $\mathcal{A}_3^{nm}$ by the integration over the zero mode ϕ_0, assuming ab initio $s \in Z_+$ and fixing the gauge $z_1 = 0, z_2 = 1, z_3 = \infty$. The only difference is that the contractions containing the matter field $X(z)$ now survive due to the presence of the (n, m) integrals over $\mathrm{d}^2 r_i$ and $\mathrm{d}^2 t_i$. Using $d_+ d_- = -\alpha_+ \alpha_- = -2$ we have:

$$\begin{aligned}\mathcal{A}_3^{nm}(k_1, k_2, k_3) = &\frac{\lambda_+^n \lambda_-^m}{n!m!} \frac{\Gamma(-s)}{\alpha_+} \bar{\mu}^s \prod_{i=1}^{n} \int \mathrm{d}^2 t_i |t_i|^{2\tilde{\alpha}} |1 - t_i|^{2\tilde{\beta}} \prod_{i<j=1}^{n} |t_i - t_j|^{4\tilde{\rho}} \\ &\times \prod_{i=1}^{m} \int \mathrm{d}^2 r_i |r_i|^{2\tilde{\alpha}'} |1 - r_i|^{2\tilde{\beta}'} \prod_{i<j}^{m} |r_i - r_j|^{4\tilde{\rho}'} \\ &\times \prod_{i=1}^{n} \prod_{j=1}^{m} |t_i - r_j|^{-4} \prod_{i=1}^{s} \int \mathrm{d}^2 z_i |z_i|^{2\alpha} |1 - z_i|^{2\beta} \prod_{i<j}^{s} |z_i - z_j|^{4\rho} ,\end{aligned} \tag{2.76}$$

where $\tilde{\alpha} = d_+ k_1$, $\tilde{\beta} = d_+ k_2$, $\tilde{\rho} = \frac{1}{2} d_+^2$, $\tilde{\alpha}' = d_- k_1$, $\tilde{\beta}' = d_- k_2$, $\tilde{\rho}' = \frac{1}{2} d_-^2$.

Except for the new momentum conservation law (2.75), the definition of s and other kinematic formulae from last section remain still valid.

The s integrals over z_i correspond to the formula $(C.39)$ as in the $n = m = 0$ case. The $n + m$ integrals over the matter sector correspond to the formula $(C.75)$. Absorbing $\frac{\pi^{n+m}}{\alpha_+}$ in the definition of the measure we obtain:

$$\begin{aligned}&\mathcal{A}_3^{nm} = \\ &= \bar{\mu}^s \Gamma(-s) \Gamma(s+1) \lambda_+^n \lambda_-^m \pi^{s+n+m} \tilde{\rho}^{-4nm} [\Delta(1-\tilde{\rho})]^n [\Delta(1-\tilde{\rho}')]^m \prod_{i=1}^{m} \Delta(i\tilde{\rho}' - n) \prod_{i=1}^{n} \Delta(i\tilde{\rho}) \\ &\times \prod_{i=0}^{m-1} \Delta(1-n+\tilde{\alpha}'+i\tilde{\rho}') \Delta(1-n+\tilde{\beta}'+i\tilde{\rho}') \Delta(-1+n-\tilde{\alpha}'-\tilde{\beta}'-(m-1+i)\tilde{\rho}') \\ &\times \prod_{i=0}^{n-1} \Delta(1+\tilde{\alpha}+i\tilde{\rho}) \Delta(1+\tilde{\beta}+i\tilde{\rho}) \Delta(-1+2m-\tilde{\alpha}-\tilde{\beta}-(n-1+i)\tilde{\rho}) \\ &\times [\Delta(1-\rho)]^s \prod_{i=1}^{s} \Delta(i\rho) \prod_{i=0}^{s-1} \Delta(1+\alpha+i\rho) \Delta(1+\beta+i\rho) \Delta(-1-\alpha-\beta-(s-1+i)\rho) .\end{aligned} \tag{2.77}$$

We use the same kinematics as in the last section ($k_1, k_3 \geq \alpha_0$, $k_2 \leq \alpha_0$), as well as the momentum conservation law in the presence of the screening charges; furthermore we eliminate $\beta, \tilde{\beta}$ and $\tilde{\beta}'$ by the expressions

$$\beta, \tilde{\beta}, \tilde{\beta}' = \begin{cases} \rho(m+1-s)+n,\ n\rho^{-1}+m-s-1,\ n+\rho(m-s-1),\ \alpha_0 > 0 \\ -1-m-(s+n)\rho,\ m-1+(s+n)\rho,\ s+n+\rho^{-1}(m-1),\ \alpha_0 < 0 \end{cases} \tag{2.78}$$

with $(\tilde{\rho}, \tilde{\rho}') = (-\rho, -\rho^{-1})$ for $\alpha_0 < 0$ and $(-\rho^{-1}, -\rho)$ for $\alpha_0 > 0$. Expressions (2.78) are obtained from their definition after (2.76), substituting k_2 in terms of k_1, a relation derived from the momentum conservation law, and definition of s, Eq. (2.55*b*). Indeed, let us take the example of $\alpha_0 > 0$; we have from (2.75)

$$k_1 + k_2 + k_3 - \alpha_+ + \alpha_- - n\alpha_- + m\alpha_+ = 0 \quad ,$$

and from (2.55*b*), with (2.5) for $\alpha_0 > 0$

$$k_1 - k_2 + k_3 + \alpha_- + s\alpha_+ = 0 \quad ,$$

from which it follows, for $\rho = -\frac{1}{2}\alpha_+^2$,

$$\beta = -\alpha_+\beta_2 = -\alpha_+(\alpha_+ - k_2) = n + \rho(m + 1 - s) \quad .$$

The other relations follow analogously. It is not difficult to see that the amplitudes (2.77) are non-trivial only for $\alpha_0 < 0$ (as in the last section). In this case we have $\tilde{\alpha} = \alpha - 2\rho$, $\tilde{\alpha}' = -2 - \rho\alpha$.

Eliminating the dependent variables and after some algebra[74,75] involving properties of Γ functions we combine the matter and gravity contributions obtaining:

$$\begin{aligned} \mathcal{A}_3^{nm} &= [\bar{\mu}\Delta(-\rho)]^s \left[\lambda_-\Delta(\rho^{-1})\right]^m [\lambda_+\Delta(\rho)]^n \\ &\times \Delta(\rho - \alpha)\Delta(\rho(s - n + 1) + \alpha - m + 1)\Delta(-m\rho^{-1} - (s + n)) \quad , \end{aligned} \tag{2.79}$$

or in a more elegant way

$$\mathcal{A}_3^{nm} = [\bar{\mu}\Delta(-\rho)]^s \left[\lambda_-\Delta(\rho^{-1})\right]^m [\lambda_+\Delta(\rho)]^n \prod_{i=1}^{3} \Delta\left(\frac{1}{2}(\beta_i^2 - k_i^2)\right) \quad ; \tag{2.80}$$

(compare with (2.60)).

The factors $[\Delta(\rho^{-1})]^m$ and $[\Delta(\rho)]^n$ appearing above can be interpreted analogously to the factor $[\Delta(-\rho)]^s$ which appears in the amplitude $\mathcal{A}_3$ in the last section. This means that each s.c. $V_+(V_-)$ plays the role of a tachyonic operator $T_{k=d_+}(T_{k=d_-})$, (or $T_{k=-\alpha_+}(T_{k=\alpha_-})$, for $\alpha_0 < 0$), and the above result suggests a correlation function of $s + n + m + 3$ tachyons:

$$\begin{aligned} \mathcal{A}_3^{nm} &= \bar{\mu}^s \frac{\lambda_+^n \lambda_-^m}{n!m!} \left\langle (T_0)^s (T_{d_+})^n (T_{d_-})^m T_{k_1} T_{k_2} T_{k_3} \right\rangle \\ &= \bar{\mu}^s \left(\frac{\lambda_+^n \lambda_-^m}{n!m!}\right) \prod_{j=1}^{s+n+m+3} \Delta\left(\frac{1}{2}(\beta_j^2 - k_j^2)\right) \quad . \end{aligned} \tag{2.81}$$

As in the case without screening charges we can, for $c < 1$, freely redefine the vertices T_k as in (2.65), as well as the cosmological constant as in (2.66). By further redefining $\lambda_\pm = 1/\Delta\left(-\frac{1}{2}d_\pm^2\right)$, we finally arrive at the expression[4)]

$$\widetilde{\mathcal{A}}_3^{nm} = \left\langle \widetilde{T}_{k_1}\widetilde{T}_{k_2}\widetilde{T}_{k_3}(\widetilde{V}_+)^n(\widetilde{V}_-)^m \right\rangle_{\bar{\mu}} = \mu^s \quad , \tag{2.82}$$

[4)] Notice that the definitions of $\lambda_\pm$ become ill defined in the limit $c \to 1$ ($d_\pm \to \pm\sqrt{2}$) due to the lack of convergence of the corresponding conformal integrals at that point.

where s is defined as before (see (2.55b)) and the momenta k_i satisfy now the new conservation law (2.75).

This new conservation, which can be written (recall that $2\alpha_0 = d_+ + d_-$) as $\sum_{i=1}^{3} k_i = (1-n)d_+ + (1-m)d_-$, allows us to calculate more general amplitudes, as the partition function:

$$\frac{\partial^3 \mathcal{Z}}{\partial \mu^3} = \mathcal{A}_3^{11}(k_i \to 0) \quad , \tag{2.83}$$

from where we get

$$\mathcal{Z} = \frac{\rho^3 \, [\mu]^{1-\rho^{-1}}}{(\rho-1)(\rho+1)} \tag{2.84}$$

(compare with (2.44)).

To calculate the amplitudes ratio in order to compare with the results obtained by Dotsenko (Eq. (2.45)), we need the 2-point function

$$\mathcal{A}_3^{\widetilde{n}\widetilde{m}}(k_1, k_2, k_3 \to 0) = \frac{\partial}{\partial \mu} \mathcal{A}_2^{\widetilde{n}\widetilde{m}}(k_1, k_2) \quad . \tag{2.85}$$

Observe that we are analyzing the minimal model case where the momenta assume discrete values given by the Kac formula $k_i = \frac{(1-r_i)}{2} d_+ + \frac{(1-r_i')}{2} d_-$. Using the momentum conservation law we have $n = \frac{1}{2}\left(\sum_{i=1}^{3} r_i - 1\right)$ and $m = \frac{1}{2}\left(\sum_{i=1}^{3} r_i' - 1\right)$, where $\widetilde{n}$, and $\widetilde{m}$ are obtained from n and m when $r_3 = r_3' = 1$. Therefore we get

$$\mathcal{A}_2^{\widetilde{n}\widetilde{m}}(k_1, k_2) = \frac{[\mu]^{\rho^{-1}\left(\frac{\alpha_+}{2}\sum_{i=1}^{2}\beta_i + \rho - 1\right)}}{\rho^{-1}\left(\frac{\alpha_+}{2}\sum_{i=1}^{2}\beta_i + \rho - 1\right)} \, , \tag{2.86}$$

and for the ratio (R) of the amplitudes, we arrive at the expression

$$R = \frac{(\mathcal{A}_3^{nm}(k_1, k_2, k_3))^2 \mathcal{Z}}{\mathcal{A}_2^{\widetilde{n}_1\widetilde{m}_1}(k_1, k_1)\mathcal{A}_2^{\widetilde{n}_2\widetilde{m}_2}(k_2, k_2)\mathcal{A}_2^{\widetilde{n}_3\widetilde{m}_3}(k_3, k_3)} = \frac{\prod_{i=1}^{3}(\alpha_+\beta_i + \rho - 1)}{(\rho-1)(1+\rho)} \, . \tag{2.87}$$

Using the formula (2.50) for the dressing β_i we obtain exactly Dotsenko's result[77], which agrees with those of top Eq. (2.40) for $r_i = r_i'$, and generalize them to $r_i \neq r_i'$ as anticipated in (2.45). The agreement among these results indicates that although the analytic continuation for $s \notin Z_+$ has no rigorous justification, the results do not seem to depend on the details of the continuation procedure.

2.4. The N-Point Function with (n,m) Screening Charges

In this section we present the calculation of the N-tachyons correlation function ($N \geq 4$) with (n,m) s.c. in the matter sector[79,80]. The total action is the same as in the previous section $(S_L + S_M^\lambda)$. Using the zero mode technique we obtain:

$$\langle T_{k_1} \cdots T_{k_N} \rangle_{S_L+S_M^\lambda} = 2\pi\delta\left(\sum_{i=1}^{N} k_i + nd_+ + md_- - 2\alpha_0\right) \mathcal{A}_N^{nm}(k_1 \cdots k_N) \quad (2.88)$$

with

$$\mathcal{A}_N^{nm} = \bar{\mu}^s \frac{\lambda_+^n}{n!}\frac{\lambda_-^m}{m!}\Gamma(-s) \prod_{i=1}^{N} \int \mathrm{d}^2 z_i \prod_{j=1}^{n} \int \mathrm{d}^2 t_j \prod_{k=1}^{m} \int \mathrm{d}^2 r_k$$
$$\times \prod_{l=1}^{s} \int \mathrm{d}^2 w_l \left\langle \mathrm{e}^{ik_i X(z_i)} \mathrm{e}^{id_+ X(t_j)} \mathrm{e}^{id_- X(r_k)} \right\rangle_{S_M} \left\langle \mathrm{e}^{\beta_i \phi(z_i)} \mathrm{e}^{\alpha_+ \phi(w_l)} \right\rangle_{\bar{\mu}=0} \quad , \quad (2.89)$$

where $s = -\frac{1}{\alpha_+}\left(Q + \sum_{i=1}^{N} \beta_i\right)$. Fixing the gauge $z_1 = 0, z_2 = 1, z_3 = \infty$ and performing all the possible contractions we get

$$\mathcal{A}_n^{nm} = \frac{\lambda_+^n}{n!}\frac{\lambda_-^m}{m!}\bar{\mu}^s \Gamma(-s) I_N^{nm} \quad . \quad (2.90)$$

The contractions in (2.89) lead to the following integral

$$I_N^{nm} = \int \prod_{j=4}^{N} \mathrm{d}^2 z_j |z_j|^{2\alpha'_j} |1 - z_j|^{2\beta'_j} \prod_{i<j=4}^{N} |z_i - z_j|^{4\rho'_{ij}}$$
$$\times \int \prod_{i=1}^{s} \mathrm{d}^2 w_i |w_i|^{2\alpha} |1 - w_i|^{2\beta} \prod_{i<j=1}^{s} |w_i - w_j|^{4\rho} \prod_{i=1}^{s}\prod_{j=4}^{N} |w_i - z_j|^{2p_j}$$
$$\times \int \prod_{i=1}^{n} \mathrm{d}^2 t_i |t_i|^{2\widetilde{\alpha}} |1 - t_i|^{2\widetilde{\beta}} \prod_{i<j}^{n} |t_i - t_j|^{4\widetilde{\rho}} \prod_{i=1}^{n}\prod_{j=4}^{N} |z_j - t_i|^{2\widetilde{\alpha}_j} \quad (2.91)$$
$$\times \int \prod_{i=1}^{m} \mathrm{d}^2 r_i |r_i|^{2\widetilde{\alpha}'} |1 - r_i|^{2\widetilde{\beta}'} \prod_{i<j=1}^{m} |r_i - r_j|^{4\widetilde{\rho}'} \prod_{i=1}^{m}\prod_{j=4}^{N} |z_j - r_i|^{2\widetilde{\alpha}'_j}$$
$$\times \prod_{i=1}^{n}\prod_{j=1}^{m} |t_i - r_j|^{-4} \quad ,$$

where besides the parameters $\alpha, \beta, \rho, \widetilde{\alpha}, \widetilde{\beta}, \widetilde{\rho}, \widetilde{\alpha}', \widetilde{\beta}', \widetilde{\rho}'$ already defined after (2.32) and (2.76), we have also $\alpha_j = k_1 k_j - \beta_1 \beta_j$, $\widetilde{\alpha}_j = d_+ k_j$, $\beta_j = k_2 k_j - \beta_2 \beta_j$, $\widetilde{\beta}_j = d_- k_j$, $\rho'_{lj} = \frac{1}{2}(k_l k_j - \beta_l \beta_j)$, $p_j = -\alpha_+ \beta_j$, with $4 \leq j, l \leq N$.

Unfortunately, the integral (2.91) is very difficult to be calculated in general (even in the case $n = m = 0$). However we notice that it possesses a set of symmetries: it is invariant under $x_i \to 1 - x_i$ once we exchange variables as $\alpha \leftrightarrow \beta$, $\alpha'_j \leftrightarrow \beta'_j$, $\widetilde{\alpha} \leftrightarrow \widetilde{\beta}$ and $\widetilde{\alpha}' \leftrightarrow \widetilde{\beta}'$. Analogously, the symmetry under inversion, $x_i \to \frac{1}{x_i}$, is guaranteed if

$$\begin{aligned} \alpha &\to -2 - \alpha - \beta - 2\rho(s-1) - \sum_{j=4}^{N} p_j \quad , \\ \widetilde{\alpha} &\to -2 - \widetilde{\alpha} - \widetilde{\beta} - 2\widetilde{\rho}(n-1) + 4m \quad , \\ \widetilde{\alpha}' &\to -2 - \widetilde{\alpha}' - \widetilde{\beta}' - \widetilde{\rho}'(m-1) + 2n \quad , \\ \alpha'_j &\to -2 - \alpha'_0 - \beta'_0 - 2 \sum_{\substack{l=4 \\ (j<l)}}^{N} \rho'_{jl} - n\widetilde{\alpha}_j - m\widetilde{\alpha}'_j \quad . \end{aligned} \tag{2.92}$$

The above symmetries still do not permit to calculate I_N^{nm}. It is very important to know its asymptotic behaviour. Using the technique presented in Appendix C it is possible to obtain the behaviour of I_N^{nm} for $\alpha \to \infty$ (or $\beta \to \infty$), which is polynomial, that is

$$I_N^{nm}(\alpha \to \infty) = \alpha^a \quad , \tag{2.93}$$

where

$$\begin{aligned} a = &-2(N-3+s+n+m) - 2s\beta - 2\rho s(s-1) - 2s\sum_4^N p_j - 2\sum_4^N \beta'_j - 4\sum_{i<j=4}^{N} \rho'_{ij} \\ &- 2n\sum_{j=4}^{N} \widetilde{\alpha}_j - 2m\sum_4^N \widetilde{\alpha}'_j - 2n\widetilde{\beta} - \widetilde{\rho} n(n-1) - 2m\widetilde{\beta}' - \widetilde{\rho}' m(m-1) + 4nm \quad . \end{aligned} \tag{2.94}$$

The properties (2.92) and (2.93) are not enough to compute I_N^{nm}. At this point it is important to remember that even in the simple case of the amplitude $\mathcal{A}_3^{n=m=0}$ the result of the integral (2.54) simplifies considerably, after the elimination of β as functions of s and ρ, using kinematic relations. In the present case we can eliminate most of the parameters of I_N^{nm} as a function of the independent variables α, β, ρ, s and p_j (for $4 \le j \le N-1$). After this elimination the translation and inversion symmetries imply respectively[79,80]

$$\begin{aligned} &I_N^{nm}(\alpha, \beta, p_j, \rho) = I_N^{nm}(\beta, \alpha, p_j, \rho) \quad , \\ &I_N^{nm}(\alpha, \beta, p_j, \rho) = I_N^{nm}(m - 1 - P - \alpha - \beta + \rho(N + n - 1 - s), \beta, p_j, \rho) \quad , \end{aligned} \tag{2.95}$$

where $P = \sum_4^{N-1} p_j$ and we assume again $\alpha_0 < 0$ in the region $k_1, k_2, \cdots, k_{N-1} \ge \alpha_0$, $k_N < \alpha_0$. With this condition, Eqs. (2.93) and (2.94) reduce to

$$I_N^{nm}(\alpha \to \infty) = \alpha^{2\beta + 2\rho(s-N-n+3) + 2P - 2m} \quad . \tag{2.96}$$

Based on (2.95) and (2.96) we write the following Ansatz:

$$\begin{aligned} \mathcal{A}_N^{nm} &= f_N^{nm}(\rho, p_j)\Delta(\rho - \alpha)\Delta(\rho - \beta)\Delta(1 - m + P + \alpha + \beta + \rho(s + 2 - N - n)) , \\ \mathcal{A}_N^{nm} &= f_N^{nm}(\rho, p_j) \prod_{j=1,2,N} \Delta\left(\frac{1}{2}(\beta_j^2 - k_j^2)\right) , \end{aligned} \tag{2.97}$$

where f_N^{nm} is still to be determined. The above Ansatz fulfills the symmetries given by (2.95) as well as the asymptotic behavior (2.96), which can be checked using the formula $\Delta(x+B) \sim x^{2B-1}\Gamma(x)/\Gamma(-x)$ valid in the limit $x \to \infty$.

The function f_N^{nm} can be determined from the 3-point function $\mathcal{A}_3^{nm}$ given by (2.80) and from the expression

$$\mathcal{A}_N^{nm}(k_1, k_2, k_j \to 0, k_N) = \frac{\partial^{N-3}}{\partial \bar{\mu}^{N-3}} \mathcal{A}_3^{nm}(k_1, k_2, k_N) \quad , \quad 3 \le j \le N-1 \quad , \tag{2.98}$$

In this way we obtain for the N-point amplitude $\mathcal{A}_N^{nm}$

$$\mathcal{A}_N^{nm} = (s+N-3)(s+N-4)\cdots(s+1)\left[\bar{\mu}\Delta(-\rho)\right]^s \times \left[\lambda_-\Delta(\rho^{-1})\right]^m \left[\lambda_+\Delta(\rho)\right]^n \prod_{j=1}^N \Delta\left(\frac{1}{2}(\beta_j^2 - k_j^2)\right) , \tag{2.99a}$$

or

$$\mathcal{A}_N^{nm} = \left(\frac{\partial^{N-3}}{\partial \bar{\mu}^{N-3}} \bar{\mu}^{s+N-3}\right) \left[\Delta(-\rho)\right]^s \times \left[\lambda_-\Delta(\rho^{-1})\right]^m \left[\lambda_+\Delta(\rho)\right]^n \prod_{j=1}^N \Delta\left(\frac{1}{2}(\beta_j^2 - k_j^2)\right) . \tag{2.99b}$$

After redefining the constants $\bar{\mu}$ and $\lambda_\pm$ as in the case of $\mathcal{A}_3^{nm}$ we have for $c=1$ ($\alpha_0 = 0$) the result

$$\mathcal{A}_N^{nm} = \left(\frac{\partial^{N-3}}{\partial \mu^{N-3}} \mu^{N+s-3}\right) \prod_{i=1}^N \frac{\Gamma(1-\sqrt{2}|k_i|)}{\Gamma(\sqrt{2}|k_i|)} \tag{2.100a}$$

and for $c < 1$ ($\alpha_0 \ne 0$) we obtain (after redefining T_k as in (2.65)) the simple result

$$\mathcal{A}_N^{nm} = \left(\frac{\partial^{N-3}}{\partial \mu^{N-3}} \mu^{N+s-3}\right) \quad . \tag{2.100b}$$

The above results reduce to the ones obtained before for $n = m = 0$.

2.4.1. Kinematics Dependence

It must be stressed that these results in principle only hold for $k_1, k_2, \ldots k_{N-1} \ge \alpha_0$, and $k_N \le \alpha_0$, i.e. according to the technique which has been used, we have chosen a certain kinematic region. We face now the question about whether the results are valid or not outside that region. In fact, the choice has been made in such a way that it is compatible with the momentum conservation law, leading to a non-trivial result for the amplitudes. Different choices are not useful since they lead to a non-convergent integral representation of the amplitude. However, this fact does not necessarily mean that it is impossible to analytically continue the integrals outside of the kinematic region used thus far. This is a non-trivial problem, since there is no theorem assuring that the amplitudes should be analytic. Due to the dependence of the amplitudes on $\beta(k)$, which displays a cut for $k = \alpha_0$, a simple analytic continuation is discarded. Nevertheless, one can continue the results assuming a space-time interpretation of the string. As we shall see later, this leads systematically to results in accordance with matrix models, where the analytic continuation problem does not exist.

As an example of such a continuation we first analyze the case of the four point amplitude $\mathcal{A}_4$

$$\mathcal{A}_4 = \int d^2 z \left\langle \widetilde{T}_{k_1}(0)\widetilde{T}_{k_2}(\infty)\widetilde{T}_{k_3}(1)\widetilde{T}_{k_4}(z)\right\rangle_{\mu} \quad . \tag{2.101}$$

Assuming that s is a non-negative integer we have found in the kinematic region $k_1, k_2, k_3 \geq \alpha_0$, $k_4 \leq \alpha_0$ (see (2.100*b*) for $N = 4$)

$$\mathcal{A}_4 = (s+1)\mu^s \quad . \tag{2.102}$$

In principle, the integrals involved in the calculation of the amplitude (2.101) have only poles and zeroes. For arbitrary s, however, the Liouville momentum β is not conserved and this non-conservation leads as we show below to the appearance of cuts instead of poles as it was first observed in [21,98] (see also [75]). The main contribution to the integral (2.101) in the region around $z = 0$ stems from the operator product expansion

$$\widetilde{T}_{k_1}(0)\widetilde{T}_{k_4}(z) \sim \int_{-\infty}^{\infty} d\beta_p f(\beta_p, \beta_1, \beta_4)\left(V_{\Delta}^{m} e^{\beta_p \phi}\right)(z\overline{z})^{-2+\Delta+\Delta_p} \quad , \tag{2.103}$$

where

$$V_p = V_{\Delta}^{m} e^{\beta_p \phi} \tag{2.104}$$

is the intermediate state coming from the contraction of $\widetilde{T}_{k_1}$ and $\widetilde{T}_{k_4}$. Notice that V_{Δ}^{m}, is a matter operator of conformal dimension Δ and $e^{\beta_p \phi}$ its Liouville counterpart with dimension Δ_p. The factor $(z\overline{z})^{-2}$ arises from the fact that both $\widetilde{T}_{k_1}$ and $\widetilde{T}_{k_4}$ are $(1,1)$ conformal operators. The unknown function f depends continuously on β_1, β_4, β_p and defines the fusion of $\widetilde{T}_{k_1}$ and $\widetilde{T}_{k_4}$. Due to the non-conservation of the Liouville momentum β_p we do not necessarily have $\beta_p = \beta_1 + \beta_4$ and there is a (continuous) infinite set of possible values for β_p in the right hand side of (2.103). This is taken into account by the integral over β_p.

From (2.103) we have the following contribution to $\mathcal{A}_4$

$$\begin{aligned}\int_{|z|<\epsilon} d^2 z \widetilde{T}_{k_1}(0)\widetilde{T}_{k_4}(z) &\sim \int_{-\infty}^{\infty} d\beta_p \int_{|z|<\epsilon} d^2 z\,(z\overline{z})^{-2+\Delta+\Delta_p} V_p(0) f(\beta_p, \beta_1, \beta_4) \\ &\sim \int_{-\infty}^{\infty} d\beta_p \frac{V_p(0) f(\beta_p, \beta_1, \beta_4)}{\Delta + \Delta_p - 1}(z\overline{z})^{\Delta+\Delta_p-1}\Big|_0^{\epsilon^2} \quad .\end{aligned} \tag{2.105}$$

Since the momentum k of the matter field X is conserved we have:

$$\Delta = \frac{1}{2}(k_1 + k_4 - \alpha_0)^2 - \frac{1}{2}\alpha_0^2 \quad . \tag{2.106}$$

Moreover, assuming that the intermediate state has imaginary energy ($E = ip$), i.e.,

$$\beta_p = -\frac{Q}{2} + ip \quad , \tag{2.107}$$

we have

$$\Delta_p = \Delta\left(e^{\beta_p \phi}\right) = -\frac{1}{2}\beta_p(\beta_p + Q) = \frac{1}{2}p^2 + \frac{1}{2}\alpha_0^2 + 1 \tag{2.108}$$

thus

$$\Delta + \Delta_p - 1 = \frac{1}{2}(p^2 + (k_1 + k_4 - \alpha_0)^2) \geq 0 \quad . \tag{2.109}$$

Therefore, we finally obtain, upon substituting back into (2.105) the integrated operator product expansion

$$\int_{|z|<\epsilon} d^2 z \widetilde{T}_{k_1}(0)\widetilde{T}_{k_4}(z) \sim \int_{-\infty}^{\infty} dp \frac{V_p(0) f(p, \beta, p_4) \epsilon^{p^2 + (k_1 + k_4 - \alpha_0)^2}}{(p + i|k_1 + k_4 - \alpha_0|)(p - i|k_1 + k_4 - \alpha_0|)} \quad . \tag{2.110}$$

Now we close the integration contour in the upper (or lower) complex p-plane. The final contribution comes from the residue at $p = \pm i|k_1 + k_4 - \alpha_0|$. Analogously, the regions around $z = \infty$ and $z = 1$ have their main contribution at $p = \pm i|k_1 + k_2 - \alpha_0|$ and $p = \pm i|k_1 + k_3 - \alpha_0|$ respectively. Therefore we expect $\mathcal{A}_4$ to be a function of $i|k_i + k_4 - \alpha_0|$, ($i = 1,2,3$); in order to have a clue about this function we look at the following identity:

$$\begin{aligned} \left\langle \widetilde{T}_{k_1}\widetilde{T}_{k_2}\widetilde{T}_{k_3}\widetilde{T}_{k_4=0}\right\rangle &= \frac{\partial}{\partial\mu}\left\langle \widetilde{T}_{k_1}\widetilde{T}_{k_2}\widetilde{T}_{k_3}\right\rangle \\ &= \left(-\frac{\alpha_-}{2}\sum_{i=1}^{3}|k_i - \alpha_0| - \frac{1}{2}(1 + \frac{1}{2}\alpha_-^2)\right)\left\langle \widetilde{T}_{k_1}\widetilde{T}_{k_2}\widetilde{T}_{k_3}\right\rangle \quad . \end{aligned} \tag{2.111}$$

The above formula and our previous discussion suggest that we write, for arbitrary values of k_4, the expression

$$\mathcal{A}_4 = \left\langle \widetilde{T}_{k_1}\widetilde{T}_{k_2}\widetilde{T}_{k_3}\widetilde{T}_{k_4}\right\rangle = \left(-\frac{\alpha_-}{2}\sum_{i=1}^{3}|k_i + k_4 - \alpha_0| - \frac{1}{2}(1 + \frac{\alpha_-^2}{2})\right)\left\langle \widetilde{T}_{k_1}\widetilde{T}_{k_2}\widetilde{T}_{k_3}\right\rangle \quad . \tag{2.112}$$

Thus, if we define the three-point function to be unit ($\mu = 1$), we arrive at

$$\mathcal{A}_4 = -\frac{\alpha_-}{2}\sum_{i=1}^{3}|k_i + k_4 - \alpha_0| - \frac{1}{2}(1 + \frac{1}{2}\alpha_-^2) \quad . \tag{2.113}$$

This is in complete agreement with matrix model results for $c = 1\,(\alpha_0 = 0)$. The cuts at intermediate momenta have also been seen. Formula (2.113) agrees with the known result (2.102) in the kinematic region $k_1, k_2, k_3 > 0$, $k_4 \leq \alpha_0$.

Actually, we can form a nice picture of (2.105) in terms of space-time Feynman diagrams at tree level, which can be used to generalize the analytic continuation for $N > 4$ particles scattering. To understand such picture it is necessary to look at the two-point function $\langle\widetilde{T}_k\widetilde{T}_{2\alpha_0-k}\rangle$, which can be obtained by integrating the three point function with respect to the cosmological constant

$$\frac{\partial}{\partial\mu}\langle\widetilde{T}_k\widetilde{T}_{2\alpha_0-k}\rangle = \langle\widetilde{T}_k\widetilde{T}_{2\alpha_0-k}\widetilde{T}_{k=0}\rangle \quad . \tag{2.114}$$

Using (2.55*b*) and (2.67) one finds

$$\langle\widetilde{T}_k\widetilde{T}_{2\alpha_0-k}\rangle|_{\mu=1} = -\frac{1}{\alpha_-|k-\alpha_0|} \quad . \tag{2.115}$$

If we identify the tachyon propagator with the inverse 2-point function, we find

$$\frac{1}{2}\left(\langle\widetilde{T}_k\widetilde{T}_{2\alpha_0-k}\rangle\right)^{-1} = -\frac{\alpha_-}{2}|k-\alpha_0| \quad . \tag{2.116}$$

Using now the momentum conservation law $k_1+k_2+k_3+k_4 = 2\alpha_0$ we can understand the first three terms in the r.h.s. of (2.113) as coming from intermediate tachyons in the s, t and u channels of a four-legs diagram with the propagators stripped from the external legs, i.e.,

$$-\frac{1}{2}|k_n+k_m-2\alpha_0| = -\frac{1}{2}|k_i+k_j-2\alpha_0| \quad , \tag{2.117}$$

which corresponds to the diagram

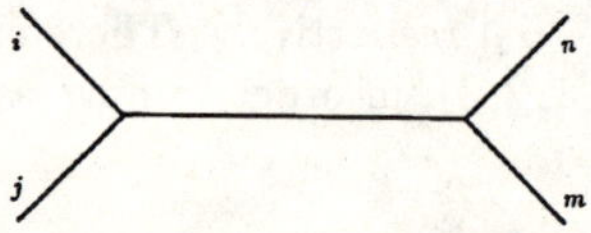

whereas the last term on the right hand side of (2.113) corresponds to a new irreducible interaction which we call $\mathcal{A}_4^{1PI}$,

$$\mathcal{A}_4^{1PI} = -\frac{1}{2}(1+\frac{1}{2}\alpha_-^2) \quad , \tag{2.118}$$

that can be depicted as

therefore, (2.113) can be depicted as in Fig. 2.1 below

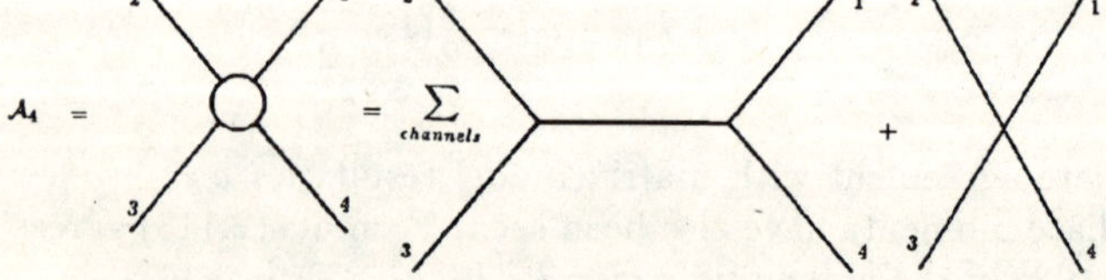

Fig. 2.1: Contributions to the $\mathcal{A}_4$ amplitude.

An analogous picture holds for $N > 4$. Actually, realizing that the insertion of the cosmological term is achieved by means of a derivative with respect to μ, we derive from (2.100b)

$$\langle T_{k_1} \cdots T_{k_N} T_{k=0} \rangle = \left[-\frac{1}{2}\alpha_- \sum_{i=1}^{N} |k_i - \alpha_0| - (\frac{N}{2} - 1)(1 + \frac{1}{2}\alpha_-^2) \right] \langle T_{k_1} \cdots T_{k_N} \rangle \ . \tag{2.119}$$

For $N = 4$ this expression suggests the diagrammatic picture of Fig. 2.2. The contribution $\mathcal{A}_5^{1PI}$ is a contact term . We complete now $\mathcal{A}_5$, which is known in the given kinematic region as before.

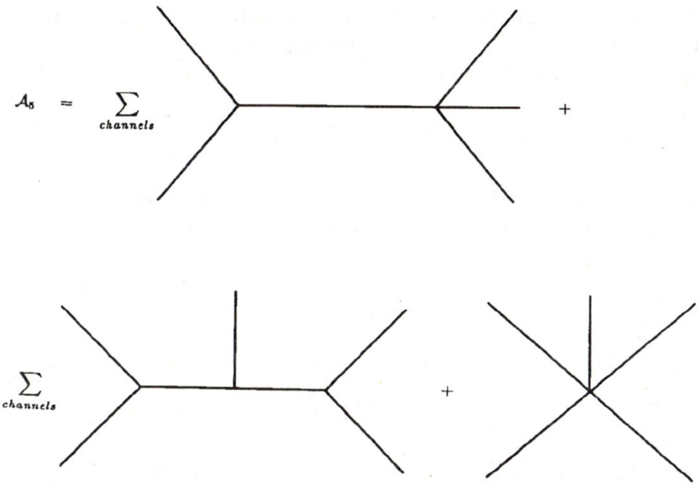

Fig. 2.2: Contributions to the five point function.

The strategy to extend $\mathcal{A}_5$ to other kinematic regions is first to compute the reducible diagrams in Fig. 2.2 through the following Feynman rules:

i) to each internal line with momentum k we associate a propagator $-\frac{\alpha_-}{2}|k - \alpha_0|$;
ii) for each three legs vertex we have a factor $\mathcal{A}_3^{1PI} = \mu^s = 1$ (for $\mu = 1$);
iii) each four legs vertex implies a factor $\mathcal{A}_4^{1PI} = -\frac{1}{2}(1 + \frac{1}{2}\alpha_-^2)$, and finally
iv) the momentum is conserved through any vertex.

We have for the five point function, the following elements: for the first diagram in Fig. 2.2 containing only one propagator and a factor $\mathcal{A}_4^{1PI}$

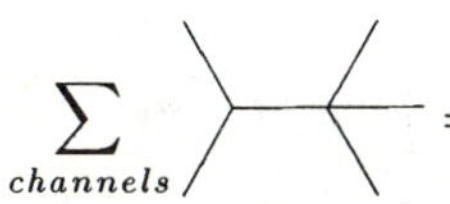

$$= \left(\sum_{i<j=1}^{5} -\frac{\alpha_-}{2}|k_i + k_j - \alpha_0| \right) (-\frac{1}{2})((1 + \frac{1}{2}\alpha_-^2) \tag{2.120}$$

while for the second set, which is a product of two propagators we obtain

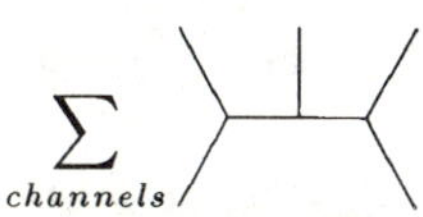

$$= \frac{\alpha_-^2}{4} [|k_2 + k_3 - \alpha_0||k_4 + k_5 - \alpha_0| + |k_2 + k_4 - \alpha_0||k_3 + k_5 - \alpha_0|$$
$$+ |k_3 + k_4 - \alpha_0||k_2 + k_5 - \alpha_0| + \text{symmetrizations}] \quad . \tag{2.121}$$

It is clear that expressions (2.120) and (2.121) hold in any kinematic region and do not need to be further examined in order to be extended.

The last step now is to calculate the irreducible part $\mathcal{A}_5^{1PI}$. We first notice that the cuts at $k_i + k_j = \alpha_0$ only appear due to zero energy intermediate particles (remember $p = \pm i|k_i + k_j - \alpha_0| = -iE$), thus they can only correspond to the massless "tachyons" T_k. The remaining massive intermediate states only appear at (non-vanishing) discrete values of the momentum. Therefore their Liouville momentum cannot be integrated over continuum values, even though it is not conserved. Thus they never produce cuts in the amplitudes and it is reasonable to associate their contribution to the irreducible part $\mathcal{A}_N^{1PI}$. Thus it is sensible to assume that $\mathcal{A}_N^{1PI}$ is analytic in the momentum. So our strategy to obtain $\mathcal{A}_5^{1PI}$ is to calculate it in a given kinematic region and then to analytically continue it. This can be done with the help of Fig. 2.2 since in the kinematic region $k_1, k_2, k_3, k_4 \geq \alpha_0$, $k_5 \leq \alpha_0$ we have (for $\mu = 1$) from (2.100b)

$$\mathcal{A}_5 = (s+2)(s+1) = \left(\alpha_- k_5 - \frac{\alpha_-^2}{2}\right)\left(\alpha_- k_5 - \frac{\alpha_-^2}{2} - 1\right) , \tag{2.122}$$

where we have used $k_N = \frac{N-3+s}{2}\alpha_+ + \frac{1}{2}\alpha_-$. Expressions (2.120) and (2.121) become respectively, in the above kinematic regions

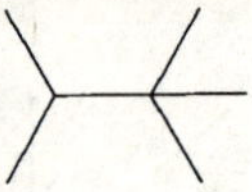

$$\sum_{channels} \quad = -\frac{1}{4}(\alpha_-^2 + 2)\left(1 - \frac{1}{2}\alpha_-^2 + 3\alpha_- k_5\right) , \tag{2.123}$$

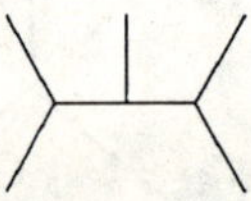

$$\sum_{channels} \quad = \frac{1}{4}\alpha_-^2 \left(\sum_{i=1}^{5} k_i^2 + 4k_5^2 - \alpha_0(2k_5 + \alpha_0)\right) . \tag{2.124}$$

Above $|k_i+k_j-\alpha_0| = k_i+k_j-\alpha_0$, $i,j = 1,2,3,4$ holds true, as well as $|k_i+k_5-\alpha_0| = \alpha_0 - k_i - k_5$, due to momentum conservation, $\sum_1^5 k_i = 2\alpha_0$. Thus, in the given kinematic region[5)] we have from Fig. 2.2, subtracting (2.123) and (2.124) from (2.122),

$$\mathcal{A}_5^{1PI} = -\frac{1}{4}\alpha_-^2 \sum_{i=1}^{5} k_i^2 + \frac{1}{4}\alpha_-^4 + 1 . \tag{2.125}$$

Since $\mathcal{A}_5^{1PI}$ is analytic and symmetric in k_i the expression (2.125) holds in any kinematic region and we finally obtain

$$\begin{aligned}\mathcal{A}_5 = &\left(\sum_{i<j=1}^{5} -\frac{\alpha_-}{2}|k_i + k_j - \alpha_0|\right)(-\frac{1}{2})\left(1 + \frac{1}{2}\alpha_-^2\right) \\ &+ \frac{\alpha_-^2}{4}\Big[|k_2 + k_3 - \alpha_0||k_4 + k_5 - \alpha_0| + |k_2 + k_4 - \alpha_0||k_3 + k_5 - \alpha_0| \\ &+ |k_3+k_4-\alpha_0||k_2+k_5-\alpha_0| + \text{symmetrizations}\Big] \\ &- \frac{\alpha_-^2}{4}\sum_{i=1}^{5} k_i^2 + \frac{\alpha_-^4}{4} + 1 .\end{aligned} \tag{2.126}$$

5) In the limit $\alpha_0 \to 0$ ($\alpha_- \to -\sqrt{2}$) expression (2.125) reduces to $\mathcal{A}_5^{1PI} = -\frac{1}{2}\sum_1^5 k_i^2 + 2$ which was calculated in Appendix B of ref. [75]

The same strategy can be clearly repeated for $\mathcal{A}_{N>5}$, see appendix B of ref. [75] for other examples in the case $c = 1$ ($\alpha_0 = 0$).

We have shown how to continue to any kinematic region the amplitudes obtained via integral representation valid in the specific region $(N-1,1)$, where $k_1, \cdots k_{N-1} \geq \alpha_0$, $k_N \leq \alpha_0$. At this point one could inquire about what happens if the integral is computed directly in another kinematic region. Suppose that we have l momenta k_i running faster than α_0 (right-runners), and $N-l$ slower (left-runners), that is

$$\begin{aligned} k_1 \cdots k_l \geq \alpha_0 \quad , \\ k_{l+1} \cdots k_N \leq \alpha_0 \end{aligned} \quad , \tag{2.127}$$

one finds

$$\begin{aligned} \beta_i &= -\frac{Q}{2} + k_i - \alpha_0 \quad , \quad i = 1, \cdots, l \\ \beta_i &= -\frac{Q}{2} + \alpha_0 - k_i \quad , \quad i = l+1, \cdots, N \quad . \end{aligned} \tag{2.128}$$

Supposing for simplicity, without loss of generality that $s = 0$. From energy and momentum conservation we have

$$\begin{aligned} &\sum_{i=1}^{N} k_i = 2\alpha_0 \quad , \\ &\sum_{i=1}^{N} \beta_i = -\frac{N}{2}Q + \alpha_0(N-2l) + \sum_{i=1}^{l} k_i - \sum_{i=l+1}^{N} k_i = -Q \quad ; \end{aligned} \tag{2.129}$$

therefore we find for the left-runners

$$2 \sum_{i=l+1}^{N} k_i = \left(-\frac{N}{2} + 1\right) Q + \alpha_0(N - 2l + 2) \quad . \tag{2.130}$$

Now using the definitions

$$-Q = \alpha_+ + \alpha_- \, ; \, Q = d_+ - d_- \, ; \, \alpha_0 = \frac{1}{2}(d_+ + d_-) \quad , \tag{2.131}$$

we have, for (2.130) the expression

$$\sum_{i=l+1}^{N} k_i = \frac{-l+2}{2} d_+ + \frac{N-l}{2} d_- \quad . \tag{2.132}$$

Moreover we find for the corresponding dressings the following expression for their sum

$$\sum_{i=l+1}^{N} \beta_i = \sum_{i=l+1}^{N} (d_- - k_i) = \frac{l-2}{2} d_+ + \frac{N-l}{2} d_- \quad . \tag{2.133}$$

Therefore if we add the "energy-momentum relations" we find

$$\sum_{i=l+1}^{N} (k^2 - \beta^2) = (l-2)d_+d_- \quad . \tag{2.134}$$

For the left-runners we finally arrive at the expression

$$-(\sum_{i=l+1}^{N} k)^2 + (\sum_{i=l+1}^{N} \beta)^2 + \sum_{i=l+1}^{N} (k^2 - \beta^2) = (l-2)\left(1 - \frac{N-l}{2}\right) d_+d_- \quad . \tag{2.135}$$

Since we know that

$$d_+d_- = (\alpha_0 + \sqrt{2+\alpha_0^2})(\alpha_0 - \sqrt{2=\alpha_0^2}) = -2 \quad , \tag{2.136}$$

we find

$$\sum_{\text{left runers}} (k_ik_j - \beta_i\beta_j) = \text{positive integer, unless } l = 1 \text{ or } l = N-1 \quad . \tag{2.137}$$

Since we expect a factor $\Delta\left(\sum\limits_{leftrun} (k_ik_j - \beta_i\beta_j)\right)$ in the final result, we must have a vanishing amplitude, unless we are in the given kinematic region of the integral. This explains, as a corollary, why we have obtained a vanishing result for the integral when $\alpha_0 \geq 0$. In this case we have $(N-1,1) \equiv (N-1, s+1)$, that is, the s factors coming from the zero mode integration can be seen as left moving components, and the integral is thus outside of the convergence region.

2.5. c=1 Theory

2.5.1. Discrete States

We have seen that the correlators display a large number of new states characterized by the poles in the amplitudes. These new states, which we call from now on "special states", are in fact an important issue in the theory of non-critical strings. We shall see in the following how they happen to show up in general and what are their consequences.

It is possible to construct an $SU(2)$ current algebra out of a $c=1$ conformal field theory, represented by vertex operators defined in terms of canonical massless fields. Indeed, let $X(z)$ be the holomorphic part of the afore mentioned field, namely

$$\underbrace{X(z)X(w)} = -\ln(z-w) \tag{2.138}$$

and the vertex operator

$$V_k = e^{ik\sqrt{2}X} \quad , \tag{2.139}$$

with $k \in Z$. The realization of $SU(2)$ is given by[85,86,99]

$$\begin{aligned} J_\pm &= \mathrm{e}^{\pm i\sqrt{2}X} \quad , \\ J_3 &= i\sqrt{2}\partial X \quad . \end{aligned} \tag{2.140}$$

The action of $J_\pm$ and J_3 on Hilbert space operators, is given, as usually, by the short distance expansions, and their realization is obtained by the quantities[85,86]

$$\hat{J}_\pm(z) = \oint \frac{\mathrm{d}w}{2\pi i} J_\pm(w+z) \quad , \tag{2.141a}$$

$$\hat{J}_3(z) = \oint \frac{\mathrm{d}w}{2\pi} \sqrt{2}\partial X(w+z) \quad . \tag{2.141b}$$

In this way the operator $V_{s/2}\mathrm{e}^{\frac{i}{\sqrt{2}}sX}$ of conformal dimension $(s/2)^2$ transforms as the highest weight of an $SU(2)$ representation of spin $s/2$. The operator J_- lowers the spin by the unit, such that

$$V_{j,m} = (J_-)^{j-m} V_j \tag{2.142}$$

has spin (j,m) as implicit in the notation. The above operators become polynomials in derivatives of X multiplied by the exponential.

If we couple this system to gravity, by means of the introduction of the Liouville field, we have to consider now

$$W_{j,n} = V_{j,n}\mathrm{e}^{\beta\phi} \quad , \tag{2.143}$$

in such way that the total conformal dimension is one; therefore, since $c=1$ and the stress-tensor of the ϕ field is

$$T_\phi = -\frac{1}{2}(\partial\phi)^2 + \sqrt{2}\phi \quad , \tag{2.144}$$

we obtain

$$\beta = (-1\pm j)\sqrt{2} \quad , \quad W^\pm_{j,j} = \mathrm{e}^{ij\sqrt{2}X+(-1\pm j)\sqrt{2}\phi} \quad . \tag{2.145}$$

Note that $W_{0,0}$ is the cosmological operator. The above "ring" of operators leads to the existence of symmetries of the conformal field theory under study. The states give rise to poles at discrete values of momenta, in the Koba-Nielsen amplitudes, as we see from the amplitudes (2.100a) obtained using the Coulomb gas formalism.

First, we analyze the structure of the OPE, and show how to compute the structure constants. If we consider the operator product expansion

$$W^{(+)}_{j_1,n_1}(z)W^{(+)}_{j_2,n_2}(w) \sim \frac{1}{z-w}\sum f^{j_3,n_3}_{j_1,n_1,j_2,n_2} W^{(+)}_{j_3,n_3}(w) \quad , \tag{2.146}$$

we aim at the computation of f, which is related to the 3-string coupling . We follow now the argumentation of Klebanov and Polyakov[85]. First one notices that due to the zero mode structure, namely $e^{i(n_1+n_2)\sqrt{2}X_0+(j_1+j_2-2)\sqrt{2}\varphi_0}$, the r.h.s. may contain only the term $j_3 = j_1 + j_2 - 1$, $n_3 = n_1 + n_2$.

Thus we write

$$f^{j_1+j_2-1,n_1+n_2}_{j_1,n_1,j_2,n_2} = C^{j_1+j_2-1,n_1+n_2}_{j_1,n_1,j_2,n_2} g(j_1,j_2) \quad , \tag{2.147}$$

where C are the Clebsh-Gordan coefficients and $g(j_1, j_2)$ an arbitrary function; in this way the n dependence is entirely contained in the Clebsh-Gordan, as a result of the $SU(2)$ invariance

$$C^{j_1+j_2-1,n_1+n_2}_{j_1,n_1,j_2,n_2} = \frac{N(j_3,n_3)}{N(j_1,n_1)N(j_2,n_2)} \frac{j_2n_1 - j_1n_2}{\sqrt{j_3(j_3+1)}} \quad , \tag{2.148}$$

with

$$j_3 = j_1 + j_2 - 1\,,\; n_3 = n_1 + n_2 \quad , \quad N(j,n) = \left[\frac{(j+n)!(j-n)!}{(2j-1)!}\right]^{1/2} \quad . \tag{2.149}$$

In order to compute $g(j_1, j_2)$ we can fix n_1 and n_2 to arbitrary values. Thus we put $n_1 = j_1 - 1$, $n_2 = j_2$, in which case we have

$$W^{(+)}_{j_1,j_1-1}(z)W^{(+)}_{j_2,j_2}(0) \approx z^{-1} f^{j_1+j_2-1,j_1+j_2-1}_{j_1,j_1-1,j_2,j_2} W^{(+)}_{j_1+j_2-1,j_1+j_2-1}(0) \quad . \tag{2.150}$$

We use the previously defined representation (2.142) and (2.143), finding

$$\begin{aligned} &W^{(+)}_{j_1,j_1-1}(z)W^{(+)}_{j_2,j_2}(0) = \\ &= \frac{1}{\sqrt{2j_1}}\oint\frac{dw}{2\pi i}:e^{-i\sqrt{2}X(z+w)}::e^{ij_1\sqrt{2}X(z)}::e^{ij_2\sqrt{2}X(0)}::e^{(j_1-1)\sqrt{2}\phi(z)}::e^{(j_2-1)\sqrt{2}\phi(0)}: \\ &= \frac{1}{\sqrt{2j_1}}\oint\frac{dw}{2\pi i} w^{-2j_1}(z+w)^{-2j_2} z^{2(j_1+j_2-1)} W^{(+)}_{j_1+j_2-1,j_1+j_2-1}(0) \quad , \end{aligned} \tag{2.151}$$

and finally after substituting $w = zw'$, $w' \to w$

$$W^{(+)}_{j_1,j_1-1}(z)W^{(+)}_{j_2,j_2}(0) = z^{-1}\left[\frac{1}{\sqrt{2j_1}}\oint\frac{dw}{2\pi i} w^{-2j_1}(1+w)^{-2j_2}\right] W^{(+)}_{j_1+j_2-1,j_1+j_2-1}(0) \tag{2.152}$$

The integral is readily computed expanding $(1+w)^{-2j_2}$ in a power series around zero, and the result corresponds to the $2j_1^{\text{th}}$ term. The result for the term inside brackets reads

$$[\cdots] = -\frac{(2j_1+2j_2-2)!}{\sqrt{2j_1}(2j_1-1)!(2j_2-1)!} \quad . \tag{2.153}$$

Therefore we obtain for the $g(j_1,j_2)$ defined in (2.147) the result

$$g(j_1,j_2) = -\frac{\sqrt{j_1+j_2}(2j_1+2j_2-2)!}{\sqrt{2j_1j_2}(2j_1-1)!(2j_2-1)!}\ , \tag{2.154}$$

and for the structure constant we find the expression

$$f^{j_3,m_3}_{j_1,m_1,j_2,m_2} = \delta_{j_3,j_1+j_2-1}\delta_{m_3,m_1+m_2}\frac{\widetilde{N}(j_3,m_3)}{\widetilde{N}(j_1,n_1)\widetilde{N}(j_2,n_2)}(j_2n_1-j_1n_2)\quad , \tag{2.155}$$

where

$$\widetilde{N}(j,n) = -\sqrt{\frac{j}{2}}(2j-1)!N(j,n)\quad . \tag{2.156}$$

Finally, redefining

$$W^{(+)}_{j,n} \to \frac{1}{\widetilde{N}(j,n)}W^{(+)}_{j,n}\quad , \tag{2.157}$$

we get the vertex operator algebra

$$W^{(+)}_{j_1,n_1}(z)W^{(+)}_{j_2,n_2}(0) = \frac{1}{z}(j_2n_1-j_1n_2)W^{(+)}_{j_1+j_2-1,n_1+n_2}(0)\quad . \tag{2.158}$$

Notice the similarity of the structure constant to the case of W_∞ algebra! Moreover one has[85]

$$\begin{aligned} W^{(-)}_{j_1,n_1}W^{(-)}_{j_2,n_2} &= z^{-1}\times 0 \\ W^{(-)}_{j_1,n_1}W^{(-)}_{j_2,n_2} &= z^{-1}\times 0\quad ,\quad j_1\geq j_2+1 \\ W^{(+)}_{j_1,n_1}W^{(-)}_{j_1+j_2-1,-n_1-n_2} &= z^{-1}(j_2n_1-j_1n_2)W^{(-)}_{j_2,-n_2}(0)\quad ,\quad j_1<j_2+1\quad . \end{aligned} \tag{2.159}$$

Indeed, defining now

$$Q^{(\pm)}_{j,n} = \oint\frac{dz}{2\pi i}W^{(\pm)}_{j,n}$$

one has

$$[Q^{(+)}_{j_1,n_1},Q^{(+)}_{j_2,n_2}] = (j_2n_1-j_1n_2)Q^{(+)}_{j_1+j_2-1,n_1+n_2} \tag{2.161}$$

which is the algebra of area-preserving diffeomorphisms of a plane[135].

We have also

$$[Q^{(+)}_{j_1,n_1},Q^{(-)}_{j_2,n_2}] = (j_2n_1+j_1n_2+n_1)Q^{(-)}_{j_2-j_1+1,n_1+n_2}\quad , \tag{2.162}$$

with $j_2-j_1+1>0$, $|m_1+m_2|\leq j_2-j_1+1$. However, new states may appear at other values of the ghost number. In fact, discrete states are obtained when taking a family of states $|\alpha(p)\rangle$ parameterized by p. Suppose that we have BRST invariance only at a particular value of $p=p_0$, that is

$$Q|\alpha(p)\rangle = f(p)|\beta(p)\rangle\,,\quad \text{with } f(p_0)=0\quad , \tag{2.163}$$

as it may happen in some explicit examples[96].

In such a case, $|\alpha(p_0)\rangle$ and $|\beta(p_0)\rangle$ are a pair of discrete states at adjacent values of the ghost number. If we consider the spin 0 BRST invariant fields of ghost number 1

$$Y_{j,n}^{(\pm)} = cW_{j,n}^{(\pm)} \quad , \tag{2.164}$$

we verify that for $|n| < j$ there are partners of ghost number 0 and 2. The ghost number 0 partners have been defined by Witten[86] as $\mathcal{O}_{l,n}$, $l = j-1$. The spin zero ghost number zero operators are

$$\mathcal{O}_{\frac{1}{2},\frac{1}{2}} = \left(cb + \frac{i}{\sqrt{2}}(\partial X - i\partial\phi) \right) \mathrm{e}^{\frac{i}{\sqrt{2}}(X+i\phi)} \quad , \tag{2.165a}$$

$$\mathcal{O}_{\frac{1}{2},-\frac{1}{2}} = \left(cb - \frac{i}{\sqrt{2}}(\partial X + i\partial\phi) \right) \mathrm{e}^{-\frac{i}{\sqrt{2}}(X-i\phi)} \quad . \tag{2.165b}$$

Such spin zero ghost number zero operators lead to new symmetries of the conformal field theory, and to the existence of the "ground ring". The spin $(1,0)$ current is

$$\mathcal{J}_{j,n,n'} = W_{j,n}^{+} \overline{\mathcal{O}}_{j-1,n'} \quad . \tag{2.166}$$

We shall come back to this point later on. Results of the previous sections indicate that discrete states appear also naturally as poles in the tachyon scattering amplitudes.

For $c<1$ matter, a similar discussion is also possible, since $c<1$ can be understood as a rotation of the $c=1$ theory. The latter is the simplest one, and should be the first to be discussed. As we have seen, the 4-point scattering is proportional to $\frac{\Gamma(1-\sqrt{2}|k|)}{\Gamma(\sqrt{2}|k|)}$ displaying poles at $k = (N-1)/\sqrt{2}$, which are the momenta where extra physical states arise. Extra physical states show up at discrete values of momenta satisfying the requirement of a physical state, due to their small statistical weight in the partition function, but for $d \leq 1$, they turn out to be essential. As remarked by Polyakov[96], such a phenomenon occurs already in ferromagnetic phase transition, where constraints disappear at zero momentum, enlarging the number of degrees of freedom; see (2.163) in such a context.

A better understanding of the problem is obtained by means of the method developed recently by several authors, based on the BRST cohomology of two-dimensional gravity. Although a full discussion of such matters is beyond the scope of this book, we present here a summary of results, which may be useful for the interested reader, who wants to compare the results previously obtained with more formal developments in terms of well defined objects in the BRST approach, which shows, in an elegant way, an abstract formulation of those results. One considers the BRST operator[100,101]

$$Q = \int \frac{\mathrm{d}z}{2\pi i} \colon\! \left(T(z) + \frac{1}{2} T^G(z) \right) c(z) \colon \quad , \tag{2.167}$$

and looks for the relative cohomology $H^{\mathrm{rel}}V$ on the Virasoro module V (as e.g., to simplify matters the Liouville sector - up to a positive energy condition) defined in terms of the relative BRST complex

$$C_n^{\mathrm{rel}} V = \{ \omega \in V \otimes L(c_{p,q}\Delta_{r,s}) \otimes \Lambda^{bc} \,, (L_0 - 1)\omega = 0 = b_0 \omega \,, \omega \text{ has antighost number } n \} \tag{2.168}$$

where $L(c_{p,q}, \Delta_{r,s})$ is the irreducible Virasoro representation with central charge $c_{p,q}$ and highest weight $\Delta_{r,s}$, and Λ^{bc} the ghost Fock space.

The solution of the BRST cohomology has been obtained upon consideration of the embedding diagram $E_{r,s}(p,q)$ of Verma modules[102]

$$\begin{array}{ccccccccc} b_0 & \longrightarrow & a_{-1} & \longrightarrow & b_{-1} & \longrightarrow & a_{-2} & \cdots \\ & \searrow & & \times & & \times & & \cdots \\ & & a_0 & \longrightarrow & b_1 & \longrightarrow & a_1 & \cdots \end{array} \tag{2.169}$$

where the above symbols represent the Kac-elements

$$\begin{aligned} a_t &= \frac{(2pqt + pr + qs)^2 - (p-q)^2}{4pq} \quad , \\ b_t &= \frac{(2pqt + pr - qs)^2 - (p-q)^2}{4pq} \quad , \end{aligned} \tag{2.170}$$

and we have

$$\begin{aligned} a_t &= \Delta_{r,2pt-s} = \Delta_{2qt-r,s} \quad , \\ b_t &= \Delta_{r,s-2pt} = \Delta_{r-2qt,s} \quad . \end{aligned} \tag{2.171}$$

Lian and Zuckerman[100] considered the dual diagram $\widetilde{E}_{r,s}(p,q)$ obtained replacing Δ by $1-\Delta$, in order to consider the gravity sector, since the dressed vertex is such that $\Delta_M + \Delta_G = 1$ This is also an embedding diagram, but it is clear that the central charge is $26 - c$ in order to match the total vanishing central charge $(c_{\mathrm{Matt.}} + c_{\mathrm{Liouv.}} - 26 = 0)$. Using these facts, they showed that the required homology is obtained from the dual diagram, such that the number of arrows $d_\Delta =$ number of arrows leading from Δ to $1 - \Delta_{r,s}$ in $\widetilde{E}_{r,s}(p,q)$ fixes the cohomology by means of the theorem

$$H^{\mathrm{rel}} L(26-c, \Delta) \neq 0 \quad \text{iff} \quad \Delta \in \widetilde{E}_{r,s}(p,q) \quad , \tag{2.172}$$

and if the latter is valid

$$\begin{aligned} \dim H_n^{\mathrm{rel}} L(26-c,\Delta) &= \begin{cases} 1 + (-1)^{n+d} - \delta_{|n|,d_\Delta} & \text{if} |n| < d_\Delta \\ 0 & \text{otherwise} \end{cases} \\ \dim H_n^{\mathrm{rel}} M(26-c,\Delta) &= \delta_{n+d_\Delta,0} \quad , \end{aligned} \tag{2.173}$$

where M is the Virasoro module, and L are the irreducible coefficients. This means that the ring structure defined by[103]

$$H_{LZ} = \bigoplus_{n \in Z} H_n \quad , \tag{2.174}$$

contains *an infinite number of physical states at different ghost numbers, appearing at the conformal weights where singular vectors arise in the Verma modules built on the matter primary states*[104]. In this language, the infinite tower of physical states, corresponding to BRST cohomologies for each conformal block, is such that there is a physical state with any given ghost number. It is exactly this set of observables that give rise to an associative ring. The ring structure is given by

$$H_n \times H_m \to H_{n+m} \quad , \tag{2.175}$$

which is defined by the OPE modulo BRST exact terms

$$\mathcal{O}_1(z)\mathcal{O}_2(w) \sim \mathcal{O}_3(w) + [Q_p . \star] \quad , \tag{2.176}$$

or $\mathcal{O}_1 \circ \mathcal{O}_2 = \mathcal{O}_3$. Indeed, using the free fields representation for the Liouville fields one has,

$$\mathrm{e}^{\alpha\phi(z)}\mathrm{e}^{\beta\phi(w)} = (z-w)^{-\alpha\beta}\mathrm{e}^{\alpha\phi(z)+\beta\phi(w)} \quad , \tag{2.177}$$

which is the equivalent to say that

$$\mathcal{O}_1[\beta_1]\mathcal{O}_2[\beta_2] = \mathcal{O}_3[\beta_1+\beta_2] \quad . \tag{2.178}$$

The Lian Zuckerman analysis of the problem of physical states starts with the irreducible matter modules, that is, the null states are removed from the beginning, and the physical states involve ghost oscillators and have non-trivial ghosts numbers. We call the physical states constructed in this way LZ states. There is one physical state with non-trivial ghost number for each null state over the primaries. The Liouville dressing of these states of non-zero ghost number is the same as the dressing of the corresponding null states.

A second construction, more natural from the point of view of the Coulomb gas formulation, is obtained representing the matter modules, realized as bosonic free field modules, in terms of Felder's resolution[105] in order to achieve irreducibility. In this latter case, the physical states of the theory with gravity are dressed tachyon states outside of the basic grid, and Lian Zuckermans ghost number grading is replaced by Felder's resolution grading. States outside the minimal conformal grid (Kac table), when provided with a suitable Liouville dressing will be called DK states[77,78] (Dotsenko-Kitazawa). They have ghost number zero.

Indeed, in the Coulomb gas formulation, the Hilbert space of the model may be taken as a direct sum of the charged bosonic Fock spaces, defined as[105]

$$F_k = \bigoplus_{k=0} \; \bigoplus_{1\geq n_1\cdots\geq n_k} a_{-n_1}\cdots a_{-n_k}|0\rangle_k \quad , \tag{2.179}$$

where a_{-n} is the creation operator of the string variable obeying $[a_n, a_m] = 2n\delta_{n,-m}$; the vertex operator can thus be written as

$$V_k(z) = z^{ika_0}\mathrm{e}^{ik\sum_1^\infty \frac{a_n}{n}z^n}\,\mathrm{e}^{-ik\sum_1^\infty \frac{a_n}{n}z^{-n}} \quad . \tag{2.180}$$

As we already know, the k's are quantized according to (compare with B.103)

$$k_{n',n} = \frac{1}{2}(1-n')d_- + \frac{1}{2}(1-n)d_+ \quad , \tag{2.181}$$

with $d_\pm = \alpha_0 \pm \sqrt{1+\alpha_0^2}$ and for rational conformal field theories $k_{n',n} = k_{p'+n',p+n}$.

The screened vertex operators are defined as

$$V_{n',n}^{r',r} = \int \prod_{i=1}^{r} \mathrm{d}u_i \prod_{j=1}^{r} \mathrm{d}v_j V_{k_{n',n}}(z) \prod_i V_{d_-}(u_i) \prod_j V_{d_+}(v_j) \quad , \tag{2.182}$$

with the integration contour $|z| > |u_1| \cdots > |v_r|$. The BRST charge Q_m is the following

$$Q_m = \frac{1}{m} \int V_{d_+}(v_1) \cdots V_{d_+}(v_m) \prod_{i=1}^{m} \mathrm{d}v_i \quad . \tag{2.183}$$

It maps $F_{m',m}$ into $F_{m',-m}$ and has the properties[105] 1) $Q_{m+n-2r-1} V_{n',n}^{r',r}(z) = \mathrm{e}^{2\pi i k_{n'n} d_+ (m+n-2r-1)} V_{n',n}^{r',n-r-1} Q_m$, where we assumed that $1 \le m' \le p'-1 \quad , \quad 1 \le m \le p-1$;

2) The operation $F_{m',2p-m} \xrightarrow{Q_{p-m}} F_{m',m} \xrightarrow{Q_m} F_{m',-n}$ enjoys the BRST property $Q_m Q_{p-m} = 0$.

3) The space of BRST states $B_{m',m} = \mathrm{Ker}\, Q_m \Im Q_{p-m}$ is isomorphic to the irreducible highest weight module $H_{m',m}$ of weight $h_{m',m}$ and central charge $c = 1 - \frac{6(p'-p)^2}{p'p}$.

The BRST property, follows, very roughly speaking, from the fact that fusing successively an enough number of $V_{d_+}(z_i)$ to $\mathrm{e}^{\frac{i}{2}(1-n)d_+ \phi(w)}$ and integrating over z_i, we eventually arrive at a positive coefficient for the operator, giving a non-singular OPE and as a consequence, vanishing upon integration. Therefore, we can define a Felder cohomology. Since we have found an infinite number of physical states in the Lian Zuckerman formulation, which actually showed up in the computation of higher correlation functions computed by means of the Dotsenko Kitazawa states, it is natural to ask how they are related. In fact, there are two types of cohomology, the LZ and the Felder cohomology (defined by Q_F obtained from $\mathrm{Ker} Q_m / \Im Q_{p-m}$). If we include now the Liouville sector, the Fock space is $\mathcal{F}(\alpha) \otimes \mathcal{F}(\beta) \otimes \mathcal{F}(gh)$, where $\mathcal{F}(\beta)$ is generated by the Liouville operators $\mathrm{e}^{\beta\phi}$ while $\mathcal{F}(\alpha)$ is the previously defined (matter) Fock space.

The Lian-Zuckerman BRST charge is

$$Q_B = \oint c(z) \left(T^M + T^c + \frac{1}{2} T_{gh} \right) \quad . \tag{2.184}$$

Let us consider the double cohomology defined by both Q_B and Q_F. The first charge Q_B implies a label defined by the usual ghost number, which in turn leads us to the LZ states. The other grading defined by Q_F is related to the Felder cohomology, and is given by the distance to the given Fock tower from the central tower in the Felder complex.

The Dotsenko-Kitazawa states

$$|k_{m',m\pm 2np}\rangle_M \otimes |\beta\rangle_L \otimes c_1 |0\rangle_{gh} \quad , \tag{2.185}$$

with β obeying the usual gravitational dressing condition, from representatives of the cohomology class[106]

$$H^{(n)} \left((H_{rel}^{(0)}(\mathcal{F}(\alpha)_M \otimes \mathcal{F}(\beta)_L \otimes \mathcal{F}_{gh}), Q_B), Q_F \right) \quad , \tag{2.186}$$

which in turn, is isomorphic to

$$H_{rel}^{(n)}\Big((H^{(0)}(\mathcal{F}(\alpha)_M\otimes\mathcal{F}(\beta)_L\otimes\mathcal{F}_{gh}),Q_F),Q_B\Big)\sim$$
$$\sim H_{rel}^{(0)}(L(m',m)\otimes\mathcal{F}(\beta)_L\otimes\mathcal{F}_{gh}),Q_B) \qquad (2.187)$$

which defines the LZ states. Since (2.185) are nothing but the DK states, there is a direct relation between LZ and DK states defined by the above isomorphism. Examples can be found in ref. [106].

Thus we find that the DK states also define an infinite set of discrete states, which are, in this case, dressed states

$$e^{ik_{n'n}X+\beta_{n'n}\phi} \quad , \qquad (2.188)$$

with $n'n$ outside the basic grid. Therefore, the correlators computed before generate the whole spectrum of the theory. The $c=1$ case is in fact rather simple, the tachyons being defined in in DK language as

$$V_j=e^{i(j\sqrt{2})X+(1-j)\sqrt{2}\phi} \qquad (2.189)$$

and $V_{j,m}$ defined as before. On the other hand, the energy momentum tensor of the general theory is given by

$$T=T_M+T_L \quad , \qquad (2.190a)$$

$$T_M=-\frac{1}{2}(\partial X)^2+i\alpha_0\partial^2X \quad , \qquad (2.190b)$$

$$T_L=-\frac{1}{2}(\partial\phi)^2+\frac{Q}{2}\partial^2\phi \quad , \qquad (2.190c)$$

with

$$\langle X(z,\bar z)X(w,\bar w)\rangle=\langle\phi(z,\bar z)\phi(w,\bar w)\rangle=-2\ln|z-w| \quad . \qquad (2.191)$$

If we make the linear transformation

$$X=\frac{1}{2\sqrt{2}}(-2i\alpha_0\varphi_1+Q\varphi_2) \quad , \qquad (2.192a)$$

$$\phi=\frac{1}{2\sqrt{2}}(-2i\alpha_0\varphi_2-Q\varphi_1) \quad , \qquad (2.192b)$$

the stress tensor takes the form

$$T=-\frac{1}{2}(\partial\varphi_1)^2-\frac{1}{2}(\partial\varphi_2)^2-\sqrt{2}\partial^2\varphi_1=T_1+T_2 \quad , \qquad (2.193)$$

with

$$\langle\varphi_1(z)\varphi_1(w)\rangle=\langle\varphi_2(z)\varphi_2(w)\rangle=-2\ln|z-w| \quad , \qquad (2.194)$$

which represents an effective $c=1$ "matter" realized by the φ_2 field ($\alpha_0=0$), and a $c=25$ "Liouville" field realized by φ_1 ($\beta_0=-1$). The vertex operators are thus

$$V_{n'n}^{(-)} = \mathrm{e}^{ij_{n'n}^{(-)}\sqrt{2}\varphi_2 + (-1-j_{n'n}^{(-)})\sqrt{2}\varphi_1} \quad , \tag{2.195}$$

with $j_{n'n}^{(-)} = \frac{n'}{2}\rho' - \frac{n}{2}$, $\rho' = \frac{\alpha_-^2}{2} = \frac{p}{p'}$,

$$V_{n'n}^{(+)} = \mathrm{e}^{ij_{n'n}^{(+)}\sqrt{2}\varphi_2 + (-1+j_{n'n}^{(+)})\sqrt{2}\varphi_1} \quad , \tag{2.196}$$

with $j_{n'n}^{(+)} = \frac{n'}{2} + \frac{n}{2}\rho$, $\rho = -\frac{\alpha_+^2}{2} = -\frac{p'}{p}$.

Therefore the $(n', 0)$ states in $j^{(+)}$ are mapped by the linear transformation (2.192) into the $SU(2)$ discrete tachyons of the $c = 1$ theory. This fact hints to the use of mixed matter Liouville screenings for the $c \leq 1$ theory as a consequence of the mixing (2.192). In fact this has been discussed in [77,107] in the framework of fusion rules of the theory coupled to gravity, in order to streamline a confused situation concerning those rules in the case of gravity interacting conformal theories. The infinite number of extra states produce new non-vanishing fusion rules mixing operators inside and outside the minimal grid[78]. Although the fusion rules have vanishing coefficients for the matter sector, the Liouville contribution may be infinite, and the previous rules may get destroyed, as suggested in [78].

We finally have to mention that in the case $c = 1$, we can think as if the basic grid became infinite, and the submodule structure became different, as proposed by Dotsenko in ref. [108]. In fact, the submodules obey an $SU(2)$ structure, by means of the screening operators

$$Q = \oint du\, \mathrm{e}^{\pm i\sqrt{2}X} \quad . \tag{2.197}$$

The screening operators are really needed, according to our previous approach in order to fulfill momentum conservation relations. However, they survive the limit $c \to 1$, and as it turns out, they are nothing but the operators we found in this section. Thus, they have a different role, namely defining new states by local action.

2.5.2. Infinite Symmetries as a Consequence of the Existence of Discrete States

Discrete physical states are remnants of the transverse string excitations, and have rather strong implications in string models. We have seen that the discrete states imply an infinite dimensional (W_∞) algebra. Therefore, the existence of the rings and modules leads to non-trivial constraints for the correlation functions. Polyakov[96] has given arguments to show that non-vanishing correlation functions must have exactly one tachyon of negative energy and the remaining one must have positive energy. This is exactly the result we have found previously. Actually, by inspection of the relevant integrals, we can convince ourselves that this fact is generally true. Indeed, the charge algebra (2.161) imply highly non-trivial Ward identities. First one notices[110] the presence of two separate Virasoro subalgebras

$$\begin{aligned} [Q_{n+1,n}, Q_{m+1,m}] &= 2(m-n)Q_{m+n+1,m+n}\,, \qquad n,m = 0, \frac{1}{2}, 1, \cdots \\ [Q_{-n+1,n}, Q_{-m+1,m}] &= 2(m-n)Q_{-m-n+1,m+n}\,, \quad n,m = 0, \frac{1}{2}, 1, \cdots \end{aligned} \tag{2.198}$$

The operator $W_{m+1,m}$ acts on a tachyon state as

$$W_{m+1,m}(z)\, c\bar{c}\, T_k^+(0) = \frac{1}{z} F_m(k)\, c\bar{c}\, T_{k+m}^+(0) + \cdots \quad , \tag{2.199a}$$

$$T_k^{(+)}(z) = \mathrm{e}^{ik\sqrt{2}X(z)+\sqrt{2}(k-1)\phi(z)} \quad , \tag{2.199b}$$

where notice that k is integer in this subsection and is equal to the momentum divided by $\sqrt{2}$. The function $F_m(k)$ can be computed acting successively with $\overline{\mathcal{O}}_{\frac{1}{2},\frac{1}{2}}$ on the tachyon $c\bar{c}e^{ik\sqrt{2}X+(-1+k)\sqrt{2}\phi}$. One finds

$$F_m(k) = (-1)^{2m}(2k)^2(2k+1)^2\cdots(2k+2m-1)^2(2k+2m) \quad . \tag{2.200}$$

On the other hand, the charge $Q_{m+n+1,m}$ while acting on a $n+1$ tachyon state, reduces the number of tachyons by n. This can be proven as follows. One first computes the action of $Q_{\frac{1}{2},-\frac{1}{2}}$ on a two tachyon state, using the explicit form of it, namely

$$W_{\frac{3}{2},-\frac{1}{2}} = \left[2(\partial X)^2 - i\sqrt{2}\partial^2 X\right]\left[\bar{c}\bar{b} - \frac{i\sqrt{2}}{2}\bar{\partial}X - \frac{\sqrt{2}}{2}\bar{\partial}\phi\right]\mathrm{e}^{\frac{\sqrt{2}}{2}(-iX+\phi)} \quad , \tag{2.201}$$

we are able to compute

$$Q_{\frac{3}{2},-\frac{1}{2}}\, c\bar{c}T_{k_1}^+(0)\int T_{k_2}^+(w,\overline{w}) = I(k_1,k_2)\, c\bar{c}\, T_{k_1+k_2-\frac{1}{2}}^+(0) + \cdots \quad , \tag{2.202}$$

where we have the contractions of the fields $\left[2(\partial X)^2 - i\sqrt{2}\partial^2 X\right]$ with the exponentials giving rise to the factor

$$-\frac{2k_1(2k_1-1)}{z^2} - \frac{2k_2(2k_2-1)}{(z-w)^2} - \frac{8k_1k_2}{z(z-w)} \quad , \tag{2.203}$$

the contraction of the term $-\frac{i\sqrt{2}}{2}\bar{\partial}X - \frac{\sqrt{2}}{2}\bar{\partial}\phi$ with the exponentials leads to $-\frac{1}{\bar{z}-\overline{w}}$; Finally the contraction of the exponentials between themselves leads to

$$|w|^{4(k_1+k_2-1)}|w-z|^{-4k_2+2}|z|^{-4k_1+2} \quad . \tag{2.204}$$

In the w integral we change the variables as $w \to zu$, and verify that the (unwanted) $\bar{z}$ piece cancels completely, and we are left with

$$\begin{aligned} I(k_1,k_2) &= \int \mathrm{d}^2u\,|u|^{4(k_1+k_2-1)}|1-u|^{-4k_2}\left(8k_1k_2 + 2k_1(2k_1-1)(1-u) + \frac{2k_2(2k_2-1)}{1-u}\right) \\ &= 2\pi(2k_1+2k_2-1)\frac{\Delta(2k_1+2k_2-1)}{\Delta(2k_1)\Delta(2k_2)} \quad . \end{aligned} \tag{2.205}$$

Absorbing the Δ factors in the tachyon wave functions, we find

$$Q_{\frac{3}{2},-\frac{1}{2}}|k_1,k_2\rangle = 2\pi(2k_1+2k_2-1)|k_1+k_2-\frac{1}{2}\rangle \quad . \tag{2.206}$$

Now one uses the commutation relation (2.161) to find

$$Q_{m+2,m}|k_1,k_2\rangle = \frac{1}{4m+3}[Q_{\frac{3}{2},-\frac{1}{2}},Q_{m+\frac{3}{2},m+\frac{1}{2}}]|k_1,k_2\rangle = 4\pi(k_1+k_2+m)|k_1+k_2+m\rangle \,. \tag{2.207}$$

One obtains, recursively, upon use of

$$[Q_{n+2,n},Q_{m+2,m}]|k_1k_2k_3\rangle = 4(m-n)Q_{m+n+3,n+m}|k_1k_2k_3\rangle \quad , \tag{2.208}$$

the expression

$$Q_{m+n+1,m}|k_1,k_2,\cdots k_{n+1}\rangle = 2\pi^n(n+1)!(m+\sum_{i=1}^{n+1}k_i)|m+\sum_{i=1}^{n+1}k_i\rangle \quad . \tag{2.209}$$

Analogous formulae can be obtained for the action of $Q_{-m+1,m}$ on the left moving tachyons ($k \le Q_0$). In fact, one can introduce annihilation (creation) operators for the (+) - tachyons $a(k)(a^+(k))$, as well as for the (−) - tachyons $b(k)(b^+(k))$, and find the representation for the charge $Q_{J,m}$

$$\begin{aligned} Q_{J,m} &= 2\pi^{J-m-1}\int \mathrm{d}k \prod_{i=1}^{J-m}\int \mathrm{d}k_1 k a^+(k)\prod_{j=1}^{J-m}a(k_j)\delta(\sum_{l=1}^{J-m}k_l+m-k) \\ &+(-1)^{2m}2\pi^{J+m-1}\int \mathrm{d}p\prod_{i=1}^{J+m}\int \mathrm{d}p_i p b^+(p)\prod_{j=1}^{J+m}b(p_j)\delta(\sum_{l=1}^{J+m}p_l+m-p) \quad . \end{aligned} \tag{2.210}$$

Thus we conveniently choose $Q_{\frac{1}{2},-\frac{1}{2}}$ to act on the 3+1 point function

$$\langle 0|S|k_1,k_2,k_3,k_4^-\rangle \quad ,$$

where S means the S-matrix operator, and the last momentum represents a left-moving tachyon. Using the rules we found in the previous sections, we have $k_1 + k_2 + k_3 = 1$, $k_4 = -\frac{1}{2}$, and the constraint on the 3+1 point function reads

$$\begin{aligned} &-2\langle 0|S|k_1,k_2,k_3;-1\rangle + 2\pi(2k_1+2k_2-1)\langle 0|S|k_1+k_2-\frac{1}{2},k_3;-\frac{1}{2}\rangle \\ &+2\pi(2k_1+2k_3-1)\langle 0|S|k_1+k_3-\frac{1}{2},k_2;-\frac{1}{2}\rangle + 2\pi(2k_2+2k_3-1)\langle 0|S|k_1,k_2+k_3-\frac{1}{2};-\frac{1}{2}\rangle = 0 \end{aligned} \tag{2.211}$$

Therefore, assuming the value unity for the 2+1 point function we find

$$\mathcal{A}_{3,1}(k_1,k_2,k_3) = \pi \quad . \tag{2.212}$$

Or, in general, iterating the above procedure, we have

$$\mathcal{A}_{N,1}(k_i) = \frac{\pi^{N-2}}{(N-2)!} \quad . \tag{2.213}$$

Up to normalization, this is the result found in (2.100b) for $s = 0$. The computation of correlation functions as a direct consequence of the use of symmetries indicates that the constraints they lead to, determine completely the dynamics in a way analogous to the situation of higher-conservation laws for integrable models[96], see [11] and references therein.

2.6. Conclusions

We have computed the dressed correlators involving the tachyon vertex of non-critical string theory. The space-time interpretation is rather clear. If we consider the tachyon $T_k = e^{ik\cdot X+\beta(k)\phi}$ we find, from the BRST invariance that $k^2-\beta(\beta+Q) = 2$ or, from $c_L = 1 + 3Q^2 = 26 - d$,

$$E^2 - k^2 = \left(\beta + \frac{Q}{2}\right)^2 - k^2 = \frac{2-d}{12} = m^2 \quad , \qquad (2.214)$$

which is the dispersion relation for the tachyon in an arbitrary $d > 2$ dimensional theory. For $d = 2$ it is massless.

There are some further different points in our analysis. The first is the existence of background charges for $c < 1$, leading to the modified momentum conservation law, $\sum k_i = 2\alpha_0$. However, the above interpretation permits us to write the wave function corresponding to the tachyon as

$$\psi = e^{ikX+(\beta+\frac{Q}{2})\phi} = g_{st}^{-1} T_k \quad , \qquad (2.215)$$

which differs from the tachyon vertex by the string coupling constant (1.76). In higher dimensions, the tachyon corresponds to the first state from an infinite tower. Below two dimensions, these further states only show up as the so-called discrete states, as discussed in the last sections of this Chapter.

When we have no Liouville self-interaction, we say that we are computing "bulk" amplitudes. This part can be computed also in higher dimensions, and we can already understand their inherent difficulties. Indeed, the four-point amplitude is a relation of Γ functions of $k_i\cdot k_j-\beta_i\beta_j$, and has poles when $(k_i+k_j, \beta_i+\beta_j+Q/2)$ is a physical momentum, $k^2-E^2 = \frac{d-2}{12}+2n$, with n a positive integer. Thus, besides the tachyons one has an infinite tower of states. As discussed in [75] this means that there is a non-trivial back reaction of the string to the propagation of tachyons, and the effect of other modes must unavoidably be taken into account. One has several poles in all possible channels, corresponding to different ways to cut the space time diagrams.

In two dimensions, as we discussed before, the higher modes appear simply as discrete states. The bulk amplitude corresponds to interactions that happened before the Liouville self interaction was important, namely beyond what one calls the "Liouville wall". Expanding the cosmological term, one realizes the presence of a (non-universal) divergence, and a $\ln 1/\mu$ term, which can be interpreted as the volume of the Liouville zero mode. The wall can be seen as a boundary effect. After switching on the Liouville self interaction one can, nevertheless, still compute exactly the amplitudes, and the scaling properties become clear.

After computing the correlators we see that we could perform a wave function renormalization including the pole factors $\Delta\left(\frac{1}{2}(\beta^2 - k^2)\right)$ into the definition of the vertices, namely

$$\widetilde{T} = -\frac{T}{\Delta\left(\frac{1}{2}(\beta^2 - k^2)\right)} \tag{2.216}$$

and the 3-point amplitudes containing $\widetilde{T}$ are constants times a power of μ. Therefore, we find a simpler expression for the correlators involving the tachyons. Although both are simply related, it is clear that the usual string theory amplitude should be related to the correlators involving T, which are the ones displaying the expected singularities. Correlators involving $\widetilde{T}$ appear in the matrix model approach. The fact[75] is that gravity in two dimensions is very mild, and almost completely decouples from the matter contribution, and the simplicity of the $\widetilde{T}$ correlators can be traced back to the possibility of (at least partially) switching off the gravitational physics in two dimensions.

3. Hermitian Matrix Model

3.1. Introduction

3.1.1. Geometrical Origins

In the previous Chapters we have reviewed some approaches in the continuum to conformal theories coupled to two-dimensional gravity. In particular, the functional integration over the two-dimensional metrics involved the difficult task of solving the quantum Liouville theory. Here we shall describe a way of regularizing the sum over geometries by means of discrete dynamical triangulations of two-dimensional surfaces.

Such discretized approaches to random surfaces are not recent: inspired by Regge calculus[111], they were introduced[52,112,113] in the 80's, followed by intense studies (see for instance refs. [114 - 120]) including numerical simulations (we suggest refs. [121,122] and references therein).

In the continuum formulation we have mentioned that the partition function for fixed genus h and area A scales as

$$Z_h(A) \propto e^{\mu_c A} A^{\gamma_h - 3} W_h \quad , \tag{3.1}$$

where W_h is a constant which depends on the matter content, μ_c represents a critical value of the cosmological constant and γ_h is a critical exponent (called string susceptibility) which depends linearly on the genus h (see Eq. (1.164)) as

$$\gamma_h - 2 = (1-h)(\gamma_0 - 2) \quad . \tag{3.2}$$

The total partition function is thus obtained integrating (3.1) over all areas and summing over all topological contributions,

$$\begin{aligned} Z &= \sum_{h=0}^{\infty} \kappa_0^{2h} \int_0^{\infty} \mathrm{d}A\, \mathrm{e}^{-\mu_B A} Z_h(A) \\ &= \int_0^{\infty} \mathrm{d}A \sum_{h=0}^{\infty} \kappa_0^{2h} \mathrm{e}^{-(\mu_B - \mu_c)A} A^{(1-h)(\gamma_0 - 2) - 1} W_h \quad , \end{aligned} \tag{3.3}$$

where $\mu_B A$ is the cosmological term of the classical action (μ_B is the bare cosmological constant) and κ_0 is the bare string coupling constant which parametrizes the topological expansion. Notice that in the weak coupling limit, given by $\kappa_0 \to 0$, all contributions cancels except for the spherical (or planar) one ($h = 0$).

Before looking for a discretization of the partition function (3.3), notice that the integral over areas can be explicitly calculated,

$$
\begin{aligned}
Z &= \sum_{h=0}^{\infty} \kappa_0^{2h} W_h \int_0^{\infty} \mathrm{d}A\, \mathrm{e}^{-(\mu_B-\mu_c)A} A^{(1-h)(\gamma_0-2)-1} \\
&= (\mu_B-\mu_c)^{2-\gamma_0} \sum_{h=0}^{\infty} \left[\frac{\kappa_0^2}{(\mu_B-\mu_c)^{2-\gamma_0}} \right]^h \Gamma((h-1)(2-\gamma_0)-1)\, W_h \\
&= (\mu_B-\mu_c)^{2-\gamma_0} \sum_{h=0}^{\infty} \kappa^{2h} \widetilde{W}_h \quad .
\end{aligned} \tag{3.4}
$$

The Γ factor has been absorbed in the constant $\widetilde{W}_h$ and we have defined

$$
\kappa = \frac{\kappa_0}{(\mu_B-\mu_c)^{(2-\gamma_0)/2}} \quad . \tag{3.5}
$$

This particular dependence on the couplings implies that the partition function satisfies a double scaling relation:

$$
Z\left[\kappa_0, (\mu_B-\mu_c)\right] = \mathrm{e}^{-(2-\gamma_0)\rho} Z\left[\mathrm{e}^{-(2-\gamma_0)\rho}\kappa_0, \mathrm{e}^{-\rho}(\mu_B-\mu_c)\right] \quad . \tag{3.6}
$$

The overall factor $\mathrm{e}^{-(2-\gamma_0)\rho}$ represents a simple constant shift in the free energy and thus can be absorbed by a normalization constant. The importance of the scaling of the parameters in (3.6) is that no information is lost if we take the limit[1)] $\rho \to \infty$: this is equivalent to approaching the critical point ($\mu_B - \mu_c \to 0$) and the weak coupling limit ($\kappa_0 \to 0$) but keeping fixed the effective genus-counting parameter κ.

This is the scaling behavior we want to reproduce in the discretized model: the weak coupling limit allows one to evaluate the partition function exactly, using perturbation theory techniques; on the other hand, when the cosmological constant approaches its critical value, the mean area $\langle A \rangle$ of the surface diverges and one can rescale the lattice length a, taking $a \to 0$, to define continuum surfaces with finite areas; and yet, keeping fixed the renormalized coupling κ, one may sum over all orders in the genus expansion. This is the idea[70] of the *double scaling limit*, which originated so much progress in the studies of non-perturbative two-dimensional gravity. We shall have a first contact with this special limit along this Chapter and review it later in Chapter 8.

Let us consider a discretized two-dimensional surface built up from elementary triangles, as illustrated by Fig. 3.1. Given the number of faces F, edges E and vertices V, its genus h is determined by Euler's theorem,

$$
2(1-h) = F - E + V \quad . \tag{3.7}
$$

---

1) We suppose that $\gamma_0 < 2$.

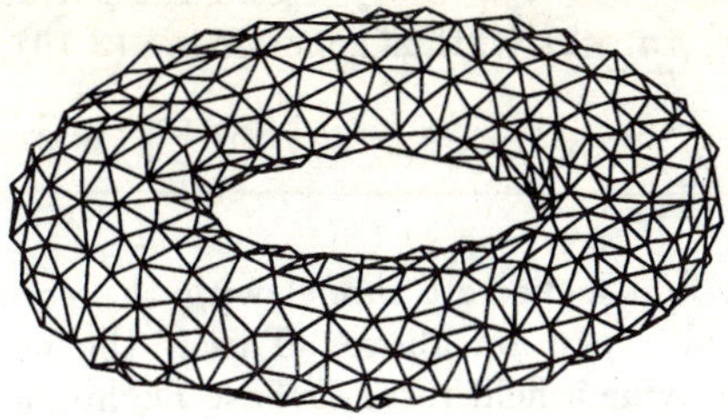

Fig. 3.1: A triangulated torus.

Taking equilateral triangles with area 1/2, the total area is given by

$$A = \frac{1}{2}F \quad . \tag{3.8}$$

One can also define local quantities, such as the intrinsic curvature R_i concentrated at the vertex i: if n_i is the total number of triangles meeting at the vertex i, the curvature is proportional to the deficit angle

$$R_i = \pi \frac{6 - n_i}{n_i} \quad . \tag{3.9}$$

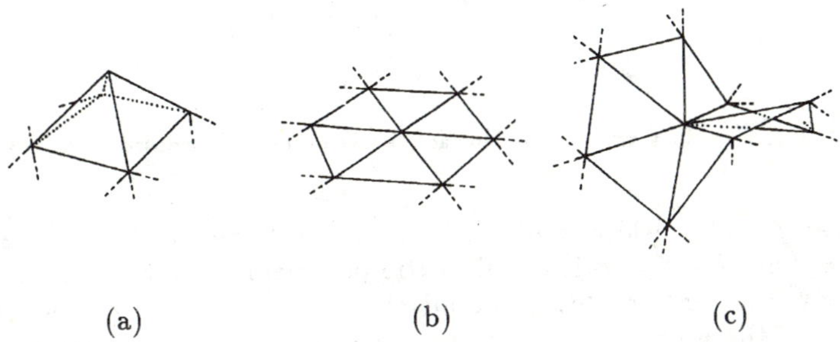

Fig. 3.2: Examples of sites with (a) positive, (b) zero and (c) negative local curvatures.

Fig. 3.2 exemplifies sites with positive, zero and negative local curvatures. We refer to [123] and references therein for a review on simplicial geometry and dynamical triangulations.

Using relations (3.7) and (3.8) we could write the discretized partition function (3.3) as

$$Z \propto \sum_{\text{triangulations}} \kappa_0^{E-F-V} \mathrm{e}^{-(\mu_B - \mu_c)\frac{1}{2}F} W_t \quad , \tag{3.10}$$

where the sum over areas and topologies was replaced by a sum over all triangulations, and W_t represents the remaining statistical weight (including contributions from the matter content).

We could have discretized the surfaces with general n-gons. The genus would still be given by (3.7). Since an n-gon can be built up from $(n-2)$ triangles we assign an area $(n-2)/2$ to each elementary n-gon. If a surface contains F_n n-gons, its total area is $A = \frac{1}{2}\sum_n (n-2)F_n$, thus generalizing (3.10) into

$$Z \propto \sum_{\text{discretizations}} \kappa_0^{E-F-V} \mathrm{e}^{-(\mu_B-\mu_c)\frac{1}{2}\sum_n (n-2)F_n} W_t \quad . \tag{3.11}$$

Therefore we must find a discrete model with the correct weight factors to reproduce the topological expansion above. This is the point where the matrix models show up: they provide a field theory whose Feynman graphs correspond to the discretizations in (3.11). The basic idea comes from the 70's with the works on the large N limit in QCD [53] and the planar approximation [124].

Consider Fig. 3.3, where broken lines represent part of a triangulated surface and the solid lines make the dual discretization. The dual graph is a typical Feynman graph of a (zero-dimensional) ϕ^3-field theory. If one uses triangles, squares, $\cdots$, n-gons to discretize the surface, its graph corresponds to a Feynman graph with vertices of types $\phi^3, \phi^4, \cdots, \phi^n$ respectively.

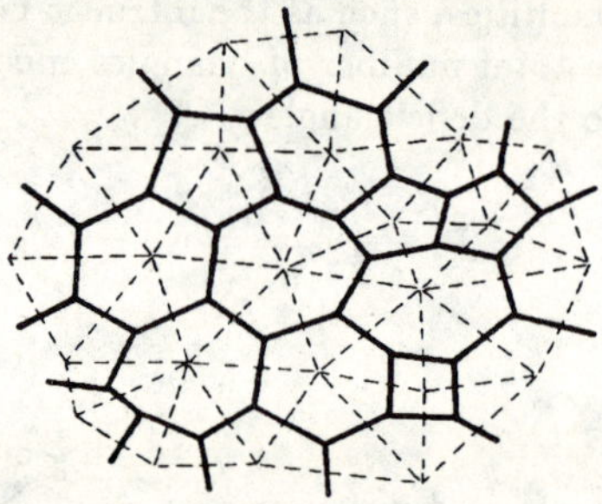

Fig. 3.3: A piece of a triangulated surface (broken lines) superposed with its dual (solid lines).

However, a scalar field ϕ does not have enough "structure" for our purposes: taking instead an $N \times N$ matrix ϕ_{ij}, the extra parameter N (the size of the matrix) will appear as the genus-counting parameter, as we show later on; moreover, the double lines in the Feynman rules for matrix fields allow one to characterize the orientability of the surface[125].

Let us take the Hermitian matrix partition function

$$Z = \int \mathcal{D}\phi \, \mathrm{e}^{-\frac{N}{\Lambda}\mathrm{tr}\, V(\phi)} \quad , \quad \phi^\dagger = \phi \quad , \tag{3.12}$$

where Λ is a constant (related to the cosmological constant, as we shall discuss later on), the measure is the flat one,

$$\mathcal{D}\phi = \prod_i \mathcal{D}\phi_{ii} \prod_{i<j} \mathcal{D}(\Re e\phi_{ij})\mathcal{D}(\Im m\phi_{ij}) \tag{3.13}$$

and the potential is taken as

$$V(\phi) = \frac{1}{2}\phi^2 + \sum_k g_k \phi^k \quad . \tag{3.14}$$

We can expand the interaction term in powers of g_k around the Gaussian point and compute Z using the Feynman rules defined below.

To reproduce the behavior (3.11) it is convenient to rescale the matrix variable as $\phi \to \tilde{\phi} = \Lambda^{-1/2}\phi$. Thus the potential term is rewritten as

$$\frac{N}{\Lambda}V(\phi) = \frac{1}{2}N\tilde{\phi}^2 + \sum_k \left(g_k N \Lambda^{(k-2)/2}\right)\tilde{\phi}^k \quad . \tag{3.15}$$

The propagator is represented by a double line, and reads

$$\langle \tilde{\phi}_{ij}\tilde{\phi}_{kl}\rangle = \frac{1}{N}\delta_{il}\delta_{jk} \quad = \tag{3.16}$$

The lines are oriented because $\tilde{\phi}$ is Hermitian and the absence of twisted propagators means that the corresponding original surface is orientable. To every propagator corresponds an edge on the original surface, so that we expect a factor N^{-E} for each graph.

A $\tilde{\phi}^3$-like vertex is given by

$$g_3 N \Lambda^{1/2}\tilde{\phi}_{ij}\tilde{\phi}_{jk}\tilde{\phi}_{ki} \quad = \tag{3.17}$$

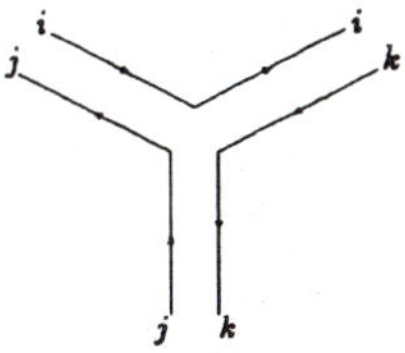

while, for a general $\tilde{\phi}^n$ interaction, we have a factor $N\Lambda^{(n-2)/2}$. According to figure 3.3 a $\tilde{\phi}^n$ vertex corresponds to an n-gonal face of the original surface, thus we find a factor $N^F \Lambda^{\sum_n F_n (n-2)/2}$ coming from each discretization.

A generic graph for Z has no external lines: all lines close in loops and for every loop we expect a factor N due to the trace in the matrix indices. We see in figure 3.3 that a loop corresponds to a vertex in the original discretization, which has V vertices. Therefore we find a factor N^V for a given graph.

Taking all factors into account we get as a result

$$Z \propto \sum_{\text{graphs}} N^{F+V-E}\Lambda^{\frac{1}{2}\sum_n (n-2)F_n} W(g_k) \quad , \tag{3.18}$$

where $W(g_k)$ depends on the coupling constants and symmetry factors[121]. Comparing the above result with (3.11) we find the expected dependence: $1/N$ plays the role of the bare string coupling constant κ_0 and therefore the weak coupling limit is given by $N \to \infty$, and corresponds to the planar (or spherical) limit; the constant Λ can be associated to $\mathrm{e}^{-\mu_B}$ and we shall also call it cosmological constant (we hope to clarify the confusing abundance of "cosmological constants" in Chapter 4). The double scaling limit therefore becomes $N \to \infty$, $\Lambda \to \Lambda_c = \mathrm{e}^{-\mu_c}$, keeping fixed the combination

$$\kappa = \frac{1}{N}\left[\ln\left(\frac{\Lambda_c}{\Lambda}\right)\right]^{(\gamma_0-2)/2} \sim N^{-1}\left[1 - \frac{\Lambda}{\Lambda_c}\right]^{(\gamma_0-2)/2} \quad , \tag{3.19}$$

where the last approximation of the logarithm holds near the critical point. We shall analyze this limit in detail later on.

Concerning the functional (3.12) we remark that it generates both connected and non-connected diagrams: indeed if we want only connected vacuum graphs to compare with the connected surfaces of the continuum approach, we must consider the free energy $F = \ln Z$ instead.

To finish this introduction we mention that there are other interesting matrix models: at the end of this Chapter we will analyze a 1-dimensional matrix model related to $c = 1$ conformal matter coupled to gravity; other matrix ensembles are discussed in refs. [125,126]; we shall also briefly describe multi-matrix models in Chapt. 8.

3.1.2. Analytical Transformations, Virasoro Constraints and Loop Equations

In this section we shall observe how analytical transformations of the Hermitian matrix variable are translated[127,128] into an infinite set of constraints satisfied by the partition function of the Hermitian matrix model. The planar limit ($N \to \infty$) of such constraints will be converted into algebraic equations between 1-point functions of the so-called *microscopic loop* operators in the next section, leading us to the planar solution of the model.

Given a set of couplings $\{g_k, k \geq 0\}$ and the bare cosmological constant Λ, we construct the partition function

$$Z = \int \mathcal{D}\phi \, e^{-\frac{N}{\Lambda}\mathrm{tr}\, V(\phi)} \quad , \quad V(\phi) = \sum_{k\geq 0} g_k \phi^k \quad . \tag{3.20}$$

Among the conceivable transformations of the matrix-valued variable, $\phi \to \phi' = f(\phi)$, we concentrate ourselves on the analytical ones of the form $\phi' = \sum_{n\geq 0} a_n \phi^n$, more precisely on those infinitesimally connected to the identity transformation, which we parametrize as

$$\phi \longrightarrow \phi + \varepsilon_n \phi^{n+1} \quad , \quad n \geq -1 \quad , \quad \varepsilon_n \ \text{ infinitesimal.} \tag{3.21}$$

Taking into account the contributions from the measure, that transforms as

$$\mathcal{D}\phi \longrightarrow \begin{cases} \mathcal{D}\phi & , n = -1, \\ \mathcal{D}\phi(1 + \varepsilon_o N^2) & , n = 0, \\ \mathcal{D}\phi \left(1 + \varepsilon_n \sum_{k=0}^{n} (\mathrm{tr}\phi^k)(\mathrm{tr}\phi^{n-k})\right) & , n \geq 1, \end{cases} \tag{3.22}$$

and the transformed potential term

$$e^{-\frac{N}{\Lambda}\mathrm{tr} V(\phi)} \longrightarrow e^{-\frac{N}{\Lambda}\mathrm{tr} V(\phi)} \left(1 - \varepsilon_n \frac{N}{\Lambda} \sum_{k\geq 1} k g_k \mathrm{tr}\phi^{n+k} \right) \quad , \tag{3.23}$$

one derives the following set of equations

$$\int \mathcal{D}\phi\, \mathrm{e}^{-\frac{N}{\Lambda}\mathrm{tr}V(\phi)} \left\{ \sum_{k\geq 1} k g_k \left(-\frac{N}{\Lambda}\mathrm{tr}\phi^{k-1} \right) \right\} = 0 \quad , \tag{3.24a}$$

$$\int \mathcal{D}\phi\, \mathrm{e}^{-\frac{N}{\Lambda}\mathrm{tr}V(\phi)} \left\{ \sum_{k\geq 1} k g_k \left(-\frac{N}{\Lambda}\mathrm{tr}\phi^{k} \right) + N^2 \right\} = 0 \quad , \tag{3.24b}$$

$$\int \mathcal{D}\phi\, \mathrm{e}^{-\frac{N}{\Lambda}\mathrm{tr}V(\phi)} \left\{ \sum_{k\geq 1} k g_k \left(-\frac{N}{\Lambda}\mathrm{tr}\phi^{k+n} \right) \right.$$
$$\left. + \frac{\Lambda^2}{N^2} \sum_{k=0}^{n} \left(-\frac{N}{\Lambda}\mathrm{tr}\phi^{k} \right) \left(-\frac{N}{\Lambda}\mathrm{tr}\phi^{n-k} \right) \right\} = 0 \quad , \quad n \geq 1 \tag{3.24c}$$

which can be seen as identities obeyed by 1- and 2-point functions of the $O(N)$-invariant quantities $\hat{w} \equiv \frac{\Lambda}{N}\mathrm{tr}\,\phi^n$, named microscopic loops (as stated in the previous section, a ϕ^n vertex in the matrix model corresponds to an n-gon on the original discretization, which tends to a microscopic loop insertion on the two-dimensional surface as one takes the continuum limit). Equations (3.24) can be written in a perhaps more familiar way, in terms of a loop generator $\hat{w}(l)$ defined as

$$\hat{w}(l) = \frac{\Lambda}{N}\mathrm{tr}\,\mathrm{e}^{l\phi} = \sum_{n\geq 0} \frac{l^n}{n!}\hat{w}^{(n)} \quad , \tag{3.25}$$

satisfying, as follows from (3.24) after some algebra, the equation

$$V'\left(\frac{\partial}{\partial l}\right) \langle \hat{w}(l) \rangle = \sum_{k\geq 1} k g_k \left(\frac{\partial}{\partial l}\right)^{k-1} \langle \hat{w}(l) \rangle = \int_0^l \mathrm{d}l' \, \langle \hat{w}(l-l')\hat{w}(l') \rangle \quad . \tag{3.26}$$

Above we recognize a typical *loop equation* – we refer to [129] for a review on the subject. It has a nice geometrical interpretation[130] as a Ward identity for a theory of interacting loop functions of *length* l. It is related to an infinitesimal change of coupling constants $\{g_k\}$, the quadratic interaction representing contact terms. We also mention that the continuum limit of (3.25) provides a possible definition[130] for macroscopic (i.e. finite length) loops. We shall refer to both, (3.24) and (3.26) as loop equations.

Notice that the potential itself can be regarded as a source term of microscopic loops, thus allowing one to rewrite equations (3.24) as a set of constraints satisfied by the partition function

$$L_n Z = 0, \quad , \quad n \geq -1 \quad , \tag{3.27}$$

$$L_{-1} = \sum_{k\geq 1} k g_k \frac{\partial}{\partial g_{k-1}} \quad , \tag{3.28a}$$

$$L_0 = \sum_{k\geq 1} k g_k \frac{\partial}{\partial g_k} + N^2 \quad , \tag{3.28b}$$

$$L_n = \sum_{k\geq 1} k g_k \frac{\partial}{\partial g_{k+n}} + \frac{\Lambda^2}{N^2} \sum_{k=0}^{n} \frac{\partial^2}{\partial g_k \partial g_{n-k}} \quad , \quad n \geq 1 \quad . \tag{3.28c}$$

It is straightforward to verify that the set of generators above obey a closed subalgebra of the Virasoro algebra, namely

$$[L_n, L_m] = (n-m) L_{n+m} \quad , \quad n, m \geq -1 \quad . \tag{3.29}$$

The presence of such Virasoro constraints was primarily detected [131,132] in the continuum (double-scaling) limit of the Hermitian matrix model, connected with the KdV hierarchic structure that lays behind the non-perturbative definition of 2-dimensional quantum gravity. Later they were also found at the original discrete model, in connection with (discrete) integrable hierarchies (see comparisons in refs. [133,134]). We shall return to the role of the Virasoro constraints in Chapter 6, as they will guide us in the supersymmetrization of the discrete model, and in Chapter 8 to comment on their connection with the integrability of the system. For the time being we shall concentrate on their planar limit – more precisely, on the planar limit of the loop equations – for the sake of comparison with the results in Chapter 2.

3.2. Loop Equations in the Planar Limit

In this section we shall approximate the loop equations to their leading order in the $1/N$ expansion. A non-linear system of equations for connected 1-point functions will come out, parametrized by a set of undetermined coefficients whose number depends on the polynomial degree of the potential.

As long as we are concerned with connected correlators, it is convenient to work directly with the *free energy* functional F, defined from

$$Z = \mathrm{e}^{N^2 F} = \int \mathcal{D}\phi \, \mathrm{e}^{-\frac{N}{\Lambda}(N g_0 + \sum_{k\geq 1} g_k \operatorname{tr} \phi^k)} \quad , \tag{3.30}$$

$$F = -\frac{g_0}{\Lambda} + \frac{1}{N^2}\left[\ln \int \mathcal{D}\phi \, \mathrm{e}^{\frac{-N}{\Lambda}\sum_{k\geq 1} g_k \operatorname{tr} \phi^k}\right] = F_0 + \frac{1}{N^2} F_1 + \cdots \tag{3.31}$$

The factor N^2 in (3.30) is just the convenient choice to guarantee a finite zeroth-order free energy, F_0, as we verify below.

The Virasoro constraints satisfied by the partition function become non-linear differential equations for the free energy,

$$\sum_{k\geq 1} kg_k \frac{\partial F}{\partial g_{k+n}} + \Lambda^2 \sum_{k=0}^{n} \left(\frac{\partial F}{\partial g_k} \frac{\partial F}{\partial g_{n-k}} + \frac{1}{N^2} \frac{\partial^2 F}{\partial g_k \partial g_{n-k}} \right) = 0 \quad , \tag{3.32}$$

but only first-order terms survive the planar limit:

$$\sum_{k\geq 1} kg_k \frac{\partial F_0}{\partial g_{k+n}} + \Lambda^2 \sum_{k=0}^{n} \left(\frac{\partial F_0}{\partial g_k} \frac{\partial F_0}{\partial g_{n-k}} \right) = 0 \quad . \tag{3.33}$$

These first-order derivatives of the free energy are proportional to the connected 1-point functions of microscopic loops,

$$w^{(n)} \equiv \frac{\Lambda}{N} \langle \mathrm{tr}\, \phi^n \rangle_c = -\frac{\Lambda^2}{N^2} \frac{1}{Z} \frac{\partial}{\partial g_n} Z = -\Lambda^2 \frac{\partial}{\partial g_n} F \quad , \tag{3.34}$$

$$w^{(0)} = \Lambda \quad , \tag{3.35}$$

which therefore satisfy, in the planar limit, the following non-linear equations

$$\begin{aligned} &\sum_{k\geq 1} kg_k w^{(k-1)} = 0 \quad , \\ &\sum_{k\geq 1} kg_k w^{(k)} - \Lambda^2 = 0 \quad , \\ &\sum_{k\geq 1} kg_k w^{k+n} - \sum_{k=0}^{n} w^{(k)} w^{(n-k)} = 0 \quad , \quad n \geq 1 \quad , \end{aligned} \tag{3.36}$$

with (3.35) as an initial condition.

We can alternatively understand the planar limit from Eq. (3.26): it is the approximation where the correlators factorize as

$$V'\left(\frac{\partial}{\partial l}\right) \langle \hat{w}(l) \rangle_c = \int_0^l \mathrm{d}l' \, \langle \hat{w}(l-l') \rangle_c \langle w(l') \rangle_c \quad . \tag{3.37}$$

a well known property of $1/N$ expansion techniques[129].

Now we must solve the non-linear system (3.36). It is shown in Appendix D that, for a polynomial potential $V(\phi) = \sum_{n=0}^{a} g_n \phi^n$ of degree a, equations (3.36) allow us to determine all but $(a-2)$ loops, which we take to be $w^{(1)}, \cdots, w^{(a-2)}$. This loss of uniqueness is a consequence of the planar approximation and we need additional criteria (like asymptotic behavior) to select the correct solution. This shall be done in the next section. Before that, it is convenient[130] to define the *Laplace transformed loop generator*

$$w(p) \equiv \int_0^\infty \mathrm{d}l\, \mathrm{e}^{-lp} \langle w(l) \rangle = \sum_{n\geq 0} \frac{w^{(n)}}{p^{n+1}} = \frac{\Lambda}{p} + \frac{w^{(1)}}{p^2} + \cdots \quad , \tag{3.38}$$

for which equations (3.36) imply

$$(w(p))^2 - V'(p)w(p) + Q(p) = 0 \quad , \tag{3.39}$$

where

$$V'(p) = \sum_{k\geq 1} k g_k p^{k-1} \quad , \tag{3.40}$$

$$Q(p) = \sum_{k\geq 2} k g_k \sum_{j=1}^{k-1} w^{(k-j-1)} p^{j-1} \quad . \tag{3.41}$$

We shall refer to (3.39) as the Laplace transformed loop equation[130].

Notice that $Q(p)$ is a polynomial in p with coefficients depending on the initial condition $w^{(0)}$ and the unknown loops $w^{(1)}, \cdots, w^{(a-2)}$. Given the potential and once we choose those undetermined coefficients, (3.39) becomes a simple quadratic equation for $w(p)$.

3.3. One-Cut Solution of the Loop Equations

We shall dedicate this section to a detailed exposition of the solution of the planar loop equations for even polynomial potentials. Such solution will be parametrized by one single function (the *cut*), which will be determined as a function of the couplings by the so-called *string equation.* Polynomial potentials of undefined parity are discussed in Appendix D.

Let us take an even potential of degree $2b$,

$$V(p) = \sum_{k=0}^{b} g_{2k} p^{2k} \quad , \quad g_{2b} \neq 0 \quad . \tag{3.42}$$

Consequently all 1-point functions of odd loops vanish, $w^{(2k+1)} = 0$, so that the loop $w(p)$ and the undetermined function $Q(p)$ have their expansions in p reduced to

$$w(p) = \frac{\Lambda}{p} + \sum_{k\geq 1} \frac{w^{(2k)}}{p^{2k+1}} \quad , \tag{3.43}$$

$$Q(p) = \sum_{k=1}^{b} 2k g_{2k} \sum_{j=0}^{k-1} w^{(2k-2-2j)} p^{2j} \quad . \tag{3.44}$$

Equation (3.39) has in principle two roots, but only one of them is compatible with the asymptotic behavior (recall (3.38)) of the loop $w(p)$:

$$w_{\pm}(p) = \frac{1}{2}\left(V'(p) \pm \sqrt{(V'(p))^2 - 4Q(p)}\right) \xrightarrow{p\to\infty} \frac{2}{\left(1 \mp \frac{|g_{2b}|}{g_{2b}}\right)} \frac{\Lambda}{p} \quad . \tag{3.45}$$

The correct choice depends on the sign of the leading coupling constant g_{2b}. Once this choice is made, we can face the problem of fixing the $(b-1)$ unknown coefficients: they show up through the square-rooted polynomial $V'(p)^2 - 4Q(p)$, which is in fact a $(2b-1)$-th order polynomial in $y = p^2$ due to our parity hypothesis. There are enough free coefficients to tune the roots of such polynomial into a set of $(b-1)$ double roots and one single root (which we shall call R), as is indicated in the following equation

$$V'(p)^2 - 4Q(p) = \left(M(p^2)\right)^2 (p^2 - R) \quad . \tag{3.46}$$

$M(p^2)$ is a polynomial of degree $(b-1)$ in p^2 whose coefficients, as well as R, are completely solved in terms of $\{g_k; \Lambda\}$ by Eq. (3.46): it is enough to identify the coefficients of $p^{4b-2} \cdots p^{2b-2}$ on both sides of the equation. We thus find

$$\begin{aligned} M(p^2) &= \sum_{n=0}^{b-1} m_n p^{2n} \\ m_n &= \sum_{k=0}^{b-n-1} \binom{2k}{k} \frac{R^k}{4^k} 2(k+n+1) g_{2(k+n+1)} \quad , \end{aligned} \tag{3.47}$$

while R is implicitly given by

$$\Lambda = \sum_{k \geq 1} k g_{2k} \binom{2k}{k} \frac{R^k}{4^k} \quad , \tag{3.48}$$

which we call (discrete) *string equation*. Once $M(p^2)$ and R are known, we use Eq. (3.46) once again: comparing the remaining terms in $p^{2b-4} \cdots p^0$ we find $Q(p)$, which finally means that $w^{(2)} \cdots w^{(2b-2)}$ are fixed. The correctness of this choice must be verified *a posteriori* – see also discussion in Appendix D.

Now we substitute the choice (3.46) back into the solution (3.45) and rewrite it as

$$w(p) = \frac{1}{2} \left(V'(p) - M(p^2) \sqrt{p^2 - R} \right) \quad . \tag{3.49}$$

In this parametrization, R plays the role of the single *cut* responsible for the non-analyticity of the $w(p)$ loop series, so we shall call (3.49) *one-cut solution*[130]. One can readily verify that the analytical contributions from the first and second terms on the r.h.s. of (3.49) do cancel each other.

The dependence of $w(p)$ with respect to Λ is particularly important: we shall use the scaling behavior of correlations functions with respect to the cosmological constant at some critical point to measure the critical exponents that will guide us in the comparison between the discrete and the Liouville formulations. Therefore it is useful to derive Eq. (3.49) with respect to the constant Λ,

$$\begin{aligned} \frac{\partial}{\partial \Lambda} w(p) &= \frac{\partial w(p)}{\partial R} \frac{\partial R}{\partial \Lambda} \\ &= \frac{1}{\sqrt{p^2 - R}} \frac{\left(\frac{1}{4} M(p^2) - \frac{(p^2 - R)}{2} \frac{\partial}{\partial R} M(p^2) \right)}{\frac{\partial \Lambda}{\partial R}} \quad . \end{aligned} \tag{3.50}$$

Using the explicit form of the polynomial (3.47) we easily find out that m_0 is the only coefficient contributing to Eq. (3.50):

$$\frac{1}{4}M(p^2) - \frac{(p^2-R)}{2}\frac{\partial M}{\partial R}(p^2) = \frac{1}{4}m_0 + \frac{R}{2}\frac{\partial m_0}{\partial R}$$
$$= \frac{1}{2}\sum_{k=0}^{b-1}(2k+1)(k+1)\binom{2k}{k}\frac{R^k}{4^k}g_{2(k+1)} \tag{3.51}$$
$$= \frac{\partial}{\partial R}\sum_{k=1}^{b} k g_{2k}\binom{2k}{k}\frac{R^k}{4^k} = \frac{\partial \Lambda}{\partial R}$$

and thus the solution (3.50) is reduced to

$$\frac{\partial}{\partial \Lambda}w(p) = \frac{1}{\sqrt{p^2-R}} \quad . \tag{3.52}$$

Finally expanding the above result in powers of p^{-2} we find the microscopic 1-loop functions

$$\frac{\partial}{\partial \Lambda}w^{(2k)} = \binom{2k}{k}\frac{R^k}{4^k} \quad . \tag{3.53}$$

As mentioned in the Introduction, the partition function (3.20) should be regarded as a compact representation of its series expansion in Gaussian integrals. We therefore require that the solution of the loop equation be perturbative around the Gaussian point. Indeed, if we take $g_2 = 2 + \varepsilon_2$ and $g_{2k} = \varepsilon_{2k}, k \geq 0$, we can apply Lagrange's method to solve the string equation (3.48), finding

$$\begin{aligned} R &= \frac{1}{n!}\sum_{n\geq 1}\left(\frac{\partial}{\partial \Lambda}\right)^{n-1}\left[-\sum_{k\geq 0} k\varepsilon_{2k}\binom{2k}{k}\frac{\Lambda^k}{4^k}\right]^n \\ &\sim \Lambda\left[1 - \sum_{k\geq 1} k\varepsilon_{2k}\binom{2k}{k}\frac{\Lambda^{k-1}}{4^k}\right] + \mathcal{O}(\varepsilon^2) \quad . \end{aligned} \tag{3.54}$$

Replacing the above series in (3.53) we find the correct perturbative expansion of the loops: for instance, at first order in ε, we have

$$w^{(2k)} \sim \frac{1}{4^k}\binom{2k}{k}\left\{\frac{\Lambda^{k+1}}{k+1} - \sum_{n\geq 1}\varepsilon_{2n}\frac{nk}{n+k}\frac{1}{4^n}\binom{2n}{n}\Lambda^{n+k}\right\} \quad . \tag{3.55}$$

Therefore we shall take (3.53) together with the string equation as the solution for the planar loop equations. Now we must identify the critical points of the theory – i.e. the critical values for the coupling constants – which exhibit interesting scaling behaviors, as well as a suitable set of scaling operators.

3.4. Double Scaling Limit (DSlim)

Guided by the string equation, we shall characterize the scaling regimes and the associated even potentials. Then, by defining macroscopic loops, we shall introduce the double scaling limit as the suitable prescription for the continuum limit of the model.

3.4.1. String Equation and Scaling Limit

Let us call Λ_c the critical value of the cosmological constant for which the theory enjoys scaling properties, and R_c the value of the cut function R associated to it through the string equation. We are interested in the properties of the theory close to this critical point. For instance, the scaling behavior of $R(\Lambda)$, given in general by

$$R_c - R \sim (\Lambda_c - \Lambda)^{-\gamma} + \mathcal{O}\left((\Lambda_c - \Lambda)^{-\gamma+1}\right) \tag{3.56}$$

determines one of the critical exponents (the *string susceptibility* γ) that characterize our model. Therefore it is convenient to expand the solution, found in the previous section, around the points $\Lambda_c(R_c)$. We start from the string equation, rewriting (3.48) as

$$\Lambda = R\frac{\partial}{\partial R}\left[\sum_{k\geq 0} g_{2k}\frac{1}{4^k}\binom{2k}{k}R^k\right] \quad . \tag{3.57}$$

If the expression between brackets is analytical around the critical value R_c – which is certainly true for polynomial potentials – we can replace it by a series in $(R_c - R)$ in the neighbourhood of R_c. Using the transformation formulae

$$\begin{gathered}\sum_{k\geq 0} a_k R^k = \sum_{n\geq 0} b_n (R_c - R)^n \quad , \\ a_k = \sum_{n\geq k}\binom{n}{k} b_n (-1)^k R_c^{n-k} \quad \Longleftrightarrow \quad b_n = \sum_{k\geq n} a_k \binom{k}{n}(-1)^n R_c^{k-n} \quad ,\end{gathered} \tag{3.58}$$

we find

$$\sum_{k\geq 0} g_{2k}\frac{1}{4^k}\binom{2k}{k}R^k = \sum_{n\geq 0}\sum_{k\geq n}\left[g_{2k}\frac{1}{4^k}\binom{2k}{k}\binom{k}{n}(-1)^n R_c^{k-n}\right](R_c - R)^n \tag{3.59}$$

and finally

$$\Lambda = \sum_{n\geq 0} t_n^B\left[(R_c - R)^n - R_c^{-1}(R_c - R)^{n+1}\right] \quad , \tag{3.60}$$

$$t_n^B = -\sum_{k\geq n+1}\left[\frac{n+1}{4^k}\binom{2k}{k}\binom{k}{n+1}(-1)^{n+1}R_c^{k-n}\right] g_{2k} \quad , \quad t_0^B = \Lambda_c \tag{3.61}$$

$$g_{2k} = \sum_{n+1\geq k}\left[-\frac{(-4)^k}{(n+1)}\frac{\binom{n+1}{k}}{\binom{2k}{k}}R_c^{n-k}\right] t_n^B \quad . \tag{3.62}$$

Equations (3.61-62) represent a linear invertible transformation between the original set of couplings $\{g_{2k}\}$ and the set $\{t_n^B\}$ of the so called bare scaling coupling constants, in terms of which the string equation (3.60) is conveniently arranged for the scaling limit. Indeed, it is easy to see how one can tune the scaling couplings to induce different scaling regimes: for instance, taking some integer $m > 1$ and choosing

$$t_n^B = \begin{cases} \Lambda_c R_c^{-n} \quad , & n < m \quad , \\ 0 \quad , & n = m \quad , \\ \text{arbitrary finite} \quad , & n > m \quad , \end{cases} \tag{3.63}$$

equation (3.64) gives

$$\Lambda = \Lambda_c - \Lambda_c R_c^{-m}(R_c - R)^m + \mathcal{O}\left((R_c - R)^{m+1}\right) \quad ,$$

implying that, near the critical point, R scales as

$$R_c - R = R_c \Lambda_c^{-1/m}(\Lambda_c - \Lambda)^{1/m} + \mathcal{O}\left((\Lambda_c - \Lambda)^{1/m+1}\right) \quad . \tag{3.64}$$

Therefore we are able to produce scaling regimes characterized by the susceptibilities $\gamma_m = -\frac{1}{m}$, with $m = 1, 2, 3, \cdots$.

To any choice of scaling couplings corresponds a potential, and as (3.63) indicates, there are infinitely many different potentials (with different values of $t_n^B, n > m$) leading to the same scaling behavior, which is referred to as the *universality* property of the system. The particular choice $t_n^B = 0$ for $n \geq m$ generates the critical potentials originally found by Kazakov[130], given by

$$V(\phi) = \sum_{k=1}^{m} g_{2k}\phi^{2k} \quad ,$$

$$g_{2k} = -\Lambda_c R_c^{-k} \frac{(-4)^k}{\binom{2k}{k}} \sum_{n+1=k}^{m} \frac{1}{n+1} \binom{n+1}{k} = -\Lambda_c R_c^{-k} \frac{(-4)^k}{k} \frac{\binom{m}{k}}{\binom{2k}{k}} \tag{3.65}$$

corresponding to the exact scaling

$$1 - \frac{R}{R_c} = \left(1 - \frac{\Lambda}{\Lambda_c}\right)^{1/m} \quad . \tag{3.66}$$

It is useful to introduce the auxiliary parameter a to control the scaling limit[71], defining

$$\Lambda \sim \Lambda_c(1 - a^2 t) \tag{3.67}$$

near the critical point, which is therefore given by the limit $a \to 0$. If we associate the constant Λ to a bare cosmological constant μ_B through the relation $\Lambda = \mathrm{e}^{-\mu_B}$, $\Lambda_c = \mathrm{e}^{-\mu_c}$, as discussed in the Introduction of this Chapter, the scaling variable t can be interpreted as a renormalized cosmological constant: $t = \lim_{a \to 0}(\mu_B - \mu_c)/a^2$.

We must also associate a scaling variable u to the cut R,

$$R \sim R_c(1 - a^{2/m}u) \quad , \tag{3.68}$$

where the a-dependence was taken in agreement with the scaling behavior of the corresponding m-th critical regime (see (3.66)). In the next section, where a renormalized free energy will be defined, u will be given an appropriate interpretation in terms of the *specific heat* of the theory.

We also define a set of renormalized couplings $\{t_n\}$ according to

$$t_n^B = \begin{cases} \Lambda_c R_c^{-n}(1 + a^{2(1-n/m)}t_n) \quad , & n < m \quad , \\ \Lambda_c R_c^{-n} a^{2(1-n/m)} t_n \quad , & n \geq m \quad , \end{cases} \tag{3.69}$$

so that Kazakov's potentials are correctly reproduced in the $a \to 0$ limit.

Finally, if we use the scaling variables above defined and take the $a \to 0$ limit, it is easy to show that the string equation (3.60) becomes

$$t = u^m - \sum_{n \geq 0} t_n u^n \quad . \tag{3.70}$$

We call it renormalized m-th critical string equation. It can be used to solve u in terms of the renormalized cosmological constant t and the couplings $\{t_n\}$. One can define exact critical regimes: for instance, when $t_n = 0$, $u = t^{1/m}$, critical exponents are measured from the scaling with respect to t. This will be done in the next section of this Chapter, while an alternative scaling regime will be used in Chapter 4.

On the other hand, notice that one can move from an m-th critical point to another by changing the t_n-couplings. These t_n-flows can be readily calculated from (3.70):

$$\frac{\partial}{\partial t_n} u = \frac{\partial}{\partial t} R_{n+1}[u] \quad , \tag{3.71}$$

where the monomials

$$R_k[u] \equiv \frac{u^k}{k} \quad , \tag{3.72}$$

are introduced for the sake of comparison with results beyond the planar limit: as we shall see in Chapter 8, a similar equation holds non-perturbatively[132], with $R_k[u]$ substituted by the Gel'fand-Dikii polynomials; up to some simple normalization factors, one can verify that these polynomials tend to (3.72) in the $N \to \infty$ limit.

We also mention that the monomials (3.72) satisfy

$$\frac{\partial}{\partial t_i} R_j[u] = \frac{\partial}{\partial t} R_{i+j}[u] \quad , \tag{3.73}$$

a useful property in the calculation of correlation functions.

Going to the critical point ($a \to 0$) constitutes one step in the prescription for the continuum limit of the matrix model. The other one is taking $N \to \infty$. In the next section we introduce the double scaling limit as a suitable simultaneous control of these two limits.

3.4.2. Macroscopic Loops and Double Scaling Limit

Keeping the simplifying assumption of even potentials, we shall carry out the construction of loop operators with non-vanishing finite lengths, called macroscopic loops. Calculating correlation functions between macroscopic loops will fix the dimension of the a scaling parameter and constrain the limits $N \to \infty, a \to 0$.

From the discussion at the beginning of this Chapter, we understand that operators like $\mathrm{tr}\,\phi^{2k}$ in the limit $k \to \infty$ correspond to the insertion of macroscopic loops[70,71] on the discretized surface: k counts the number of edges of the polygonal loop, and therefore, given some lattice scale, measures the loop length. In order to control this limit, we will evaluate the 1-point function $\lim_{k\to\infty}\langle \mathrm{tr}\,\phi^{2k}\rangle_c$ in the planar approximation at the critical point ($a \to 0$). Taking the planar result (3.53) and using the Stirling formula to calculate

$$\frac{1}{4^k}\binom{2k}{k} \overset{k\to\infty}{\longrightarrow} \frac{1}{\sqrt{\pi k}} \quad , \tag{3.74}$$

we find

$$\begin{aligned}
\langle \mathrm{tr}\,\phi^{2k}\rangle_c &= \frac{N}{\Lambda}\partial_\Lambda^{-1} \lim_{k\to\infty} \partial_\Lambda w^{(2k)} \\
&\overset{k\to\infty}{\longrightarrow} \frac{R_c}{\sqrt{\pi k}}\frac{N}{\Lambda}\partial_\Lambda^{-1} \mathrm{e}^{k\ln(R/R_c)} \overset{a\to 0}{\longrightarrow} -\frac{R_c^k}{\sqrt{\pi k}} N a^2 \partial_t^{-1} \mathrm{e}^{-ka^{2/m}u} \\
&= -R_c^k \frac{(Na^{2+1/m})}{\sqrt{\pi(ka^{2/m})}} \partial_t^{-1} \mathrm{e}^{-(ka^{2/m})u} \quad .
\end{aligned} \tag{3.75}$$

Above the operator $-\partial_t^{-1}$ means $\int_t^\infty \mathrm{d}t'$. The R_c^k factor can be absorbed in the future definition of the loop operator. On the other hand, in order to have a non-trivial (i.e. u-dependent) answer, the $ka^{2/m}$ factor should be finite. Therefore, we associate to the discrete length k the renormalized loop length

$$l = ka^{2/m} \quad , \tag{3.76}$$

showing that $a^{2/m}$ plays the role of length scale in our lattice[70]. This allows one to determine the dimensions of the renormalized coupling constants and respective operators (see next Chapter).

Notice also that the factor $Na^{2+1/m}$ should be kept constant,

$$Na^{2+1/m} = \kappa^{-1} \quad , \tag{3.77}$$

to guarantee a finite result in (3.75), thus binding the limits $a \to 0$, $N \to \infty$. This is the double scaling outlined in the Introduction: one approaches the critical point (parametrized by a) while the matrix size (represented by N) is increased in the ratio implied by Eq. (3.77)[2)]. Remembering that $1/N$ was the parameter that

---

2) Using the definition (3.45), we can rewrite the double scaling condition as

$$N\left(1-\frac{\Lambda}{\Lambda_c}\right)^{1+1/2m} \to \tilde{\kappa} \quad \Longrightarrow \quad N \to \tilde{\kappa}(\Lambda_c - \Lambda)^{-1-1/2m} \quad ,$$

$\tilde{\kappa}$ constant, to be compared with (3.19).

controlled the genus expansion – the bare string coupling constant in the string interpretation of the sum over random surfaces – we can read Eq. (3.77) as the renormalization of the genus counting parameter, $\kappa = (1/N)a^{-2-1/m}$, which we thus call (renormalized) *string coupling constant.*

We conclude that

$$W(l) = \mathrm{DSlim}_{k \to la^{-2/m}} \frac{\mathrm{tr}\,\phi^{2k}}{R_c^k} \quad , \tag{3.78}$$

DSlim meaning Double-Scaling-limit ($a \to 0, N \to a^{-2-1/m}\kappa^{-1}$), is a suitable definition for a macroscopic loop operator of length l, and its planar 1-point function follows from (3.75)

$$\langle W(l) \rangle = -\frac{1}{\kappa\sqrt{\pi l}} \partial_t^{-1} \mathrm{e}^{-lu} \quad . \tag{3.79}$$

The role of the parameter κ is not so clear if we limit ourselves to a fixed topology: indeed, in the spherical continuum results of the previous Chapters, such parameter was absent. The same would happen if we absorbed κ in the loop definition (3.78). The full power of the double scaling limit, and thus the usefulness of the parameter κ, comes out only when we consider contributions from all genera. Nevertheless we prefer to keep the definition (3.78), leaving to Chapter 8 a more convincing justification of the limit (3.77).

We also introduce the rescaled loop operator

$$\widetilde{W}(l) \equiv \sqrt{\frac{\pi}{l}} W(l) \quad , \quad \langle \widetilde{W}(l) \rangle = -\frac{1}{\kappa l} \partial_t^{-1} \mathrm{e}^{-lu} \quad , \tag{3.80}$$

in terms of which some results will come out more transparent.

Let us close this section calculating n-point functions[31,135] of macroscopic loops in the planar limit. A 2-point function can be obtained by insertion of a loop into (3.80), as follows:

$$\begin{aligned} \langle \widetilde{W}(l_1)\widetilde{W}(l_2) \rangle &= \mathrm{DSlim}_{k_1 \to l_1 a^{-2/m}} \sqrt{\frac{\pi}{l_1}} \langle \frac{\mathrm{tr}\,\phi^{2k_1}}{R_c^{k_1}} \widetilde{W}(l_2) \rangle \\ &= \mathrm{DSlim}_{k_1 \to l_1 a^{-2/m}} \sqrt{\frac{\pi}{l_1}} \frac{1}{R_c^{k_1}} \left(-\frac{\Lambda}{N} \frac{\partial}{\partial g_{2k_1}} \right) \langle \widetilde{W}(l_2) \rangle \\ &= -\partial_t^{-1} \left[\mathrm{e}^{-l_2 u} \mathrm{DSlim}_{k_1 \to l_1 a^{-2/m}} \frac{1}{\kappa N} \sqrt{\frac{\pi}{l_1}} \frac{\Lambda_c}{R_c^{k_1}} \frac{\partial}{\partial g_{2k_1}} u \right] \quad . \end{aligned} \tag{3.81}$$

To calculate the limit inside brackets we must use the flows (3.71) and the definitions (3.62) and (3.69) that relate the renormalized couplings t_n to the g_{2k}'s. We find

$$\begin{aligned} \mathrm{DSlim}_{k \to la^{-2/m}} \frac{1}{\kappa N} \sqrt{\frac{\pi}{l}} \frac{\Lambda_c}{R_c^k} \frac{\partial}{\partial g_{2k}} u &= \mathrm{DSlim}_{k \to la^{-2/m}} \frac{1}{\kappa N} \sqrt{\frac{\pi}{l}} \frac{\Lambda_c}{R_c^k} \sum_n \left(\frac{\partial t_n}{\partial g_{2k}} \right) \frac{\partial}{\partial t_n} u \\ &= \mathrm{DSlim}_{k \to la^{-2/m}} \frac{1}{\kappa N a^2} \sqrt{\frac{\pi}{l}} \frac{k}{4^k} \binom{2k}{k} \sum_{n=0}^{k-1} \binom{k-1}{n} (-1)^n a^{2n/m} u^n \frac{\partial}{\partial t} u \\ &= \mathrm{DSlim}_{k \to la^{-2/m}} \frac{\sqrt{k}}{\kappa N a^2 \sqrt{l}} (1 - a^{2/m} u)^{k-1} \partial_t u \\ &= \mathrm{e}^{-lu} \partial_t u \end{aligned} \tag{3.82}$$

and therefore

$$\begin{aligned}\langle\widetilde{W}(l_1)\widetilde{W}(l_2)\rangle &= -\partial_t^{-1}\mathrm{e}^{-(l_1+l_2)u}\partial_t u = \frac{\mathrm{e}^{-(l_1+l_2)u}}{(l_1+l_2)} \\ &= (-\kappa\partial_t)\left(-\frac{1}{\kappa(l_1+l_2)}\partial_t^{-1}\mathrm{e}^{-(l_1+l_2)u}\right) \\ &= (-\kappa\partial_t)\langle\widetilde{W}(l_1+l_2)\rangle \end{aligned} \tag{3.83}$$

From the results above we learn that, given a 1-loop function with length l_1, the insertion of another loop of length l_2, produced by the action of the operator $I_{l_2} = \sqrt{\frac{\pi}{l_2}}\mathrm{DSlim}_{k_2}(-\frac{\Lambda}{R_c^{k_2}N}\frac{\partial}{\partial g_{2k}})$, generates another 1-loop function with total length l_1+l_2 operated on by $(-\kappa\partial_t)$. Since the operators I_l and $(-\kappa\partial_t)$ commute, we readily deduce the n-loop function formula

$$\langle\widetilde{W}(l_1)\cdots\widetilde{W}(l_n)\rangle = (-\kappa\partial_t)^{n-1}\langle\widetilde{W}(l_1+\cdots+l_n)\rangle \quad . \tag{3.84}$$

Correlation functions for $W(l)$ loops can be easily obtained from the equation above. However we remark that, written in terms of the rescaled loops $\widetilde{W}(l)$, those correlators only depend on the *total* length $l_t = \sum_i l_i$. A similar property also holds in the supersymmetric model, as we shall verify in Chapter 6.

Let us write down the 1-loop function as a series in the length variable, as follows from (3.79),

$$\langle W(l)\rangle = \frac{1}{\sqrt{\pi}\kappa}\sum_{n\geq 0}\frac{(-1)^n}{n!}l^{n+1/2}\partial_t^{-1}\left(\frac{u^{n+1}}{n+1}\right) + \mathcal{O}(l^{-1/2}) \quad , \tag{3.85}$$

since this formula is useful to relate loops to scaling operators. When $u = t^{1/m}$, the integration of Eq. (3.85) is straightforward:

$$\begin{aligned}\langle W(l)\rangle &= \frac{1}{\sqrt{\pi}\kappa}\sum_{n\geq 0}\frac{(-1)^n}{n!}l^{n+1/2}\frac{m\,t^{n/m+1/m+1}}{(n+1)(n+1+m)} + (\text{singular terms in } l) \\ &= -\frac{1}{\sqrt{\pi}\kappa}t^{1+1/2m}\sum_{s\geq 1}\frac{m}{m+s}\frac{(-1)^s}{s!}x^{s-1/2} + (\text{singular terms in } l) \quad , \end{aligned} \tag{3.86}$$

where we have defined the dimensionless variable $x \equiv lt^{1/m}$. This is indeed a particular case of the result obtained in ref. [31] for a general regime: writing the string equation in the form $t = \sum t_n u^n$ (so that the regime $u = t^{1/m}$ is obtained as $t_m \to 1$ and $t_n \to 0, n \neq m$) and performing the $t \to u$ change of variables to integrate (3.79), one finds

$$\langle W(l)\rangle = \frac{1}{\sqrt{\pi}\kappa}\sum_{n\geq 1} t_n u^{n+1/2}\left(\frac{n}{x}\right)\psi_{n-1}(x) \quad , \tag{3.87}$$

where

$$x = lu \quad , \tag{3.88a}$$

$$\begin{aligned}\left(\frac{n}{x}\right)\psi_{n-1}(x) &= n!x^{-n-1/2}\mathrm{e}^{-x}\sum_{s=0}^{n-1}\frac{x^s}{s!} \\ &= \frac{n!}{x^{n+1/2}} - \frac{1}{x^{1/2}} - \sum_{s\geq 1}\frac{n}{n+s}\frac{(-1)^s}{s!}x^{s-1/2} \quad .\end{aligned} \tag{3.88b}$$

Indeed the function $(\frac{m}{x})\psi_{m-1}(x)$ reproduces the behavior found in (3.86), but the information contained in (3.87-88) is more general and precise: it exactly provides the singular terms in l, which will be analyzed later on; moreover it is valid for any choice of the couplings $\{t_n\}$, in particular for the important conformal basis studied in Chapter 4.

Having found a suitable prescription for the continuum limit of the discrete matrix model, we proceed to the next ambitious step: to find operators corresponding to the vertex operators of the Liouville approach. Recall that the later exhibit a precise scaling behavior with respect to the cosmological constant (see Eq. (2.28)). We take this scaling property as a first guide, postponing the analysis of correlation functions (the decisive test).

3.5. Scaling Operators

Coupled to the renormalized constants presented in the previous section, a set of scaling operators[71,130] will be defined. From their correlation functions in the double scaling limit, we shall construct a renormalized free energy and identify scaling dimensions. A preliminary comparison with results from the Liouville approach will be given at the end of this section.

We learned from the previous section how the t_n^B-coupling constants appear as one expands the string equation about the critical point, and how those couplings are renormalized into the set $\{t_n\}$ to produce the renormalized string equation. Therefore the operators coupled to the t_n's are natural candidates for scaling operators[130]. The relation between them and the microscopic loops $w^{(2k)}$ is the same one connecting the operators $\frac{\partial}{\partial t_n}$ and $\frac{\partial}{\partial g_{2k}}$, which can be calculated from (3.62) and (3.69):

$$\frac{\partial}{\partial t_n} = \Lambda_c R_c^{-n} a^{2(1-n/m)}\frac{\partial}{\partial t_n^B} \quad , \tag{3.89a}$$

$$\frac{\partial}{\partial t_n^B} = \sum_k \frac{\partial g_{2k}}{\partial t_n^B}\frac{\partial}{\partial g_{2k}} = \sum_{k=0}^{n+1}\left[-\frac{(-4)^k}{(n+1)}\frac{\binom{n+1}{k}}{\binom{2k}{k}}R_c^{n-k}\right]\frac{\partial}{\partial g_{2k}} \quad . \tag{3.89b}$$

Therefore we define the operators

$$\sigma_n \equiv \sum_{k=0}^{n+1} \left[-\frac{(-4)^k}{(n+1)} \frac{\binom{n+1}{k}}{\binom{2k}{k}} \right] \frac{a^{2(1-n/m)} \Lambda_c}{R_c^k} (-\frac{N}{\Lambda} \mathrm{tr}\, \phi^{2k}) \quad , \tag{3.90}$$

for $n = 0, 1, \cdots$. Their 1-point functions, in the planar approximation, are easily calculated using (3.53):

$$\begin{aligned} \langle \sigma_n \rangle &= \sum_{k=0}^{n+1} \left[-\frac{(-4)^k}{(n+1)} \frac{\binom{n+1}{k}}{\binom{2k}{k}} \right] \frac{a^{2(1-n/m)} \Lambda_c}{R_c^k} \frac{-N^2}{\Lambda^2} \binom{2k}{k} \frac{1}{4^k} \partial_\Lambda^{-1} R^k \\ &= -\frac{N^2 a^2}{n+1} \frac{\Lambda_c}{\Lambda} (-\Lambda^{-1} \partial_\Lambda^{-1}) a^{-2n/m} \sum_{k=0}^{n+1} \binom{n+1}{k} \left(-\frac{R}{R_c} \right)^k \\ &= -\frac{N^2 a^{4+2/m}}{n+1} \frac{\Lambda_c}{\Lambda} (-a^{-2} \Lambda^{-1} \partial_\Lambda^{-1}) \left(a^{-2/m} \left(1 - \frac{R}{R_c} \right) \right)^{n+1} \quad , \end{aligned} \tag{3.91}$$

which, in the double scaling limit, become

$$\langle \sigma_n \rangle = -\frac{1}{\kappa^2} \partial_t^{-1} \frac{u^{n+1}}{(n+1)} = -\frac{1}{\kappa^2} \partial_t^{-1} R_{n+1}[u] \quad . \tag{3.92}$$

In the exact m-th critical regime where $u = t^{1/m}$, the above result implies that the operator σ_n scales as a power of the renormalized cosmological constant, $\langle \sigma_n \rangle \propto t^{1+1/m+n/m}$. Moreover we remark the reappearance of the $R_k[u]$ monomials, whose flows in the t_n-coupling space, given by Eq. (3.73), teach us how to go from 1-point functions to higher order correlators. Bearing in mind this aim, we rewrite Eq. (3.92) as

$$\begin{aligned} \langle \sigma_n \rangle &= -\frac{1}{\kappa^2} \partial_t^{-2} \frac{\partial}{\partial t} R_{n+1}[u] = -\frac{1}{\kappa^2} \partial_t^{-2} \frac{\partial}{\partial t_n} R_1[u] \\ &= \frac{\partial}{\partial t_n} \left(-\frac{1}{\kappa^2} \partial_t^{-2} R_1[u] \right) \quad . \end{aligned} \tag{3.93}$$

To complete the definition of our model in its continuum limit we must find the free energy in the double-scaling limit: it can thus be used to calculate connected correlation functions and its scaling behavior defines the susceptibility exponent (γ) that characterizes the conformal theory on the random surfaces.

3.5.1. Free Energy and N-Point Functions

Assuming the existence of a renormalized free energy function $\mathcal{F}(t;\{t_n\})$, from which 1-point functions of scaling operators can be calculated as $\langle\sigma_n\rangle = \frac{\partial}{\partial t_n}\mathcal{F}$, its t_n-dependent part can be read from Eq. (3.93) and comes implicitly through the $u(t_n)$ function. Therefore, as far as scaling operators are concerned, we can take

$$\mathcal{F} = -\frac{1}{\kappa^2}\partial_t^{-2} R_1[u] = -\frac{1}{\kappa^2}\partial_t^{-2} u \quad . \tag{3.94}$$

With the free energy and the flows (3.73) in hand, the calculation of n-point functions is straightforward: since the operators ∂_t and $\frac{\partial}{\partial t_i}$ commute, we obtain

$$\begin{aligned}\langle\sigma_i\sigma_j\rangle &= -\frac{1}{\kappa^2}\partial_t^{-2}\frac{\partial^2}{\partial t_i \partial t_j} R_1[u] = -\frac{1}{\kappa^2}\partial_t^{-2}\left(\frac{\partial}{\partial t}\right)^2 R_{i+j+1}[u] \\ &= -\frac{1}{\kappa^2}\frac{u^{i+j+1}}{i+j+1} \quad ,\end{aligned} \tag{3.95}$$

for 2-point functions, and

$$\begin{aligned}\langle\prod_{i=1}^{n}\sigma_{a_i}\rangle &= -\frac{1}{\kappa^2}\partial_t^{n-2}R_{a+1}[u] = -\frac{1}{\kappa^2}\partial_t^{n-2}\frac{u^{a+1}}{a+1} \\ a &= \sum_{i=1}^{n} a_i \quad ,\end{aligned} \tag{3.96}$$

for n-point correlators of scaling operators.

Consider equation (3.94): in its statistical mechanical interpretation t plays the role of temperature and therefore $u \propto \partial_t^2\mathcal{F}$ represents the specific heat, as we had anticipated in the discussion after (3.68). Notice also that the specific heat can be identified[132] with the 2-point function of the *puncture* operator σ_o, as follows from Eq. (3.95),

$$u = -\kappa^2\langle\sigma_o\sigma_o\rangle \tag{3.97}$$

and the insertion of another scaling operator σ_n into (3.97) is equivalent to the n-th flow of u,

$$-\kappa^2\langle\sigma_n\sigma_o\sigma_o\rangle = \frac{\partial u}{\partial t_n} = \partial_t R_{n+1}[u] \quad . \tag{3.98}$$

It can be shown that the above flows hold beyond the planar limit, with $R_n[u]$ replaced by the Gel'fand-Dikki polynomials of the KdV hierarchy. We shall comment on this property and its relation to the integrability of the model along Chapter 8.

We call the reader's attention to the simplicity of the n-point correlators of loops and scaling operators, respectively given by (3.84) and (3.96). There is a relation between these two sets of operators which will be established in the next section.

3.5.2. Macroscopic Loops Versus Scaling Operators

We have presented the planar solution in the double scaling limit by means of two types of operators: the macroscopic loops and the scaling operators. Using their correlation functions, it will be shown that a macroscopic loop can actually be expanded as a series in terms of scaling operators.

Compare the 1-point functions (3.85) and (3.92): they imply

$$\langle W(l)\rangle = -\frac{\kappa}{\sqrt{\pi}} \sum_{n\geq 0} \frac{(-1)^n}{n!} l^{n+1/2} \langle \sigma_n \rangle + \mathcal{O}(l^{-1/2}) \quad . \tag{3.99}$$

Actually the singular piece indicated by the term $\mathcal{O}(l^{-1/2})$ can be explicitly calculated: using Eqs. (3.87-88) and the string equation, we have

$$\begin{aligned}
\langle W(l)\rangle &= \frac{1}{\sqrt{\pi}\kappa} \sum_{n\geq 1} t_n u^{n+1/2} \left(\frac{n!}{x^{n+1/2}} - \frac{1}{x^{1/2}} \right) + (\text{regular in } l) \\
&= \frac{1}{\sqrt{\pi}\kappa} \left(\sum_{n\geq 1} \frac{n! t_n}{l^{n+1/2}} - \frac{1}{l^{1/2}} \sum_{n\geq 1} t_n u^n \right) + (\text{regular in } l) \\
&= \frac{1}{\sqrt{\pi}\kappa} \left(-\frac{t}{l^{1/2}} + \sum_{n\geq 0} \frac{n! t_n}{l^{n+1/2}} \right) + (\text{regular in } l) \quad ,
\end{aligned} \tag{3.100}$$

We see that singular terms are analytical in the coupling constants (t_n). In ref. [143] it is argued that such terms are in fact expected also in the continuum formulation: singular contributions are interpreted as arising from small area surfaces, and despite the analyticity in t_n, they are considered universal.

Having understood the origin of the singular terms, we use the notation of ref. [31] to write the operator expansion

$$W(l) \quad \text{“}=\text{”} \quad -\frac{\kappa}{\sqrt{\pi}} \sum_{n\geq 0} \frac{(-1)^n}{n!} l^{n+1/2} \sigma_n \quad , \tag{3.101}$$

where quotation marks remind us that this "equation" does not reproduce the singularities of the 1-point functions. This expansion will be used in Chapter 4, in the studies on wave functions of scaling operators and the mininsuperspace approximation.

3.6. Scaling Dimensions and Preliminary Comparisons

Now we have in our hands enough data to start the comparison with the results from Chapter 2.

Following the arguments of the introductory discussion of this Chapter, as well as the renormalizations performed along the double scaling limit, we have interpreted t as the renormalized cosmological constant of the model. On the other hand we recall that, in the Liouville approach, the scaling behavior of correlation functions is indeed measured in terms of the cosmological coupling (μ); therefore we must find the equivalent regime in the matrix model formulation.

Let us consider the regime defined by the limit $t_n \to 0$: as the string equation implies that $u = t^{1/m}$, we shall define some critical exponents by observing the scaling behavior of correlation functions with respect to the constant t.

The scaling of the free energy defines the string susceptibility γ_m: it can be taken from the integration of Eq. (3.94) in the $u = t^{1/m}$ regime, and we find

$$\mathcal{F}_{zc} = -\frac{1}{\kappa^2}\frac{m^2}{(m+1)(2m+1)}\, t^{2+1/m} \propto t^{2-\gamma_m} \quad , \tag{3.102}$$

where zc means zero-coupling ($t_n = 0$). We therefore have the same family of susceptibilities $\gamma_m = -\frac{1}{m}$ found at the discrete level (i.e. before the double scaling limit).

We mentioned in the Introduction that the free energy of the matrix model, being the generator of connected surfaces, is the function to be compared with the partition function Z of the Liouville description. Recalling Eq. (1.163), that provides the scaling $Z \sim \mu^{2-\gamma}$, we learn that for the $c = 0$ model (i.e. pure gravity) we must have $\gamma = -1/2$, which coincides with the susceptibility observed in (3.102) for the $m = 2$ critical matrix model. If we take minimal unitary models, whose central charges are $c = 1 - \frac{6}{m(m+1)}$, $m = 1, 2, \cdots$, we obtain from Eq. (1.163) the corresponding susceptibilities $\gamma = -1/m$, i.e. the same family found in (3.102).

This coincidence led to the conjecture[130] that the m-th critical 1-matrix models corresponded to minimal unitary theories coupled to gravity: although we have started from discretized surfaces without any matter fields, it was predicted the "possibility of the appearance of some collective excitations of the metric field g_{ab}, which are similar to conformal fields of unitary models interacting with quantum fluctuations of 2D-gravity"[130].

However, the exponent γ does not determine uniquely the model and we anticipate that 1-matrix models in fact correspond to the non-unitary $(2, 2m-1)$ family of minimal models coupled to gravity. The detailed identification will be drawn in Chapter 4.

Let us first examine the scaling behavior of correlation functions of scaling operators[71]. We can associate a scaling dimension to every scaling operator, taken from the scaling of their correlators with respect to the coupling t. Calculating the 1-point functions

$$\langle\sigma_n\rangle_{zc} = -\frac{1}{\kappa^2}\frac{m}{(n+1)(n+m+1)}\, t^{1+1/m+n/m} \propto t^{2-\gamma_m+(d_{\sigma_n}-1)} \quad , \tag{3.103}$$

we find the exponent

$$d_{\sigma_n} = \frac{n}{m} \quad , \tag{3.104}$$

$(1-d_{\sigma_n})$ being the analog of β/α_+ in Eq. (2.28). Indeed we verify that the insertion of a scaling operator σ_n implies the addition of a term $(d_{\sigma_n}-1)$ to the exponent of t, as one can see in the correlation functions below:

$$\begin{aligned}\langle\sigma_i\sigma_j\rangle_{zc} &= -\frac{1}{\kappa^2}\frac{(t^{1/m})^{i+j+1}}{i+j+1} = -\frac{1}{\kappa^2}\frac{1}{i+j+1}\, t^{2+1/m+(i/m-1)+(j/m-1)} \\ &= -\frac{1}{\kappa^2}\frac{1}{i+j+1}\, t^{2-\gamma_m+(d_i-1)+(d_j-1)} \quad ,\end{aligned} \tag{3.105}$$

$$\begin{aligned}\langle\prod_{i=1}^{n}\sigma_{a_i}\rangle_{zc} &= -\frac{1}{\kappa^2}\partial_t^{n-2}\frac{(t^{1/m})^{a+1}}{1+a} = -\frac{1}{m\kappa^2}\partial_t^{n-3}t^{-1+1/m+a/m} \\ &= -\frac{1}{m\kappa^2}\frac{\Gamma(1/m+n/m)}{\Gamma(3-n+1/m+n/m)}\, t^{2+1/m+a/m-n} \\ &= \frac{\gamma_m}{\kappa^2}\frac{\Gamma\left(n-\gamma_m+\sum(d_{a_i}-1)\right)}{\Gamma\left(3-\gamma_m+\sum(d_{a_i}-1)\right)}\, t^{2-\gamma_m+\sum(d_{a_i}-1)} \quad ,\end{aligned} \tag{3.106}$$

where the integer a was defined in (3.96).

Take the simplest model, given by $m=2$: the susceptibility $\gamma=-1/2$ suggests that it might correspond to the theory of pure gravity. In that theory the Area operator is the relevant one: within the Liouville approach, it has two representations, namely $T_{k=0}$ and $T_{k=2\alpha_0}$, and we choose either one or the other depending on the kinematics (momentum conservation); in the matrix model formulation it is identified with the scaling operator σ_0, i.e. the one coupled to the cosmological constant $t\sim t_0$. From equations (2.100*b*) and (3.106) we have

$$\langle T_{2\alpha_0}(T_0)^{n-1}\rangle = \left(\frac{\partial}{\partial\mu}\right)^{n-3}\mu^{-1/2} \propto \mu^{\frac{5}{2}-n} \quad , \tag{3.107}$$

$$\langle(\sigma_o)^n\rangle = -\frac{1}{2\kappa^2}\left(\frac{\partial}{\partial t}\right)^{n-3} t^{-1+1/2} \propto t^{\frac{5}{2}-n} \quad , \tag{3.108}$$

which means that, up to a trivial normalization, the correlation functions are identical.

The problems come out when we go to higher values of m, if we insist on identifying the scaling operators σ_n with the dressed primary operators of the minimal unitary models: neither scaling dimensions nor correlation functions are compatible.

Take $m=3$ for instance: a detailed comparison[56,61–64] between the Ising model (characterized by (p,q)=(3,4), c=1/2) on random surfaces with the $m=3$ critical matrix model finally showed that the latter corresponds in fact to the theory of the Yang-Lee edge singularity, (p,q)=(2,5) and c=-22/5, coupled to gravity.

However, even if one succeeds in identifying the respective minimal model, we point out some troublesome aspects of the correlators (3.106): namely, the 2-point functions (3.105) are not orthogonal (in fact all of them are different from zero), as opposed to the orthogonality of vertex operators (recall that the momentum conservation implies that the only non-vanishing 2-point functions are those of the form $\langle T_k T_{2\alpha_0 - k}\rangle$); moreover, the only vertex operator whose 1-point function is not zero is the area operator $T_{2\alpha_0}$ (due to momentum conservation too), while (3.103) is telling us that all scaling operators have non-vanishing expectation values.

We conclude that a naive identification of the σ_n-operators with T_k-operators is likely to fail. We shall dedicate the whole Chapter 4 to the solution of this question.

3.7. c=1 Matrix Model

These models must be treated separately. This is not a surprise, since from this point on the string theory presumably undergoes a phase transition, which is signalized by the imaginary term in (1.121) for $1 < c < 25$. The $c = 1$ matrix models have been reviewed by D. Gross[136]. We shall use several results and techniques from that work.

Moreover, the tachyon, which is the degree of freedom of the string representing its center of mass, is massless in 2 dimensions ($c = 1$); above this point it is really a tachyonic degree of freedom, in the sense of having imaginary mass as it is clear form (2.214). Indeed, a simple counting of degrees of freedom, gives for the ground state energy $E \sim m^2$ the value $\frac{(d-2)}{2}\sum n = \frac{2-d}{24} = \frac{1-c}{24}$, using the standard ζ-function regularization.

Nevertheless, the $c = 1$ theory can still be treated using the matrix model method. We consider the following action

$$S[\phi] = \int \mathrm{d}t \, \mathrm{tr} \left[\frac{1}{2}\dot{\phi}^2(t) + U(\phi(t))\right] \quad , \tag{3.109}$$

where we introduce a kinetic term, as compared with (3.20) in order to take into account time dependence[3)]. Fortunately, in one dimension the exponential propagator falls into the same universality class as the Gaussian theory, due to the mild (in fact absent) ultra violet divergencies. On the other hand, the infrared behavior is the same, that is, both e^{-p^2/μ^2} and $\frac{1}{p^2+\mu^2}$, behave as $1 - p^2/\mu^2$, for small momentum. Therefore the partition function is

$$\mathcal{Z} = \int \mathcal{D}^{N^2} \phi(t) \, \mathrm{e}^{-\beta S[\phi]} \quad , \tag{3.110}$$

where we introduce the parameter $\beta \equiv N/\Lambda$.

The Hamiltonian is given by the expression

$$H = \mathrm{tr}\left[-\frac{1}{2}\frac{\delta^2}{\delta\phi_{ij}^2} + U(\phi)\right] \quad . \tag{3.111}$$

3) From the usual discretization[136] of Polyakov action, we obtain an exponential propagator.

In the above we allow for a general interaction described by the potential $U(\phi)$. There are also some different possibilities of defining the target space, which could be the real line, a circle of radius R, or a discrete set of points[136]. In fact, the latter is equivalent to the first when the points are at distances smaller than the Planck length. We shall, in the following stay in the case of a real line. See however [94] for the case of a circle of radius R.

The most important step allowing for a solution of the problem concerns the integration over the angular degrees of freedom, reducing a problem of N^2 degrees of freedom represented by $\phi(t)$, to the N degrees of freedom represented by the eigenvalues of $\phi(t)$. We first notice that the potential is angular invariant, and can be written directly in terms of the eigenvalues. Consider the wave function ψ, eigenfunction of (3.111), and the object

$$\chi = \psi\Delta(\lambda) \quad , \tag{3.112}$$

where $\Delta(\lambda)$ is the Vandermond determinant

$$\Delta(\lambda) = \prod_{i<j}(\lambda_i - \lambda_j) \quad . \tag{3.113}$$

The ϕ integration measure can be written in terms of the above as

$$\mathcal{D}\phi = \mathcal{D}\Omega\Delta^2(\lambda)\mathcal{D}\lambda \quad , \tag{3.114}$$

since $\phi = \Omega^\dagger\Lambda\Omega$, where Ω is the angular variables matrix and Λ is the diagonal matrix of the eigenvalues. For the kinetic term of the Hamiltonian we have,

$$\int\prod_i \mathcal{D}\lambda_i\mathcal{D}\Omega\sum_{ij}\left(\frac{\partial\psi}{\partial\phi_{ij}}\right)^2\Delta^2(\lambda) = \text{angular volume} \times \int\prod_i \mathrm{d}\lambda_i\sum\left(\frac{\partial\psi}{\partial\lambda_i}\right)^2\Delta^2(\lambda) \tag{3.115}$$

where only the radial terms survived for angular symmetric wave functions. Using the definition (3.112) we have for the (normalized) kinetic term

$$\begin{aligned}\int\prod_j \mathcal{D}\lambda_j\sum\left(\frac{\partial\chi}{\partial\lambda_i} - \chi\frac{\partial\ln\Delta(\lambda)}{\partial\lambda_i}\right)^2 &= \\ = \int\prod_j \mathcal{D}\lambda_j\sum\left(\frac{\partial\chi}{\partial\lambda_i}\right)^2 &+ \int\prod_j \mathcal{D}\lambda_i\chi^2\sum\left[\frac{\partial^2\ln\Delta}{\partial\lambda_i^2} + \left(\frac{\partial\ln\Delta}{\partial\lambda_i}\right)^2\right]\end{aligned} \tag{3.116}$$

The last term vanishes identically as is easily seen from the definitions. Therefore the Hamiltonian is given by

$$H = N\sum_i\left[-\frac{1}{2}\frac{\partial^2}{\partial\lambda_i^2} + U(\lambda_i)\right] \tag{3.117}$$

in terms of the new wave functions χ as defined in (3.113), or else, using the eigenfunctions ψ and including the angular terms, we find

$$H = -\frac{1}{2\Delta(\lambda)}\sum \frac{\partial^2}{\partial\lambda_i^2}\Delta(\lambda) + \sum U(\lambda_i) + \sum_{i<j}\frac{\pi_{ij}^2 + \tilde{\pi}_{ij}^2}{(\lambda_i - \lambda_j)^2} \quad , \tag{3.118}$$

where π_{ij} and $\tilde{\pi}_{ij}$ are the generators of the left and right rotations.

The above results show that the problem boils down to a theory of N non-interacting fermions, since the wave function is antisymmetric because of the (explicit) appearance of the Vandermonde determinant. In the partition function this fact shows up due to the decomposition $\phi(t) = \Omega^\dagger(t)\Lambda(t)\Omega(t)$ where Ω is unitary and Λ is diagonal (eigenvalue) matrix; following this decomposition one computes the kinetic term of the action as[136]

$$\operatorname{tr}\dot{\phi}^2 = \operatorname{tr}(\dot{\Lambda}^2 + [\Lambda, A][\Lambda, A]) \quad , \tag{3.119}$$

where $A \equiv \dot{\Omega}(t)\Omega^\dagger(t)$. The only non-trivial appearance of Ω is through A above, and using the formula

$$\int \mathcal{D}U e^{\operatorname{tr} UAU^\dagger B} = \frac{\det e^{a_i b_j}}{\Delta(a_i)\Delta(b_j)} \quad , \tag{3.120}$$

where a_i and b_i are the eigenvalues of A and B, we can discretize the time, writing

$$\begin{aligned} A \equiv \Omega(t)\Omega^\dagger(t) &= \frac{1}{\epsilon}\left(\Omega(t_{n+1})\Omega^\dagger(t_n) - 1\right) \\ &= -\frac{1}{\epsilon}\left(\Omega(t_n)\Omega^\dagger(t_{n+1}) - 1\right) \quad , \end{aligned} \tag{3.121}$$

and integrate explicitly over $\Omega(t_n)$, since it shows up only in

$$\operatorname{tr}[\Lambda, A]^2 = \frac{2}{\epsilon^2}\operatorname{tr}\left(\Lambda\Omega(t_{n+1})\Omega^\dagger(t_n)\Lambda\Omega(t_n)\Omega^\dagger(t_{n+1}) - \Lambda^2\right) \quad . \tag{3.122}$$

Therefore, integration over Ω upon defining $U = \Omega(t_{n+1})\Omega^\dagger(t_n)$ yields the square of the Vandermonde determinant in the denominator, cancelling the factors of $\Delta(\lambda)$ arising from the change of variables

$$\mathcal{D}^{N^2}\phi = \mathcal{D}\Omega \prod \mathcal{D}\lambda_i \Delta^2(\lambda) \quad . \tag{3.123}$$

A comment is in order at this point. For a circle (or in general a non-contractible loop) the U's above obey the constraint $\prod U_i = 1$, and the Ω's can not be easily integrated. This is the case of a plaquette in lattice QCD[136].

In the large N expansion of the model, we shall be interested in the double scaling limit[4)] where $\frac{N}{\beta} \to 1$, and the theory displays critical behavior, in such a way that we are near the critical level of the potential, where the Fermi level μ_F is near $\mu_c = U(\lambda_c)$ and $U'(\lambda_c) = 0$.

[4)] We normalize $\Lambda_c = 1$.

We are interested in results which are universal, not depending on details of the given potential used to construct the triangulation. A possible choice is the well known double well potential $U(\lambda) = -g_2\lambda^2 + g^4\lambda^4$, for which $\mu_c = 0 = \lambda$.

At the critical point the energy starts to drop away from the potential well, since the Fermi level has reached the critical (top) value. In this case we shall be able to express $1 - \Lambda = 1 - \frac{N}{\beta}$ in terms of $\mu \equiv \mu_c - \mu_F$, and this relation displays a singularity.

The study of the $c = 1$ theory starts out of the definition of the density of the states

$$\rho(E) = \frac{1}{\beta}\sum \delta(E - E_i) \quad . \tag{3.124}$$

In terms of the Hamiltonian operator, this is a trace, and can be evaluated in terms of the imaginary part of the resolvent, that is

$$\rho(E) = \frac{1}{\beta}\mathrm{tr}\,\delta(H - E) = \frac{1}{\pi\beta}\Im m\,\mathrm{tr}\,\frac{1}{H - E - i\epsilon} \tag{3.125}$$

and obeys the constraint

$$\beta\int_0^{\mu_F} \rho(E)\mathrm{d}E = N \quad , \tag{3.126}$$

since it counts the number of states.

The average energy is defined as

$$E = \beta^2\int_0^{\mu_F} \rho(E)E\mathrm{d}E \quad . \tag{3.127}$$

The above definitions of E and N (or $\Lambda = N/\beta$) will lead, in the critical case to a non-analytic behavior of the relations $\Lambda(\mu)$ and $E(\Lambda)$ (or $E(\mu)$). As explained in the case $c < 1$, here also Λ will be identified with the cosmological constant.

From the above equations we find

$$\frac{\partial\Lambda}{\partial\mu} = -\rho(\mu_F) \tag{3.128}$$

and

$$\frac{\partial E}{\partial\Lambda} = \frac{-\frac{\partial E}{\partial\mu_F}}{\frac{\partial\Lambda}{\partial\mu}} = \beta^2\mu_F = -\beta^2(\mu - \mu_c) \quad . \tag{3.129}$$

The density of eigenvalues, on the other hand, may be computed integrating over the $\{\lambda, p\}$ phase space, in the semiclassical approximation, and one finds

$$\rho(E) = \int\frac{\mathrm{d}\lambda\mathrm{d}p}{2\pi}\delta(\frac{p^2}{2} + U - E) = \frac{1}{\pi}\int_{\lambda_0}\frac{\mathrm{d}\lambda}{\sqrt{2(E - U(\lambda))}} \quad . \tag{3.130}$$

Notice that this is $1/2\pi$ times the time of flight of a particle of unit mass in this potential. We expand the potential as

$$U(\lambda) = U(\lambda_c) - \frac{1}{2}(\lambda - \lambda_c)^2 = \mu_c - \frac{1}{2}(\lambda - \lambda_c)^2 \quad , \tag{3.131}$$

from which we find

$$\rho(\mu_F) = -\frac{1}{2\pi}\ln\mu + \cdots \quad , \tag{3.132}$$

where the dots contain terms depending on details of the potential. Fortunately, we are only interested in the singular term, which corresponds to the logarithmic divergence associated with the time of flight of the particle in the given potential. We thus find

$$\Lambda \approx \int^{\mu} \rho(E)\mathrm{d}E = \frac{1}{4\pi}\mu\ln\mu + \cdots \tag{3.133}$$

and for the energy

$$\frac{\partial E}{\partial \mu} = \beta^2\mu\rho \quad \text{implying} \quad E = -\frac{N^2}{4\pi}\mu^2\ln\mu + \cdots \quad . \tag{3.134}$$

Notice the logarithmic deviations which we have interpreted in the case of continuous gravity as a reminiscent from the new cosmological term in the $c = 1$ theory (see discussion prior to (3.71)).

In the full quantum theory, the energy has also been computed as a perturbative expansion[51]. We present only the result

$$E = \frac{1}{4\pi}\left\{-(\beta\mu)^2\ln\mu + \frac{1}{9}\ln\mu - \sum_1^{\infty}(2^{2n+1}-1)\frac{B_{2n+2}}{n(n+1)(2n+1)}(\beta\mu)^{-2n}\right\} \quad , \tag{3.135}$$

where B_{2n} is a Bernoulli number.

Kostov[137] considered this problem and using the WKB approximation computed the period T of the classical trajectory, supposing a symmetric potential, and found

$$T = 2\ln\mu \quad , \tag{3.136}$$

which is nothing but (3.130) up to a constant that can be traced back to the symmetric potential. The energy splitting is given in terms of the frequency w by

$$\frac{\mathrm{d}E_n}{\mathrm{d}n} = w = \frac{2\pi}{T} = \frac{\pi}{\ln\mu} \quad . \tag{3.137}$$

Gross, Klebanov and Newman[93] considered the puncture operators

$$\mathcal{O}^{(q)}(x) = \frac{1}{N}\mathrm{tr}\,\phi^q(x) \tag{3.138}$$

and computed its 2-point function. This computation uses the semi-classical approximation in quantum mechanics. Consider the connected Greens function

$$\begin{aligned} G^{(2)}(x) &= \langle \mathrm{tr}\,\phi^q(0)\mathrm{tr}\,\phi^q(x)\rangle_c = \\ &= \langle 0|\sum \lambda_k^q \sum_{\chi\neq 0} |\chi\rangle\langle\chi| \mathrm{e}^{iHx} \sum \lambda_j^q \mathrm{e}^{-iHx}|0\rangle \\ &= \sum_{\chi\neq 0} \mathrm{e}^{iN(E_l-E_0)x} |\langle 0|\sum \lambda_k^q |\chi_l\rangle|^2 \quad . \end{aligned} \tag{3.139}$$

As mentioned before, the states are given by the Slater determinant of one particle state, and the above matrix elements are non-vanishing only if χ_l is a one particle excitation.

Notice that the vacuum is the Slater determinant of the N lowest eigenstates. Therefore one of the states is excited to a state above N, and we have

$$G^{(2)}(x) = \sum_{n=1}^{\infty} \sum_{k=N-n+1}^{\infty} e^{iN(E_{k+n}-E_k)x} |\langle E_k|\lambda^q|E_{k+n}\rangle|^2 \quad . \tag{3.140}$$

The matrix element may be computed within the semi classical approximation by standard methods[138]. The matrix elements between two given states (1,2) of a physical quantity f is

$$f_{1,2} = \frac{f(\lambda)\mathrm{d}\lambda}{\{4(E_1-U)(E_2-U)\}^{1/4}} e^{iN\int \mathrm{d}x(\sqrt{2(E_1-U)}-\sqrt{2(E_2-U)})} \quad . \tag{3.141}$$

Using $\dot{\lambda} = \sqrt{2(E-U)}$, and expanding for $E_k - E_{n+k} << N$ (using the saddle point), and $E_k \sim kw$ we obtain,

$$\langle E_k|\lambda^q|E_{k+n}\rangle = \int \mathrm{d}x \lambda^q(x) e^{inwx} = f_n^{(q)} \tag{3.142}$$

and

$$G^{(q)}(x) = \sum n|f_n^{(q)}|^2 e^{inwt} \quad . \tag{3.143}$$

Going to the Euclidian space we have

$$G^{(q)}(x) = \sum_1^{\infty} n|f_n^{(q)}|^2 e^{-nw|x|} \quad . \tag{3.144}$$

We consider the simplest potential[137]

$$U(\lambda) = \frac{1}{4\alpha}(\lambda^2 - \lambda^4) \tag{3.145}$$

and obtain in the semi classical approximation

$$p(\lambda) = \frac{\mathrm{d}\lambda}{\mathrm{d}\tau} = \sqrt{2(\mu - U)} = \left[\frac{1}{2\alpha}(\lambda_+^2 - \lambda^2)(\lambda_-^2 \lambda^2)\right] \quad , \tag{3.146}$$

with λ_+, λ_- being the turning points of the potential. The above equation has been solved in terms of elliptic integrals[93]

$$\lambda(\tau) = \lambda - sn\left(\frac{\lambda_+ 2}{\sqrt{2\alpha}}, k\right) \tag{3.147}$$

and the matrix element $f_n^{(2)}$ reads

$$f_n^{(2)} = -\frac{2nw^2\alpha}{\sinh(nwT'/2)} \quad , \tag{3.148}$$

with

$$w = \frac{2\pi}{T'} = -\frac{\pi}{\sqrt{\alpha}\ln\mu} \quad , \quad T' = 2\pi\sqrt{\alpha} \quad . \tag{3.149}$$

Therefore the 2-point function for $q = 2$ is

$$\begin{aligned} G^{(2)}(p) &= 8\alpha^2 \sum_{n\,\mathrm{even}} \frac{wn^2}{p^2 + n^2w^2} \frac{w^4n^2}{\sinh^2(n\pi w\sqrt{\alpha})} \\ &= -i \oint \mathrm{d}z \cot\left(\frac{\pi z}{2w}\right) \frac{\alpha^2 z^4}{(p^2+z^2)\sinh^2(2\pi\sqrt{\alpha})} \quad , \end{aligned} \tag{3.150}$$

where the integral runs as in figure 3.4. We can distort it as in figure 3.5. The single poles above contribute as

$$\begin{aligned} -\frac{2\pi\alpha^2|p|^3}{\sin^2(p\pi\sqrt{\alpha})} \coth\frac{\pi p}{2w} &\sim -\left(1 + 2\mathrm{e}^{-\pi p/w}\right) \\ &= \left(1 + 2\mu^{p\sqrt{\alpha}}\right) \quad , \end{aligned} \tag{3.151}$$

where Eq. (2.150) has been used.

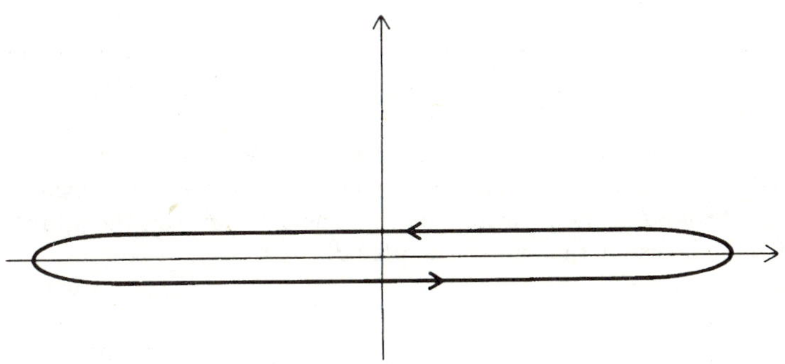

Fig. 3.4 Integration contour corresponding to Eq. (3.150).

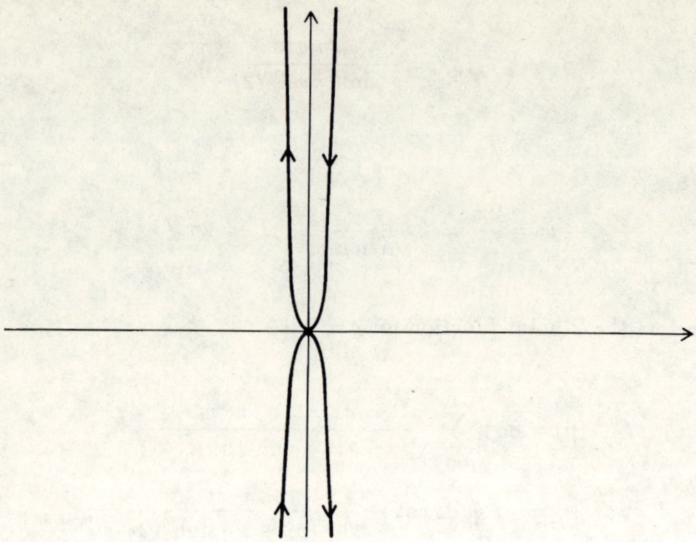

Fig. 3.5 Deformed integration contour.

The constant term is independent of μ and can be dropped. For the second term, we have a dependence $\mu^{p\sqrt{\alpha}}$, and the scaling at the double poles is $\sim \mu^n$.

After collecting the double poles contributing at $z = \pm \frac{in}{\sqrt{\alpha}}$, we have

$$G_{\text{double}} = \frac{-8n^4\mu^n}{w(\alpha p^2 - n^2)} + \frac{16\sqrt{\alpha}\mu^n n^3}{\pi}\left[\frac{2}{\alpha p^2 - n^2} + \frac{n^2}{(\alpha p^2 - n^2)^2}\right] \quad . \tag{3.152}$$

In general, for a correlator with $q = 2r$, r integer, one finds

$$f_n^{(2r)} = \frac{\alpha n w^2}{\sinh(n\pi w\sqrt{\alpha})}\sum_{m=1}^{r} a_m(\alpha n^2 w^2)^{m-1} \quad , \tag{3.153}$$

in such a way that $F^{(2r)}(p)$ is a polynomial of degree $2r - 2$ in αp^2 multiplying $F^{(2)}(p)$. This is a non-universal behavior, which depends on details of the discretization.

The universal behavior is the presence of the poles at $p = n/\sqrt{\alpha}$, which are the discrete states we found before in the continuum case, although historically they have been first found in the present context. One can compare them to those in the $c = 1$ Liouville theory at the end of Chapter 2.

4. Conformal Basis for Scaling Operators

4.1. Introduction

A preliminary attempt to identify the vertex with scaling operators in $c < 1$ minimal models, as described in the previous Chapter, turned out to be unsuccessful, since the results seemed incompatible at the level of 1- and 2-point functions already. Here we shall see how one can define alternative sets of scaling operators, related to the original ones by analytical transformations in the space of coupling constants. One of these sets, referred to as the conformal basis of scaling operators, happens to exhibit the orthogonality properties of vertex operators correlation functions, within a suitable scaling regime, and thus shall be taken as the appropriate set for comparisons with the Liouville formulation of the theory. The conformal basis was originally found in ref. [31] using the minisuperspace approximation of the Wheeler-DeWitt equation. Physical wave functions are expected to obey such equation and section 4.4. is dedicated to its discussion.

4.1.1. Dimensional Analysis of Coupling Constants

Here we shall identify the coupling constant which has the dimension of a *physical* cosmological constant, in terms of which a new critical regime will be defined.

Let us examine the dimensions of the renormalized coupling constants. From their definition in Eq. (3.69), we can relate their dimensions to the scaling parameter a:

$$[t_n] = [a^{2/m}]^{n-m} \quad . \tag{4.1}$$

On the other hand, while constructing macroscopic loops, we learned that $a^{2/m}$ represented a length scale for the discretized surface. We therefore conclude that the dimension of each coupling constant depends on m,

$$[t_n] = [\text{length}]^{n-m} \quad , \tag{4.2}$$

which thus fixes the dimensions of the scaling operators as

$$[\sigma_n] = [\text{length}]^{m-n} \quad . \tag{4.3}$$

Concerning the specific heat u and the string coupling constant κ, their dimensions can be read from Eqs. (3.68) and (3.77) as given by, respectively

$$[u] = [\text{length}]^{-1} \quad , \tag{4.4}$$

$$[\kappa] = [\text{length}]^{-(m+1/2)} \quad . \tag{4.5}$$

The "physical" cosmological constant is usually taken to be the one coupled to the area operator (the cosmological term, or the dressed identity operator in the Liouville description) and should therefore have dimension of inverse of area, or $[\text{length}]^{-2}$. Since $[t] = [t_0] = [\text{length}]^{-m}$, we conclude that t deserves the name of cosmological constant only when $m = 2$. In general t_{m-2} is the coupling with the true dimension of a cosmological constant. If one wishes to compare results with the Liouville approach, as they stand in Chapter 2, one should select a regime where the scaling properties are measured with respect to some coupling as t_{m-2} rather than t.

Therefore the issue of reaching a scaling regime involves the tuning of coupling constants, and since the dependence on the couplings is fundamentally carried by the specific heat u, as a solution of the string equation, we here rewrite the latter as

$$u^m = t + \sum_{n\geq 0} t_n u^n \quad . \tag{4.6}$$

Notice that u depends on the sum $(t_0 + t)$ only, which guarantees that $\frac{\partial}{\partial t}u = \frac{\partial}{\partial t_0}u$. Still t and t_0 can be varied independently so that we can take t_0 as the source for the σ_0 (puncture) operator: in particular we can insert puncture operators into a correlation function by successive applications of the derivative $\frac{\partial}{\partial t_0}$ and then taking the limit $t_0 \to 0$; the resulting correlator, as a function of t, would correspond to an average in the presence of a $t\sigma_0$ term.

As it stands in (4.6) the string equation seems suitably designed for this σ_0-averaged regime: one takes the t_n couplings as perturbative sources to calculate correlation functions in the limit $t_n \to 0$, when $u^m = t$. This regime was explored in the previous Chapter, but one can think of many others.

Considering that we are looking for a regime where a constant with dimension of t_{m-2} is the relevant one, it is convenient to shift the couplings in (4.6) as

$$\begin{aligned} t_0 &\to t_0 - t \quad , \\ t_{m-2} &\to t_{m-2} + \mu \quad , \\ t_n &\to t_n \quad , \quad n \neq 0, m-2 \quad , \end{aligned} \tag{4.7}$$

where μ is taken as the physical cosmological constant, so that the string equation becomes

$$u^m = \mu u^{m-2} + \sum_{n\geq 0} t_n u^n \quad . \tag{4.8}$$

Now u depends on $(t_{m-2} + \mu)$ and the couplings t_n can be used as perturbative sources for n-point functions calculated in the presence of the term $\mu\sigma_{m-2}$, in the limit $t_n \to 0$, when $u^2 \to \mu$. The resulting correlators would then be scaling functions of μ.

In this σ_{m-2}-averaged regime we have the following flows

$$\frac{\partial u}{\partial t_n} = u^{n-(m-2)}\frac{\partial u}{\partial \mu} = u^{-(m-2)}\frac{\partial}{\partial \mu}R_{n+1}[u] \quad , \tag{4.9}$$

which can be easily derived from (4.8) and are the counterparts of the t-type flows in (3.71). As before, they shall be useful in the calculation of correlation functions.

Although the term $\mu\sigma_{m-2}$ exhibits already the dimensional properties of a two-dimensional cosmological term, the shift proposed in (4.7) does not exhaust all possibilities in our search for an appropriate set of couplings/operators. As it is discussed in Appendix F, we should consider analytical transformations[139] in the space of coupling constants as corresponding to admissible changes of basis.

Therefore we shall analyze the set of couplings $\{\hat{t}_n; n = 0, 1, 2, \cdots\}$, related to the original set by

$$t_n = f_n(\hat{t}_0, \cdots, \mu + \hat{t}_{m-2}, t_{m-1}, \cdots) \quad , \tag{4.10}$$

where f_n are analytical functions.

Let us further require that every coupling $\hat{t}_n$ from the new set have a well defined dimension: we assume that $[\hat{t}_n] = [t_n]$. We also impose that the limits $t_n \to 0$ and $\hat{t}_n \to 0$ be equivalent, so that both sets of couplings can be taken as perturbations around the same point. Moreover, as long as 1- and 2-point functions are concerned, one can reduce (4.10) to its first-order approximation in the $\hat{t}_n$ couplings. Taken all these restrictions and simplifications into account, the most general transformation would be written as

$$t_n \equiv \sum_{s=0}^{\infty} A_s^{(n+2s)} \mu^s \hat{t}_{n+2s} \quad , \tag{4.11}$$

where $A_s^{(i)}$ are some dimensionless coefficients. We remark that the linear transformation (4.11) is upper triangular; moreover, if we distinguish the even- and odd-index subsets, $\{t_n, n\,\mathrm{even}\}$ and $\{t_n, n\,\mathrm{odd}\}$, Eq. (4.11) relates them to analogous subsets, respectively $\{\hat{t}_n, n\,\mathrm{even}\}$ and $\{\hat{t}_n, n\,\mathrm{odd}\}$, without mixing among even and odd indices.

In terms of the couplings $\hat{t}_n$ the string equation reads

$$\begin{aligned} u^m &= \mu u^{m-2} + \sum_{n\geq 0} \left\{ \sum_{s=0}^{\infty} A_s^{(n+2s)} \mu^s \hat{t}_{n+2s} \right\} u^n \\ &= \mu u^{m-2} + \sum_{n\geq 0} \hat{t}_n \left\{ \sum_{r=0}^{[n/2]} A_r^{(n)} \mu^r u^{n-2r} \right\} \\ &= \mu u^{m-2} + \sum_{n\geq 0} \hat{t}_n (\sqrt{\mu})^n P_n\left(u/\sqrt{\mu}\right) \quad , \end{aligned} \tag{4.12}$$

where

$$P_n(x) = \sum_{s=0}^{[n/2]} A_s^{(n)} x^{n-2s} \tag{4.13}$$

are characteristic polynomials of a given basis (at the linear approximation (4.11) a basis is determined by its coefficients $A_s^{(n)}$).

The analytical transformations (4.11) also lead us to a new set of scaling operators, $\{\widehat{\sigma}_n\}$, coupled to the constants $\widehat{t}_n$. They can be written in terms of the operators σ_n in the same way that $\frac{\partial}{\partial \widehat{t}_n}$ relates to $\frac{\partial}{\partial t_n}$:

$$\frac{\partial}{\partial \widehat{t}_n} = \sum_k \frac{\partial t_k}{\partial \widehat{t}_n} \frac{\partial}{\partial t_k} = \sum_{s=0}^{[n/2]} A_s^{(n)} \mu^s \frac{\partial}{\partial t_{n-2s}} \quad , \tag{4.14}$$

i.e., we define

$$\widehat{\sigma}_n \equiv \sum_{s=0}^{[n/2]} A_s^{(n)} \mu^s \sigma_{n-2s} \quad . \tag{4.15}$$

This formula represents a change of basis of scaling operators governed by the coefficients $A_s^{(n)}$. Notice that it preserves the dimensions ($[\widehat{\sigma}_n] = [\sigma_n]$) and is independent of the perturbative couplings $\widehat{t}_n$ (this is a consequence of the linear approximation in (4.11)).

It is interesting to calculate the $\widehat{t}_n$-flows, which follow from (4.9) and (4.14),

$$\begin{aligned} \frac{\partial u}{\partial \widehat{t}_n} &= \sum_{s=0}^{[n/2]} A_s^{(n)} \mu^s u^{n-2s-(m-2)} \frac{\partial}{\partial \mu} u \\ &= (\sqrt{\mu})^n P_n\left(\frac{u}{\sqrt{\mu}}\right) u^{-(m-2)} \frac{\partial u}{\partial \mu} \quad . \end{aligned} \tag{4.16}$$

In the zero-coupling (zc) regime, when $\widehat{t}_n \to 0$ and $u/\sqrt{\mu} \to 1$, the equation above implies in particular that

$$\frac{\partial u}{\partial \widehat{t}_{m-2}} \to P_{m-2}(1) \frac{\partial u}{\partial \mu} \quad . \tag{4.17}$$

Normalizing the polynomials $P_n(x)$ such that $P_{m-2}(1) = 1$, we conclude that deriving with respect to the cosmological constant μ corresponds to the insertion of a $\widehat{\sigma}_{m-2}$ operator, which thus is our candidate for the role of *area* operator.

The two-steps coupling redefinition parametrized by (4.7) and (4.11) have led us to a family of different basis for scaling operators parametrized by (4.15). Now we must check whether there is at least one basis in which the correlation functions coincide with those predicted by the conformal (continuum) formulation of the problem.

4.2. Conformal Basis

Here we shall identify the orthogonal set of scaling operators, which will be later interpreted as a lowest order approximation to the conformal basis we have been looking for. We shall also calculate the n-point functions of orthogonal operators.

We have adopted the orthogonality of 2-point functions of vertex operators as a guiding criterion. From the definition (4.15) and also using Eq. (3.95) which was valid in any regime, we find

$$\begin{aligned}\langle\widehat{\sigma}_i\widehat{\sigma}_j\rangle &= \sum_{s=0}^{[i/2]}\sum_{r=0}^{[j/2]} A_s^{(i)}A_r^{(j)}\mu^{s+r}\langle\sigma_{i-2s}\sigma_{j-2r}\rangle = \\ &= -\frac{1}{\kappa^2}\sum_{s=0}^{[i/2]}\sum_{r=0}^{[j/2]} A_s^{(i)}A_r^{(j)}\mu^{s+r}\frac{u^{i+j-2s-2r+1}}{(i+j-2s-2r+1)} \quad .\end{aligned} \tag{4.18}$$

In the zero-coupling limit, when $u^2 \to \mu$, the equation above reduces to

$$\begin{aligned}\langle\widehat{\sigma}_i\widehat{\sigma}_j\rangle &= -\frac{\mu^{(i+j+1)/2}}{\kappa^2}\sum_{s=0}^{[i/2]}\sum_{r=0}^{[j/2]} A_s^{(i)}A_r^{(j)}\frac{1}{(i+j-2s-2r+1)} \\ &= -\frac{\mu^{(i+j+1)/2}}{\kappa^2}g_{ij}(A) \quad ,\end{aligned} \tag{4.19}$$

where

$$g_{ij}(A) \equiv \sum_{s=0}^{[i/2]}\sum_{r=0}^{[j/2]}\frac{A_s^{(i)}A_r^{(j)}}{(i+j-2s-2r+1)} \tag{4.20}$$

can be interpreted[139] as a metric $g_{ij}(t_n)$ in the space of coupling constants: the coefficients $A_s^{(i)}$ parametrize a curve on that space and we would like to find the point of this curve where the metric is diagonal, i.e. $g_{ij} \propto \delta_{ij}$. Having this aim in mind, we remark that

$$\begin{aligned}[1-(-1)^{i+j+1}]g_{ij}(A) &\equiv \sum_{s=0}^{[i/2]}\sum_{r=0}^{[j/2]} A_s^{(i)}A_r^{(j)}\frac{[1-(-1)^{i+j+1}]}{(i+j-2s-2r+1)} \\ &= \sum_{s=0}^{[i/2]}\sum_{r=0}^{[j/2]} A_s^{(i)}A_r^{(j)}\int_{-1}^{+1}\mathrm{d}x\, x^{i+j-2s-2r} \\ &= \int_{-1}^{+1}\mathrm{d}x\, P_i(x)P_j(x) \quad ,\end{aligned} \tag{4.21}$$

where $P_i(x)$ is the i-th characteristic polynomial defined in (4.13). We see that the problem of orthogonalizing 2-point functions has now been reduced to finding orthogonal polynomials with respect to a given integral. In our case this integral is simply $\int_{-1}^{+1}\mathrm{d}x$, for which the problem has a well known solution: we must take

$P_n(x)$ as the Legendre polynomials. We have summarized in Appendix E some of their properties.

In particular, Eq. (E.2) indicates that the coefficients $A_s^{(i)}$ should be taken as

$$A_s^{(i)} = \frac{(-1)^s 2^{i-2s}}{\sqrt{\pi}} \frac{\Gamma(i-s+1/2)}{\Gamma(s+1)\Gamma(i-2s+1)} \quad , \tag{4.22}$$

to define the conformal basis. Using the orthogonality property (E.5) in Eq. (4.21) we observe that:

- when $i+j=2k, k=0,1,\cdots$, one has

$$g_{ij} = \frac{1}{(2i+1)}\delta_{ij} \quad ; \tag{4.23}$$

- for $i+j+1=2k, k=1,2,\cdots$, Eq. (4.21) is sterile. Back to the definition (4.20) one finds that in general $g_{ij} \neq 0$ (for instance, $g_{12}=1/8$). But, according to Eq. (4.19), such 2-point functions are analytical functions of μ,

$$\langle \widehat{\sigma}_i \widehat{\sigma}_{2k-1-i} \rangle = -\frac{g_{i,2k-1-i}}{\kappa^2}\mu^k \quad , \tag{4.24}$$

which, as it is discussed in Appendix F, is a non-universal neglectable term.

Therefore we can say that the scaling operator basis given by[31]

$$\widehat{\sigma}_n = \sum_{s=0}^{[n/2]} \frac{(-1)^s 2^{n-2s}}{\Gamma(1/2)} \frac{\Gamma(n-s+1/2)}{\Gamma(s+1)\Gamma(n-2s+1)}\mu^s \sigma_{n-2s} \tag{4.25a}$$

and the inversion relation

$$\sigma_n = \frac{\sqrt{\pi} n!}{2^n} \sum_{s=0}^{[n/2]} \frac{(2n-4s+1)}{s!\Gamma(n-s+3/2)}\mu^s \widehat{\sigma}_{n-2s} \quad , \tag{4.25b}$$

are so that the basis $\{\widehat{\sigma}_n\}$ is orthogonal up to analytical terms in μ:

$$\langle \widehat{\sigma}_i \widehat{\sigma}_j \rangle = \begin{cases} \text{analytical in } \mu & \text{when } i+j \text{ is odd} \quad , \\ -\frac{1}{\kappa^2}\frac{\mu^{i+1/2}}{(2i+1)}\delta_{ij} & \text{when } i+j \text{ is even} \quad , \end{cases} \tag{4.26}$$

as expected for the dressed vertex operators of the Liouville formulation. We discuss the scaling factor ($\mu^{i+1/2}$) later on, after the identification of model in section 4.3.

After successful results with 2-point correlators we must now examine 1-point functions. Using the normalization property $P_n(1)=1$ of the polynomials (E.2) and the area-operator insertion equation (4.17), we find

$$\begin{aligned} \frac{\partial}{\partial \mu}\langle \sigma_i \rangle &= \langle \widehat{\sigma}_{m-2} \sigma_i \rangle = \\ &= \sqrt{\pi}\frac{i!}{2^i} \sum_{s=0}^{[i/2]} \frac{(2i-4s+1)}{s!\Gamma(i-s+3/2)}\mu^s \langle \widehat{\sigma}_{m-2} \widehat{\sigma}_{i-2s} \rangle \quad . \end{aligned} \tag{4.27}$$

Using now the orthogonality of the conformal basis, we see that the only relevant term in (4.27) comes from $i-2s=m-2$. Since $0 \le s \le [i/2]$, this term is present only when $i=(m-2)+2k$, $k=0,1,\cdots$. In that case,

$$\begin{aligned}\frac{\partial}{\partial\mu}\langle\sigma_{(m-2)+2k}\rangle &= \frac{(m-2+2k)!}{2^{m-2+2k}}\frac{(2m-3)\sqrt{\pi}}{k!\Gamma(m+k-1/2)}\mu^k\langle\widehat{\sigma}^2_{m-2}\rangle \\ &= -\frac{\sqrt{\pi}}{\kappa^2}\mu^{m+k-3/2}\frac{1}{2^{m-2+2k}}\frac{(m-2+2k)!}{k!\Gamma(m+k-1/2)} \quad ,\end{aligned} \tag{4.28}$$

whose integration yields

$$\langle\sigma_{(m-2)+2k}\rangle = -\frac{\sqrt{\pi}}{\kappa^2}\mu^{m+k-1/2}\frac{1}{2^{m-2+2k}}\frac{(m-2+2k)!}{k!\Gamma(m+k+1/2)} \quad . \tag{4.29}$$

Now we can calculate $\langle\widehat{\sigma}_i\rangle$ recalling the definition (4.25a): the non-trivially vanishing 1-point functions are

$$\begin{aligned}&\langle\widehat{\sigma}_{(m-2)+2k}\rangle = \\ &= \sum_{s=0}^{[\frac{(m-2)}{2}]+k}\frac{(-1)^s 2^{m-2+2(k-s)}}{\Gamma(1/2)}\frac{\Gamma(m-2+2k-s+1/2)}{s!(m-2+2k-2s)!}\mu^s\langle\sigma_{m-2+2(k-s)}\rangle \quad ,\end{aligned} \tag{4.30}$$

for $k=0,1,2,\cdots$. Since $\langle\sigma_{m-2+2(k-s)}\rangle \ne 0$ only when $s \le k$, we have

$$\begin{aligned}\langle\widehat{\sigma}_{(m-2)+2k}\rangle &= \sum_{s=0}^{k}\frac{(-1)^s 2^{(m-2)+2(k-s)}}{\Gamma(1/2)}\frac{\Gamma(m-2+2k-s+1/2)}{s!(m-2+2k-2s)!}\mu^s\times \\ &\times\left(-\frac{\sqrt{\pi}}{\kappa^2}\right)\mu^{m+(k-s)-1/2}\frac{1}{2^{m-2+2(k-s)}}\frac{(m-2+2k-2s)!}{(k-s)!\Gamma(m+k-s+1/2)} \\ &= -\frac{1}{\kappa^2}\mu^{m+k-1/2}\sum_{s=0}^{k}(-1)^s\frac{1}{s!(k-s)!}\frac{\Gamma(m+2k-s-3/2)}{\Gamma(m+k-s+1/2)} \quad .\end{aligned} \tag{4.31}$$

We notice that the sum right above can be rewritten as

$$\begin{aligned}&\sum_{s=0}^{k}(-1)^s\frac{1}{s!(k-s)!}\frac{\Gamma(m+k-s-1/2+(k-2)+1)}{\Gamma(m+k-s-1/2+1)} \\ &= \sum_{s=0}^{k}(-1)^s\frac{1}{s!(k-s)!}\left\{\left(\frac{\partial}{\partial x}\right)^{k-2}x^{(m+k-s-1/2)+k-2}\right\}_{x=1} \\ &= \left\{\frac{1}{k!}\left(\frac{\partial}{\partial x}\right)^{k-2}x^{(m-1/2)+(k-2)}\sum_{s=0}^{k}\frac{k!}{s!(k-s)!}x^{k-s}(-1)^s\right\}_{x=1} \\ &= \left\{\frac{1}{k!}\left(\frac{\partial}{\partial x}\right)^{k-2}x^{(m-1/2)+k-2}(x-1)^k\right\}_{x=1} \quad ,\end{aligned} \tag{4.32}$$

which *vanishes* for $k \geq 2$ (as the expression between curled brackets is proportional to $(x-1)$ whenever $k \geq 2$). Therefore, the only non-vanishing cases are $k = 0, 1$, which we calculate separately:

$$\begin{aligned}
\langle \widehat{\sigma}_{m-2} \rangle &= -\frac{1}{\kappa^2} \mu^{m-1/2} \frac{\Gamma(m-3/2)}{\Gamma(m+1/2)} \\
&= -\frac{1}{\kappa^2} \mu^{m-1/2} \frac{1}{(m-1/2)(m-3/2)} \\
&= -\frac{4}{\kappa^2} \frac{\mu^{m-1/2}}{(2m-1)(2m-3)} \quad , \qquad (4.33)\\
\langle \widehat{\sigma}_m \rangle &= -\frac{1}{\kappa^2} \mu^{m+1/2} \left\{ \frac{\Gamma(m+1/2)}{\Gamma(m+3/2)} - \frac{\Gamma(m-1/2)}{\Gamma(m+1/2)} \right\} \\
&= -\frac{1}{\kappa^2} \mu^{m+1/2} \left\{ \frac{1}{(m+1/2)} - \frac{1}{(m-1/2)} \right\} \\
&= -\frac{1}{\kappa^2} \mu^{m+1/2} \frac{(-1)}{(m+1/2)(m-1/2)} \\
&= \frac{4}{\kappa^2} \frac{\mu^{m+1/2}}{(2m+1)(2m-1)} \quad . \qquad (4.34)
\end{aligned}$$

Some comments are in order: equation (4.34) exemplifies the fact that, in general,

$$\frac{\partial}{\partial \mu} \langle \widehat{\sigma}_i \rangle \neq \langle \widehat{\sigma}_{m-2} \widehat{\sigma}_i \rangle \quad , \qquad (4.35)$$

because the $\widehat{\sigma}$-operators have an explicit μ-dependence in their definition (4.25). That is why we had better calculate $\langle \sigma_i \rangle$ first, since the operators σ_i do satisfy

$$\frac{\partial}{\partial \mu} \langle \sigma_i \rangle = \langle \widehat{\sigma}_{m-2} \sigma_i \rangle \quad . \qquad (4.36)$$

The results (4.33,34) are very encouraging: the area operator $\hat{\sigma}_{m-2}$ has a non-vanishing 1-point function, as predicted in the Liouville approach; the operator $\hat{\sigma}_m$ was interpreted in ref. [31] as corresponding to the surface term in the Liouville equation of motion, and therefore it is not written as a vertex operator (the interpretation of special operators will be given latter on); all the remaining 1-point functions vanish, in agreement with the results in the continuum.

Integrating the 1-point function (4.33) we can also determine the scaling of the free energy

$$\mathcal{F} = -\frac{8}{\kappa^2} \frac{\mu^{m+1/2}}{(2m+1)(2m-1)(2m-3)} \propto \mu^{2-\widehat{\gamma}} \quad . \qquad (4.37)$$

The string susceptibility $\widehat{\gamma}$ in the new regime can be taken either from Eq. (4.37) or from the 2-point function of the area operator $\widehat{\sigma}_{m-2}$ and reads $\widehat{\gamma}_m = -m + 3/2$.

Now we can calculate $(N+2)$-point functions by inserting N scaling operators into the correlator (4.18), as follows:

$$\begin{aligned}
\langle \left(\prod_{i=1}^{N} \widehat{\sigma}_{n_i}\right) \widehat{\sigma}_a \widehat{\sigma}_b \rangle &= \prod_{i=0}^{N} \frac{\partial}{\partial \widehat{t}_{n_i}} \langle \widehat{\sigma}_a \widehat{\sigma}_b \rangle = \\
&= \frac{-1}{\kappa^2} \sum_{s=0}^{[a/2]} \sum_{r=0}^{[b/2]} A_s^{(a)} A_r^{(b)} \mu^{s+r} \prod_{i=1}^{N} \frac{\partial}{\partial \widehat{t}_{n_i}} \left\{ \frac{u^{a+b+1-2s-2r}}{a+b+1-2s-2r} \right\} \\
&= \frac{-1}{\kappa^2} \sum_{s=0}^{[a/2]} \sum_{r=0}^{[b/2]} A_s^{(a)} A_r^{(b)} \mu^{s+r} \prod_{i=1}^{N} \sum_{q_i=0}^{[n_i/2]} A_{q_i}^{(n_i)} \mu^{q_i} \frac{\partial}{\partial t_{n_i - 2q_i}} \left\{ \frac{u^{a+b+1-2s-2r}}{a+b+1-2s-2r} \right\} \\
&= \frac{-1}{\kappa^2} \sum_{s,r,q_i} A_s^{(a)} A_r^{(b)} A_{q_1}^{(n_1)} \cdots A_{q_N}^{(n_N)} \mu^{s+r+\sum_i q_i} \times \\
&\qquad \times \frac{\partial^N}{\partial t_{n_1 - 2q_1} \cdots \partial t_{n_N - 2q_N}} \left\{ \frac{u^{a+b+1-2s-2r}}{a+b+1-2s-2r} \right\} \quad . \qquad (4.38)
\end{aligned}$$

Using Eq. (4.9) one can show that

$$\begin{aligned}
\frac{\partial}{\partial t_k} \left(\frac{u^j}{j} \right) &= u^{j-1} \frac{\partial u}{\partial t_k} = u^{j-1+k-(m-2)} \frac{\partial u}{\partial \mu} \\
&= \frac{\partial}{\partial \mu} \left(\frac{u^{j+k-(m-2)}}{j+k-(m-2)} \right) , \qquad (4.39) \\
\frac{\partial^2}{\partial t_{k_1} \partial t_{k_2}} \left(\frac{u^j}{j} \right) &= \frac{\partial}{\partial \mu} \left[\frac{\partial}{\partial t_{k_2}} \left(\frac{u^{j+k_1-(m-2)}}{j+k_1-(m-2)} \right) \right] \\
&= \frac{\partial}{\partial \mu} \left[u^{j+k_1-(m-2)-1} \frac{\partial u}{\partial t_{k_2}} \right] \\
&= \frac{\partial}{\partial \mu} \left[u^{j+k_1+k_2-2(m-2)-1} \frac{\partial u}{\partial \mu} \right] \\
&= \frac{\partial^2}{\partial \mu^2} \left[\frac{u^{j+k_1+k_2-2(m-2)}}{j+k_1+k_2-2(m-2)} \right] , \qquad (4.40)
\end{aligned}$$

and by induction,

$$\frac{\partial^N}{\partial t_{k_1} \cdots \partial t_{k_N}} \left(\frac{u^j}{j} \right) = \left(\frac{\partial}{\partial \mu} \right)^N \left[\frac{u^{j+\sum_i k_i - N(m-2)}}{j+\sum_i k_i - N(m-2)} \right] . \qquad (4.41)$$

Finally substituting Eq. (4.41) into (4.38) we find, for $N \geq 2$,

$$\langle \prod_{i=1}^{N} \widehat{\sigma}_{n_i} \rangle = \frac{-1}{\kappa^2} \sum_{s_i=0}^{[n_i/2]} A_{s_1}^{(n_1)} \dots A_{s_N}^{(n_N)} \mu^{\sum_i s_i} \times \left(\frac{\partial}{\partial\mu}\right)^{N-2} \left[\frac{u^{1+\sum_i (n_i-2s_i)-(N-2)(m-2)}}{1+\sum_i (n_i-2s_i)-(N-2)(m-2)} \right] . \tag{4.42}$$

Notice that the equation above is valid in any regime. In the zero-coupling point ($u \to \sqrt{\mu}$) it reduces to:

$$\begin{aligned}\langle \prod_{i=1}^{N} \widehat{\sigma}_{n_i} \rangle &= \\ &= \frac{-1}{\kappa^2} \sum_{s_i=0}^{[n_i/2]} A_{s_1}^{(n_1)} \dots A_{s_N}^{(n_N)} \mu^{\sum_i s_i} \left(\frac{\partial}{\partial\mu}\right)^{N-2} \left[\frac{\mu^{(1+\sum_i (n_i-2s_i)-(N-2)(m-2))/2}}{1+\sum_i (n_i-2s_i)-(N-2)(m-2)} \right] \\ &= \frac{-1}{2\kappa^2} \sum_{s_i=0}^{[n_i/2]} A_{s_1}^{(n_1)} \dots A_{s_N}^{(n_N)} \mu^{\sum_i s_i} \left(\frac{\partial}{\partial\mu}\right)^{N-3} \mu^{(\sum_i (n_i-2s_i)-(N-2)(m-2)-1)/2} \\ &= \frac{-1}{2\kappa^2} \mu^{(\sum_i n_i-(N-2)(m-2)-1)/2-(N-3)} f_N(n_i) \quad , \end{aligned} \tag{4.43}$$

where f_N can be written as

$$f_N(n_i) = \left(\frac{\partial}{\partial x}\right)^{N-3} \left\{ (\sqrt{x})^{-(N-2)(m-2)-1} \prod_{i=1}^{N} P_{n_i}(\sqrt{x}) \right\}_{x=1} \quad , \tag{4.44}$$

and is a μ-independent coefficient. As regards the scaling factor, we can rearrange it as

$$\begin{aligned}\mu^{(\sum_i n_i-(N-2)(m-2)-1)/2-(N-3)} &= \mu^{(\sum_i (n_i-m)+2(N-3)+2m+1)/2-(N-3)} \\ &= \mu^{(m+1/2)+\frac{1}{2}\sum_i (n_i-m)} \quad . \end{aligned} \tag{4.45}$$

The final result is therefore given by

$$\langle \prod_{i=1}^{N} \widehat{\sigma}_{n_i} \rangle = -\frac{1}{2\kappa^2} \mu^{(m+1/2)+\frac{1}{2}\sum_{i=1}^{N} (n_i-m)} f_N(n_i) \quad . \tag{4.46}$$

These correlation functions should be compared with the continuum result, Eq. (2.28): in particular the scaling exponent of μ should be identified with the parameter $s = -(Q + \sum_i \beta_i)/\alpha_+$. This allows one to determine the central charge and the spectrum of conformal dimensions of the theory which corresponds to the matrix model, as explained in the next section.

4.3. Comparison with Continuum Results

After calculating the correlation functions of the scaling operators $\widehat{\sigma}_i$ in terms of the cosmological constant μ, we are ready to compare results with those obtained in the continuum. The first thing we have to know is which is the model coupled to gravity described by the Hermitian matrix model. For this purpose it is enough to look at the scaling behaviour of the free energy (4.37)

$$\begin{aligned} \mathcal{F} &\sim \mu^{m+1/2} \quad \text{(matrix model)} \quad , \\ \mathcal{Z} &\sim \mu^{-Q/\alpha_+} \quad \text{(continuum approach)} \quad , \end{aligned} \tag{4.47}$$

where $Q = 2\sqrt{2+\alpha_0^2}$, $c = 1 - 12\alpha_0^2$ and $\alpha_\pm = -\frac{Q}{2} \pm |\alpha_0|$.

In terms of the coprimes q, p ($q > p$ by assumption), which characterize a minimal model, we have $\alpha_0 = \frac{(p-q)}{\sqrt{2pq}}$ and therefore we conclude that

$$-\frac{Q}{\alpha_+} = 1 + \frac{q}{p} = m + \frac{1}{2} \quad . \tag{4.48}$$

Since p and q are coprimes, the equation above implies that $(p,q) = (2, 2m-1)$, as already mentioned in Chapter 3. Excluding the case $m = 2$, these values correspond to *non-unitary* minimal models. The mistaken identification with the $(m, m+1)$ series can be understood as follows.

The parameter α_+ that shows up in Eqs. (4.47-48) is the dressing $\beta(k)$ of the identity (or area) operator, given by e^{ikx} with $k = 0$ or $k = 2\alpha_0$. Its conformal weight is $\Delta = 0$ and is indeed the minimal weight in unitary models (in general one has $\Delta_{min} = \frac{1-(p-q)^2}{8pq}$). But if the theory is non-unitary, there exists an operator given by $e^{ik_{min}x}$ with a smaller conformal dimension $\Delta_{min} = \frac{1}{2}k_{min}(2\alpha_0 - k_{min}) < 0$, whose dressing reads $\beta_{min} = -Q/2 + |k_{min} - \alpha_0|$. If t is the coupling constant of this minimal weight operator, one expects[64] the following scaling of the partition function

$$\mathcal{Z} \sim t^{-Q/\beta_{min}} \quad , \tag{4.49}$$

where the exponent can be written in terms of the (p,q) indices as

$$-\frac{Q}{\beta_{min}} = \frac{2(p+q)}{(p+q)-1} \quad . \tag{4.50}$$

In particular for the series $(2, 2m-1)$ one has $\frac{2(p+q)}{(p+q)-1} = 2 + \frac{1}{m}$. This is precisely the scaling observed in (3.102), indicating that σ_0 is the minimal weight operator and t ($\sim t_0$) the corresponding coupling constant.

Observe that if one measures the scaling of the partition function (or free energy in the matrix model approach) with respect to the minimal weight coupling t, Eqs. (49-50) say that one will be able to determine only the sum $p+q$: for instance, Eq. (3.102) gives $p + q = 2m + 1$, which has many solutions (including the misleading unitary case $(m, m+1)$). In order to determine (p,q) uniquely one needs extra information, as the ratio q/p given by Eq. (4.48).

Briefly, we may uniquely determine the indices (p, q) of the minimal theory by measuring the scalings of the free energy with respect to the minimal weight coupling t and the physical cosmological constant μ.

In our case the solution is:

$$q = 2m - 1 \quad ; \quad p = 2 \quad ;$$
$$c = 1 - 3\frac{(2m-3)^2}{(2m-1)} = 0, -\frac{22}{5}, -\frac{72}{7}, \cdots \quad . \tag{4.51}$$

Therefore for $m = 2$ we have $c = 0$ which corresponds to the case of pure gravity. For $m = 3$, it corresponds to the Lee-Yang singularity $c = -22/5$, which is a non-unitary model ($c < 0$) like all other models with $m > 2$.

In order to finish the identification of the models we have to relate the scaling operators $\widehat{\sigma}_n$, $n = 0, 1, \cdots, m-1$ with the minimal operators $\mathcal{O}_{rr'}$ properly dressed. There are $\frac{1}{2}(q-1)(p-1)$ minimal operators $\mathcal{O}_{rr'}$ for each (p, q) minimal model, thus we should have $(m-1)$ scaling operators for $(q, p) = (2m-1, 2)$. From the continuum point of view the conformal weight $\Delta_{nn'}$ of such primary operator is given by the formula:

$$\Delta(\mathcal{O}_{rr'}) = \Delta_{rr'} = \frac{(rp - r'q)^2 - (q-p)^2}{4pq} \quad , \tag{4.52}$$

with $1 \leq r' \leq p - 1 = 1 (\Rightarrow r' = 1)$ and $1 \leq r \leq q - 1 = 2(m-1)$, i.e.,

$$\Delta_{r,1} = \frac{(m-1-r)(m-r) - (m-2)(m-1)}{2(2m-1)} \quad . \tag{4.53}$$

From the matrix model point of view we deduce from (4.46) that the scaling contribution from an operator $\widehat{\sigma}_n$ is given by:

$$\widehat{\sigma}_n \sim \mu^{\frac{n-m}{2}} \quad , \tag{4.54}$$

whereas in the continuum we have

$$\widehat{\sigma}_n \sim \mu^{-\beta/\alpha_+} \quad , \tag{4.55}$$

where β is a function of the conformal weight Δ_n of the operator $\widehat{\sigma}_n$ as follows:

$$\Delta_n - \frac{\beta(\beta + Q)}{2} = 1 \quad . \tag{4.56}$$

Comparing (4.54) with (4.55) we fix the dressing β of $\widehat{\sigma}_n$ and plugging the result in (4.56) we get:

$$\Delta(\widehat{\sigma}_n) = \Delta_n = \frac{n(n+1) - (m-2)(m-1)}{2(2m-1)} \quad . \tag{4.57}$$

Defining $n = m - 1 - r$ we recover formula (4.53) of the continuum and when we vary r from 1 to $m-1$ the index n runs from $m-2$ to 0. Therefore the set of scaling operators $\{\widehat{\sigma}_n$ with $n = 0, 1, \cdots, m-2\}$ spans the set of dressed primary operators

$\mathcal{O}_{r,1}$ with $r = 1, 2, \cdots, (m-1)$. These are actually all primary operators that we have, since for $m \le r \le 2(m-1)$ we just have the well known doubling of such operators.

Here we have a couple of examples:

a) $m = 2$ (Pure gravity):
In this case $r = 1, 2$ and $\Delta_{11} = \Delta_{21} = 0$. From the matrix model point of view we only have the operator $\widehat{\sigma}_0 = \widehat{\sigma}_{m-2}$ and $\Delta(\widehat{\sigma}_{n=0}) = 0$ when $n = 2$. The operators $\mathcal{O}_{11}$ and $\mathcal{O}_{21}$ correspond to the identity operator which becomes, after the dressing, the $\widehat{\sigma}_0$ puncture (or area) operator.

b) $m = 3$ (Lee-Yang singularity):
From (4.53) and (4.57) we have

$$\begin{aligned} \Delta_{r,1} &= \frac{(r-1)(r-4)}{10} = 0, -1/5, -1/5, 0 \quad ; \quad r = 1, 2, 3, 4 \\ \Delta_n &= \frac{n(n+1)-2}{10} = -1/5, 0, 2/5, 1 \quad ; \quad n = 0, 1, 2, 3 \quad . \end{aligned} \tag{4.58}$$

In the above table we varied r from 1 to $2(m-1)$ and n from 0 to m.

To conclude, the agreement is complete between the conformal dimension of $\widehat{\sigma}_0$, $\widehat{\sigma}_1, \cdots, \widehat{\sigma}_{m-2}$ and those of $\mathcal{O}_{m-1,1}, \mathcal{O}_{m-2,1}, \cdots, \mathcal{O}_{1,1}$.

We shall discuss the remaining scaling operators after studying the properties of macroscopic loops and wave functions in the new scaling regime.

4.4. Macroscopic Loops: Bessel Equation and Minisuperspace Approximation

In the μ-scaling regime we shall calculate the expectation value of macroscopic loops and define the wave functions of scaling operators. The results will be compared with the minisuperspace approximation of the theory in the continuum and some special operators will be identified.

If we begin by neglecting singular terms in the loop length l, the calculation of a macroscopic loop 1-point function is straightforward: we simply take the expansion (3.99) in terms of scaling operators and then use Eq. (4.29), finding as a result

$$\begin{aligned} \langle W(l) \rangle &= -\frac{\kappa}{\sqrt{\pi}} \sum_{s \ge 0} \frac{(-1)^m l^{m-3/2+2s}}{(m-2+2s)!} \langle \sigma_{(m-2)+2s} \rangle + \cdots \\ &= (-1)^m \frac{2^{3/2}}{\kappa l} (\sqrt{\mu})^{m-1/2} \sum_{s \ge 0} \frac{1}{s!\Gamma(s+(m-1/2)+1)} \left(\frac{\sqrt{\mu} l}{2} \right)^{m-1/2+2s} + \cdots \\ &= (-1)^m \frac{2^{3/2}}{\kappa l} (\sqrt{\mu})^{m-1/2} I_{m-1/2}(\sqrt{\mu} l) + \cdots \quad , \end{aligned} \tag{4.59}$$

where the dots mean "singular terms in l" and we have recognized one of the modified Bessel functions, given in general by

$$I_\alpha(x) = \sum_{s=0}^{\infty} \frac{1}{s!\Gamma(s+\alpha+1)} \left(\frac{x}{2}\right)^{\alpha+2s} \quad . \tag{4.60}$$

The reader can find in Appendix E some properties of Bessel functions which we found useful along the calculations.

However we could have started from the general result given by Eq. (3.87): taking the transformation (4.11) and the property (E.28), we find

$$\langle W(l)\rangle = \frac{2^{5/2}}{\pi\kappa l}(\sqrt{\mu})^{m-1/2} K_{m-1/2}(\sqrt{\mu}l) \quad , \tag{4.61}$$

where $K_\alpha(x)$ are the modified Bessel functions of second kind. The consistency between the results (4.59) and (4.61) can be understood if we take Eq. (E.27) to write

$$K_{m-1/2} = \frac{\pi}{2}(-1)^m \left[I_{m-1/2}(x) - I_{-m+1/2}(x)\right] \quad , \tag{4.62}$$

and since

$$\begin{aligned} &(-1)^m(\sqrt{\mu})^{m-1/2} I_{-m+1/2}(\sqrt{\mu}l) \\ &\quad = (-1)^m \sum_{s=0}^{\infty} \frac{1}{s!\Gamma(s-m+3/2)}(\sqrt{\mu})^{m-1/2}\left(\frac{\sqrt{\mu}l}{2}\right)^{-m+1/2+2s} \\ &\quad = (-1)^m \sum_{s=0}^{\infty} \frac{1}{s!\Gamma(s-m+3/2)}\mu^s \left(\frac{l}{2}\right)^{2s-m+1/2} \end{aligned} \tag{4.63}$$

is analytical in μ, we conclude that a replacement like

$$(-1)^m I_{m-1/2}(\sqrt{\mu}l) \longleftrightarrow \frac{2}{\pi} K_{m-1/2}(\sqrt{\mu}l) \quad , \tag{4.64}$$

in (4.59) only amounts to analytical terms in μ. Some of them are singular for $l \to 0$, corresponding to the universal analytical terms discussed in Chapter 3. In either case, we conclude that the macroscopic loop operator obeys[31]

$$\left[-\left(l\frac{\partial}{\partial l}\right)^2 + \mu l^2 + (m-\frac{1}{2})^2\right] \langle lW(l)\rangle = 0 \quad . \tag{4.65}$$

The Bessel function $K_{m-1/2}(x)$ plays here the role that the function $\psi_{m-1}(x)$, defined by Eq. (3.88), played in the t-coupling regime. This analogy can be further extended if we expand the loop operator in the conformal basis: using the inversion formula (4.25b) to write the σ-operators in terms of the orthogonal basis, one can rewrite the loop expansion (3.101) as

$$\widetilde{W}(l) = \sqrt{2}\kappa \sum_{k=0}^{\infty} \widehat{\sigma}_k \frac{(-1)^{k+1}(2k+1)}{(\sqrt{\mu})^{k+1/2}} I_{k+1/2}(\sqrt{\mu}l) + \cdots \quad , \tag{4.66}$$

where singular terms in l have been omitted. We see that the family of Bessel functions $\{I_{k+1/2}\}$ has replaced the basis $\{l^{k+1/2}\}$ in the expansion in the loop length l; in this sense we can say that the Bessel functions are characteristic of the conformal basis[31].

We can also associate wave functions to scaling operators: they are defined as the 2-point functions $\Psi_i = \langle \widehat{\sigma}_i W(l) \rangle$ and can be interpreted as correlation functions between two loops where one loop has been shrunk into a microscopic loop (or puncture). From the expansion (4.66) and the orthogonality (4.26) we find

$$\begin{aligned}
\Psi_i &\equiv \langle \widehat{\sigma}_i \widetilde{W}(l) \rangle \\
&= \sqrt{2}\kappa \sum_{k=0}^{\infty} \langle \widehat{\sigma}_i \widehat{\sigma}_k \rangle \frac{(-1)^{k+1}(2k+1)}{(\sqrt{\mu})^{k+1/2}} I_{k+1/2}(\sqrt{\mu} l) \\
&= \frac{\sqrt{2}}{\kappa} \frac{\mu^{i+1/2}}{(2i+1)} \frac{(-1)^i (2i+1)}{(\sqrt{\mu})^{i+1/2}} I_{i+1/2}(\sqrt{\mu} l) \\
&= \frac{\sqrt{2}}{\kappa} (-1)^i (\sqrt{\mu})^{i+1/2} I_{i+1/2}(\sqrt{\mu} l) \quad .
\end{aligned} \tag{4.67}$$

The above result should be regarded as valid up to singular terms in l. The correct answer[31] is again obtained by replacing the function $I_{i+1/2}(x)$ by the second kind Bessel function $K_{i+1/2}$, so that $\Psi_i = \frac{2^{3/2}}{\pi\kappa} (\sqrt{\mu})^{i+1/2} K_{i+1/2}(\sqrt{\mu} l)$. The important result is that these wave functions satisfy the Bessel equation

$$\left[-\left(l \frac{\partial}{\partial l} \right)^2 + \mu l^2 + \left(i + \frac{1}{2} \right)^2 \right] \Psi_i = 0 \quad . \tag{4.68}$$

From the discussion in Chapter 1 (see Eqs. (1.57-61)) we expect that wave functions of vertex operators satisfy the Wheeler-DeWitt equation

$$[H - \Delta_{\mathcal{O}}] \Psi(\mathcal{O}) = 0 \quad , \tag{4.69}$$

where H is the Hamiltonian (1.59) and $\Delta_{\mathcal{O}}$ the conformal weight of the operator $\mathcal{O}$. In the minisuperspace approximation[31], one ignores the space dependence and drops the derivative terms in (1.59). Moreover, since the boundary of the surface where the macroscopic loop was inserted is measured by the contour integral

$$l = \oint d\xi \, e^{\frac{1}{2}\alpha_+ \phi} \quad , \tag{4.70}$$

we can associate, within this space-independent approximation,

$$\begin{aligned}
p \sim \frac{\partial}{\partial \phi} \quad &\leftrightarrow \quad l \frac{\partial}{\partial l} \quad , \\
\mu e^{\alpha_+ \phi} \quad &\leftrightarrow \quad \mu l^2 \quad ,
\end{aligned} \tag{4.71}$$

and also for the operators $\widehat{\sigma}_i \sim \mathcal{O}_{m-1-i,1}$ we find $\left(Q^2/8 - \Delta\right)/\alpha_+^2 = (i+1/2)^2$ (recall Eq. (4.52)). Thus we reproduce Eq. (4.68). The remarkable fact is[31] that (4.68) was obtained from the matrix model formulation without any approximation, suggesting that in this case the minisuperspace approximation leads to an exact result.

The wave functions (4.67) can also be used to give a more precise interpretation to some special operators.

Take Ψ_{m-1} for instance: comparing the loop expectation value (4.59) and Eq. (4.67) for $i = m-1$ it is easy to see that

$$\langle \widehat{\sigma}_{m-1} W(l) \rangle = -\frac{1}{2} l \langle W(l) \rangle \quad . \tag{4.72}$$

This means that the operator $\widehat{\sigma}_{m-1}$ just measures the length of the loop, which is the boundary of the surface where that loop was inserted. Therefore we call $\widehat{\sigma}_{m-1}$ the *boundary operator*, and up to some trivial normalization we associate it to (4.70),

$$\widehat{\sigma}_{m-1} \quad \leftrightarrow \quad \oint e^{\alpha + \phi/2} \quad . \tag{4.73}$$

We also remark the presence of the operator $l\frac{\partial}{\partial l}$ in (4.68) and (4.71): we might look for a scaling operator ε_m satisfying

$$\langle \varepsilon_m W(l) \rangle = l \frac{\partial}{\partial l} \langle W(l) \rangle \quad . \tag{4.74}$$

Indeed, using the properties (E.30) and (E.31) of Bessel functions we find

$$\begin{aligned} l\frac{\partial}{\partial l}\langle W(l)\rangle &= \\ &= \frac{\sqrt{2}\,(\sqrt{\mu})^{m+1/2}(-1)^m}{\kappa \quad (m-1/2)} \left[(m-3/2) I_{m-3/2}(\sqrt{\mu} l) - (m+1/2) I_{m+1/2}(\sqrt{\mu} l) \right] \\ &= \frac{(2m-3)}{(2m-1)} \mu \langle \widehat{\sigma}_{m-2} W(l) \rangle - \frac{(2m+1)}{(2m-1)} \langle \widehat{\sigma}_m W(l) \rangle \\ &= \left\langle \left[-\frac{(2m+1)}{(2m-1)} \widehat{\sigma}_m + \frac{(2m-3)}{(2m-1)} \mu \widehat{\sigma}_{m-2} \right] W(l) \right\rangle \quad . \end{aligned} \tag{4.75}$$

Therefore we identify

$$\varepsilon_m = -\frac{(2m-3)}{(2m-1)} \left\{ \frac{(2m+1)}{(2m-3)} \widehat{\sigma}_m - \mu \widehat{\sigma}_{m-2} \right\} \quad . \tag{4.76}$$

From the 1-point functions (4.33-34) we can calculate the expectation value

$$\langle \varepsilon_m \rangle = -\frac{8}{\kappa^2} \frac{\mu^{m+1/2}}{(2m-1)^2} \quad . \tag{4.77}$$

Comparing the scalings in (4.37) and (4.77) we conclude that the operator ε_m has the same dimension of the free energy $\mathcal{F}$. We therefore call it *energy operator*.

From the interpretation above, together with the non-vanishing 1-point function (4.34), we conclude that the operator $\widehat{\sigma}_m$ does not correspond to a vertex operator: it is rather related to operators involving derivatives of the Liouville field, as pointed in ref. [31].

4.5. Conclusions

Having found the conformal basis of operators $\{\widehat{\sigma}_n\}$ (in its lowest order approximation) we conclude that the m-th critical Hermitian 1-matrix model corresponds to the $(2, 2m-1)$ minimal conformal theory coupled to two-dimensional gravity, and we can draw the associations[31]

$$\widehat{\sigma}_n \quad \leftrightarrow \quad \int d^2\xi \, e^{\beta_n \phi} \mathcal{O}_{m-1-n,1} \, , \, n = 0, 1, \cdots, m-2 \tag{4.78}$$

$$\widehat{\sigma}_{m-1} \quad \leftrightarrow \quad \oint d\xi \, e^{\alpha_+ \phi/2} \quad , \tag{4.79}$$

$$\widehat{\sigma}_m - \frac{(2m-3)}{(2m+1)} \mu \widehat{\sigma}_{m-2} \leftrightarrow \text{energy operator} \quad . \tag{4.80}$$

Concerning the remaining operators, $\{\widehat{\sigma}_n \, , \, n > m\}$, we notice that their conformal weights differ from the weights of $\{\widehat{\sigma}_n \, , \, n \leq m\}$ by integers, and thus correspond to secondary operators[44]. This can be easily verified using Eq. (4.57) (see also the example (4.58)).

The 1- and 2-point functions found in the discrete and continuum approaches are in perfect agreement. Higher order correlators involve the issue of fusion rules and lie beyond the approximation assumed in Eq. (4.11). Nevertheless the scaling factor in (4.46) agrees with the predictions of the continuum. We shall comment on fusion rules along Chapter 8.

5. Correlation Functions for the N=1 Super Liouville Theory

5.1. Introduction

The successful comparison between the results of the matrix models of Chapter s 3, 4 and the ones obtained in the continuum formulation of the non-critical strings in the Chapter 2, motivates us to generalize such computations for the non-critical superstring. The technique is similar to the bosonic case. The results of this section have been obtained in refs. [79,80] where we have used the formulation of refs. [13,50] in which the measure is invariant under translations.

In the superconformal gauge the super gravity sector of the superstring is represented by the super Liouville action[140] S_{SL}. Assuming that we are dealing with a translation invariant measure we have the following action for S_{SL},

$$S_{SL} = \frac{1}{4\pi} \int \mathrm{d}^2\mathbf{z} E \left(\frac{1}{2} D_\alpha \Phi_{SL} D^\alpha \Phi_{SL} - \frac{Q}{2} Y \Phi_{SL} - 4i\bar{\mu} \mathrm{e}^{\alpha_+ \Phi_{SL}} \right) \quad , \tag{5.1}$$

where Q and α_+ will be determined analogously to the bosonic case and $\Phi_{SL} = \phi + \theta\psi + \bar{\theta}\bar{\psi} + \theta\bar{\theta}F$ is the Liouville superfield with ψ and $\bar{\psi}$ the partners of ϕ and F is an auxiliary field.

The field E is the super determinant of the super zweibein E_{ab}, while Y is the curvature superfield[69,141,142]; it contains the gravitational field, the gravitino, and further auxiliary fields, but one can fix the gauge where one is left only with the graviton (Liouville) field, after the Grassmann integration. The symbol D_α stands for the covariant super derivative, and the infinitesimal superspace volume element is given by

$$\mathrm{d}^2\mathbf{z} = \mathrm{d}^2 z \mathrm{d}\theta \mathrm{d}\bar{\theta} \quad . \tag{5.2}$$

When we couple S_{SL} to a super matter action S_M with central charge c the critical value of the central charge becomes[47,50] $c = 3/2$, which corresponds to only one superfield ($c = 1 + 1/2$). Thus, the non-critical superstring is in principle defined for $c \leq 3/2$. Analogously to the bosonic case we can have a super Coulomb gas formulation where we introduce a background charge at the infinity without breaking supersymmetry. This corresponds to the following action for the supermatter,

$$S_M = \frac{1}{4\pi} \int \mathrm{d}^2 z \mathrm{d}^2\theta E \left(\frac{1}{2} D_\alpha \Phi_M D^\alpha \Phi_M - i\alpha_0 Y \Phi_M \right) \quad , \tag{5.3}$$

where $\Phi_M = X + \theta\zeta + \overline{\theta\zeta} + \theta\bar{\theta}G$ is the matter superfield with $\zeta, \bar{\zeta}$ partners of X, and G another auxiliary field. Therefore the total action we are going to work with is given by

$$S = S_{SL} + S_M + S_{gh} \quad , \tag{5.4}$$

where S_{gh} stands for the ghost action whose explicit form we do not need here (see ref. [2]).

In order to compute Q and α_+ we have to impose respectively that the total central charge c_T vanishes and $e^{\alpha_+\Phi_{SL}}$ be a (1/2,1/2) conformal operator. For this computation we need the bosonic components of the super energy momentum tensor which are given, in components by

$$\begin{aligned} T_{SL} &= -\frac{1}{2}{:}\partial\phi\partial\phi{:} - \frac{1}{2}{:}\psi\partial\psi{:} + \frac{Q}{2}\partial^2\phi \quad , \\ T_M &= -\frac{1}{2}{:}\partial X\partial X{:} - \frac{1}{2}{:}\zeta\partial\zeta{:} - i\alpha_0\partial^2 X \quad , \\ T_{gh} &= T_{bc} + T_{\beta\gamma} = {:}c\partial b{:} + {:}2(\partial c)b{:} - \frac{3}{2}{:}(\partial\gamma)\beta{:} - \frac{1}{2}{:}\gamma\partial\beta{:} \quad , \end{aligned} \tag{5.5}$$

where $b, c(\beta, \gamma)$ are the reparametrization (supersymmetry) ghosts. Analogous expressions hold for the antiholomorphic part ($\overline{T}$). Using the free propagators

$$\begin{aligned} \langle\phi(z)\phi(w)\rangle &= \langle X(z)X(w)\rangle = -\ln(z-w) \quad , \\ \langle\zeta(z)\zeta(w)\rangle &= \langle\psi(z)\psi(w)\rangle = (z-w)^{-1} \quad , \\ \langle c(z)b(w)\rangle &= \langle\gamma(z)\beta(w)\rangle = (z-w)^{-1} \quad , \end{aligned} \tag{5.6}$$

and remembering that $\beta, \gamma(b, c)$ are (anti)-commuting quantities we can compute the central charges for super Liouville, supermatter and ghosts, which are given, respectively by

$$\begin{aligned} c_{SL} &= \frac{3}{2}(1+2Q^2) \quad , \\ c &= \frac{3}{2}(1-8\alpha_0^2) \quad , \\ c_{gh} &= c_{bc} + c_{\beta\gamma} = -26 + 11 = -15 \quad . \end{aligned} \tag{5.7}$$

Imposing that the total central charge vanishes $c_T = c_{SL} + c + c_{gh} = 0$ we deduce that the parameter Q is now given by

$$Q = 2\sqrt{\alpha_0^2 + 1} \quad . \tag{5.8}$$

Calculating the conformal weight Δ of the operator $e^{\alpha\Phi}$ using T_{SL} we have a similar expression as before,

$$\Delta(e^{\alpha\Phi}) = -\frac{\alpha(\alpha+Q)}{2} \quad , \tag{5.9}$$

therefore, requiring that $\Delta = 1/2$ as explained above, we obtain for the parameter α the value

$$\alpha = \alpha_\pm = -\frac{Q}{2} \pm \frac{1}{2}\sqrt{Q^2-4} = -\frac{Q}{2} \pm |\alpha_0| \quad , \quad \alpha_+\alpha_- = 1 \quad . \tag{5.10}$$

Analogously to the bosonic case we choose $\alpha = \alpha_+$ for the same reasons presented in Chapter 1. Having defined the action we consider in the next two sections the definition of the physically relevant vertex operators and the computation of the corresponding correlators.

5.2. The Neveu-Schwarz Vertex and Its 3-Point Function

The particle content of the non-critical superstring consists of a scalar (NS) and a spinor (R) particle in space-time. Both are massless. The Ramond vertex (or spinor emission vertex) is more complicated and will be studied in the next section, while the Neveu-Schwarz vertex (or scalar emission vertex) is simply the supersymmetric extension of a planar wave (the "tachyon" of the bosonic string) being defined by

$$\begin{aligned}\Psi_{NS}(k) &= \int \mathrm{d}^2 z \mathrm{d}^2\theta \mathrm{e}^{ik\Phi_M+\beta\Phi_{SL}} \\ &= \int \mathrm{d}^2 z \mathrm{e}^{ikx+\beta\phi}[(ik\zeta+\beta\psi)(ik\overline{\zeta}+\beta\overline{\psi})+\beta F+ikG] \quad .\end{aligned} \tag{5.11}$$

Analogously to the cosmological term of the action we impose that $\mathrm{e}^{ik\Phi_M+\beta\Phi_{SL}}$ be a $(1/2,1/2)$ conformal operator, that is,

$$\Delta(\mathrm{e}^{ik\Phi_M+\beta\Phi_{SL}}) = \frac{1}{2}k(k-2\alpha_0) - \frac{\beta}{2}(\beta+Q) = \frac{1}{2} \quad . \tag{5.12}$$

Thus, choosing the positive energy solution once again, we have the usual dispersion relation given below. This determines the dressing $\beta(k)$,

$$E = \beta + \frac{Q}{2} = |k-\alpha_0| \quad . \tag{5.13}$$

Indeed, expression (5.13) is exactly of the same form as the bosonic dispersion relation, Eq. (2.50), and this is one of the points which lead to the similarities between the two cases.

Having defined the action and the NS-vertex Ψ_{NS} we can start calculating its amplitudes. As in the bosonic case the simplest correlator is the 3-point amplitude $\mathcal{A}_3$. The exponential interaction in S_{SL} allows us to use the zero mode technique again. Integrating over ϕ_0 and X_0 we have

$$\left\langle \prod_{i=1}^{3} \int \Psi_{NS}(\tilde{\mathbf{z}}_i, k_i) \right\rangle_S = 2\pi\delta\left(\sum_{i=1}^{3} k_i - 2\alpha_0\right) \mathcal{A}_3(k_1,k_2,k_3) \quad , \tag{5.14a}$$

with

$$\begin{aligned}&\mathcal{A}_3(k_1,k_2,k_3) = \Gamma(-s)(i\bar{\mu})^s \\ &\quad \left\langle \int \prod_{i=1}^{3} \mathrm{d}^2\tilde{\mathbf{z}}_i \mathrm{e}^{ik_i\Phi_M(\tilde{\mathbf{z}}_i)} \mathrm{e}^{\beta_i\Phi_{SL}(\tilde{\mathbf{z}}_i)} \left(\int \mathrm{d}^2\mathbf{z}\mathrm{e}^{\alpha_+\Phi_{SL}(\mathbf{z})}\right)^s \right\rangle_{S(\bar{\mu}=0)} \quad .\end{aligned} \tag{5.14b}$$

As in the bosonic case we define

$$s = -\frac{1}{\alpha_+}\left(Q + \sum_{i=1}^{3}\beta_i\right) \quad . \tag{5.15}$$

Rewriting in components and assuming again that s is a non-negative integer we arrive at the following expression for the 3-point function in terms of free field correlators,

$$\mathcal{A}_3 = (i\bar{\mu})^s \Gamma(-s) \Big\langle \int \prod_{i=1}^{s} \mathrm{d}^2 z_i \mathrm{e}^{\alpha_+ \phi} (\alpha_+^2 \psi \overline{\psi} + \alpha_+ F) \times$$
$$\times \int \prod_{j=1}^{3} \mathrm{d}^2 \tilde{z}_j \mathrm{e}^{ik_j x + \beta_j \phi} [(ik_j \zeta + \beta \psi)(ik_j \overline{\zeta} + \beta_j \overline{\psi}) + \beta_j F + ik_j G] \Big\rangle_{S(\bar{\mu}=0)} \tag{5.16}$$

with the action computed with zero cosmological constant, given by

$$\begin{aligned} S(\bar{\mu}=0) = \frac{1}{8\pi} \int \mathrm{d}^2 z \sqrt{\hat{g}} [\partial\phi\overline{\partial}\phi - Q\hat{R}\phi - \psi\overline{\partial}\psi - \overline{\psi}\partial\overline{\psi} + F^2 \\ + \partial X \overline{\partial} X + 2i\alpha_0 \hat{R} X - \zeta\partial\zeta - \overline{\zeta}\partial\overline{\zeta} + G^2] \quad . \end{aligned} \tag{5.17}$$

The auxiliary fields F and G appear in a trivial way in the action which means that their propagators consist of delta functions, that is

$$\langle F(z_i) F(z_j) \rangle = \langle G(z_i) G(z_j) \rangle \sim \delta^{(2)}(z_i - z_j) \quad . \tag{5.18}$$

Taking into account the contractions of other fields in (5.16), from which we get typically $\underbrace{\mathrm{e}^{a\phi(z_i)} \mathrm{e}^{b\phi(z_j)}} \sim |z_i - z_j|^{-2ab}$, we see that if we maintain the auxiliary fields G and F we shall have ill defined expressions such as $\delta^{(2)}(z_i - z_j)|z_i - z_j|^{-2ab}$ in the correlation function. This problem already appeared in the context of superstrings[143] and in our case it will be circumvented by discarding such fields[1)], i.e., we fix $F = G = 0$. This amounts to assume that $ab < 0$.

After such simplification the NS-vertex given in (5.11) becomes simply:

$$\begin{aligned} \Psi_{NS} &= \int \mathrm{d}^2 z \mathrm{d}^2 \theta \mathrm{e}^{ik\Phi_M + \beta\Phi_{SL}} \\ &= \int \mathrm{d}z (ik\zeta + \beta\psi) \int \mathrm{d}\overline{z} (ik\overline{\zeta} + \beta\overline{\psi}) \mathrm{e}^{ikx + \beta\phi} \quad . \end{aligned} \tag{5.19}$$

The next step in calculating (5.16) is to notice that the group of residual symmetry of the super conformal gauge is the $OSP(2,1)$ which contains besides the three bosonic generators $L_0, L_\pm$ also two fermionic ones $G_{\pm 1/2}$ which allow us to fix three bosonic $\tilde{z}_j$ and two fermionic coordinates $\tilde{\theta}_j$ which we choose to be[144]

$$\tilde{z}_1 = 0 \quad , \quad \tilde{z}_2 = 1 \quad , \quad \tilde{z}_3 = \infty \quad , \quad \tilde{\theta}_2 = \tilde{\theta}_3 = 0 \quad . \tag{5.20}$$

Therefore using the propagators given in (5.6) we get

$$\begin{aligned} \mathcal{A}_3(k_1, k_2, k_3) = \Gamma(-s)(i\bar{\mu})^s \alpha_+^{2s} \prod_{i=1}^{s} \int \mathrm{d}^2 z_i |z_i|^{2\alpha} |1 - z_i|^{2\beta} \prod_{i<j}^{s} |z_i - z_j|^{4\rho} \\ \times \left\langle (ik_1 \zeta(0) + \beta_1 \psi(0))(ik_1 \overline{\zeta}(0) + \beta_1 \overline{\psi}(0)) \prod_{i=1}^{s} \overline{\psi}\psi(z_i) \right\rangle_{\bar{\mu}=0} \quad . \end{aligned} \tag{5.21}$$

---

1) For a more rigorous treatment of such terms (called contact terms) including the auxiliary fields see ref. [143]

Due to the symmetries $\psi \to -\psi\,(\overline{\psi} \to -\overline{\psi})$ and $\zeta \to -\zeta\,(\overline{\zeta} \to -\overline{\zeta})$ of $S(\bar{\mu}=0)$ the terms of (5.21) involving the matter fermionic partner $\zeta(\overline{\zeta})$ drop out of $\mathcal{A}_3$ and in order to have a non-vanishing non-critical amplitude we must assume that s is odd:

$$s = 2l+1 \quad , \quad l = 0,1,2\cdots \quad . \tag{5.22}$$

We return to (5.21) using the free fermion propagator (5.6), i.e., $\langle\overline{\psi}\psi(z)\overline{\psi}\psi(w)\rangle = |z-w|^{-2}$ and get for odd s:

$$\mathcal{A}_3(k_1,k_2,k_3) = \Gamma(-s)\frac{1}{\alpha_+^2}\left(i\alpha_+^2\bar{\mu}\right)^s \alpha^2(-1)^{\frac{s+1}{2}}\, s!!\, I_G^s \quad , \tag{5.23}$$

with

$$I_G^s = \prod_{i=1}^{s}\int d^2 z_i |z_i|^{2\alpha}|1-z_i|^{2\beta}\prod_{i<j=1}^{s}|z_i-z_j|^{4\rho}\prod_{i=1}^{\frac{s-1}{2}}|z_{2i-1}-z_{2i}|^{-2}|z_s|^{-2} \quad . \tag{5.24a}$$

where we have used the same definitions of Chapter 2 for α,β, and ρ, that is, $\alpha = -\alpha_+\beta_1$, $\beta = -\alpha_+\beta_2$, $\rho = -\frac{1}{2}\alpha_+^2$.

The above integral corresponds to the supersymmetric generalization of Eq. (C.22) and was given in ref. [79],

$$I_{susy} = \int\prod_{i=1}^{3}d^2\tilde{\mathbf{z}}_i\prod_{i=1}^{s}d^2\mathbf{z}_i\prod_{i<j}^{3}|\tilde{\mathbf{z}}_{ij}|^{2k_ik_j-2\beta_i\beta_j}\prod_{i=1}^{3}\prod_{j=1}^{s}|\tilde{z}_i - z_j - \tilde{\theta}_i\theta_j|^{-2\alpha_i\beta_i}\prod_{i<j}^{s}|\mathbf{z}_{ij}|^{-2\alpha_+^2} \tag{5.24b}$$

Recently the calculations of refs. [76-78] have been generalized in refs. [79-82] for the case of $N=1$ superconformal theories. In [79] making use of a set of Ansätze compatible with the symmetries and the asymptotic behavior for $(\alpha,\beta\to\infty)$ of the integral we have obtained the result (see derivation in Appendix C, formula (C.64))

$$\begin{aligned}\mathcal{A}(k_1,k_2,k_3) = {}& \Gamma(-s)\frac{\Gamma(s+1)}{\alpha_+^2}\left(\frac{\alpha_+^2\bar{\mu}}{2}\right)^{2l+1}\left[\Delta(\tfrac{1}{2}-\rho)\right]^{2l+1}\prod_{i=1}^{l}\Delta(2i\rho)\\ &\times\prod_{i=1}^{l}\Delta(\tfrac{1}{2}+(2i+1)\rho)\prod_{i=0}^{l}\Delta(1+\alpha+2i\rho)\Delta(1+\beta+2i\rho)\Delta(-\alpha-\beta+(2i-4l)\rho)\\ &\times\prod_{i=1}^{l}\Delta(\tfrac{1}{2}+\alpha+(2i-1)\rho)\Delta(\tfrac{1}{2}+\beta+(2i-1)\rho)\Delta(-\tfrac{1}{2}-\alpha-\beta+(2i-4l-1)\rho).\end{aligned} \tag{5.25}$$

As in the bosonic case the above expression simplifies after eliminating β. For the kinematics $k_1,k_3 \ge \alpha_0$, $k_2 \le \alpha_0$ one finds, upon definitions below (5.24a), (5.15) and the definition of β_2 (see e.g. (5.28)),

$$\beta = \begin{cases}\rho(1-s) & (\alpha_0 > 0)\\ -\frac{1}{2}-\rho s & (\alpha_0 < 0)\end{cases} \quad . \tag{5.26}$$

Also in this case, we only arrive at a non-trivial result for $\mathcal{A}_3$ for $\alpha_0 < 0$, in which case the amplitude results in

$$\mathcal{A}_3(k_1, k_2, k_3) = \left[\bar{\mu}\Delta\left(\frac{1}{2}-\rho\right)\right]^s \Delta\left(\frac{1}{2}-\frac{s}{2}\right)\Delta\left(1+\alpha-(s-1)\rho\right)\Delta\left(\frac{1}{2}-\alpha+\rho\right) \quad . \tag{5.27}$$

Since $k_1, k_3 \geq \alpha_0$, $k_2 \leq \alpha_0$, we have

$$\begin{aligned} \beta_{1,3} &= \alpha_+ + k_{1,3} \quad , \\ \beta_2 &= \alpha_- - k_2 \end{aligned} \tag{5.28}$$

and we can obtain, upon substitution of (5.28), the relations

$$\begin{aligned} \frac{1}{2} + \frac{1}{2}(\beta_1^2 - k_1^2) &= \frac{1}{2} - \alpha + \rho \quad , \\ \frac{1}{2} + \frac{1}{2}(\beta_2^2 - k_2^2) &= \frac{1}{2} - \frac{1}{2}s \quad , \\ \frac{1}{2} + \frac{1}{2}(\beta_3^2 - k_3^2) &= 1 + \alpha + (s-1)\rho \quad , \end{aligned} \tag{5.29}$$

which permit to rewrite (5.27) as

$$\mathcal{A}_3(k_1, k_2, k_3) = \left[\bar{\mu}\Delta\left(\frac{1}{2}-\rho\right)\right]^s \prod_{j=1}^{3} \Delta\left(\frac{1}{2}+\frac{1}{2}(\beta_j^2 - k_j^2)\right) \quad . \tag{5.30}$$

The above result is completely parallel to the bosonic one (see (2.60)). It is convenient once again to separate the two cases $\hat{c} < 1\,(\alpha_0 \neq 0)$ and $\hat{c} = 1\,(\alpha_0 = 0)$. For $\hat{c} < 1$ the factors $\Delta(\frac{1}{2} + \frac{\beta^2 - k^2}{2})$ never vanish or diverge and we renormalize the cosmological constant and the NS-vertices according to

$$\bar{\mu} = \frac{\mu}{\Delta\left(\frac{1}{2}-\rho\right)} \quad , \quad \widetilde{\Psi}_{NS}(k_j) = \frac{\Psi_{NS}(k_j)}{\Delta\left(\frac{1}{2}+\frac{1}{2}[\beta_j^2 - k_j^2]\right)} \quad , \tag{5.31}$$

after which we have for the three-point function of the renormalized vertices the expression

$$\mathcal{A}_3(k_1, k_2, k_3) = (\mu)^s \quad . \tag{5.32}$$

For $\hat{c} = 1\,(\alpha_0 = 0)$ we are not allowed to redefine the vertices; thus we get

$$\mathcal{A}_3(\hat{c} = 1) = (\mu)^{(\sum_{i=1}^{3} |k_i| - 1)} \prod_{i=1}^{3} \frac{\Gamma(1 - |k_i|)}{\Gamma(|k_i|)} \quad . \tag{5.33}$$

Expression (5.33) deserves some comments. First, we have the presence of poles at $|k_i| = 1, 2, 3 \cdots$ indicating the existence of an infinity set of discrete states as in the bosonic case. These discrete states have also been seen from the BRST point of view in ref. [104]. Another interesting feature of (5.33) is the appearance of the renormalized cosmological constant μ which can be interpreted as the cosmological constant associated with the cosmological term $\Phi e^{-\Phi}$ which plays the role of the term $\phi e^{-\sqrt{2}\phi}$ of the bosonic case. To check this interpretation notice that for $\alpha_0 = \epsilon(\epsilon \to 0)$ we have $\alpha_+ \approx -1 + \epsilon\,(\rho \sim -1/2 + \epsilon)$ and therefore the cosmological term may be written as

$$\frac{\mu}{\Delta\left(\frac{1}{2}-\rho\right)} e^{\alpha_+\Phi} \sim \mu\Phi e^{-\Phi} \quad + \quad \text{(divergent term)} \quad . \tag{5.34}$$

As before we discard the divergent term relying on the argument of non-universality. All such similarities between the bosonic and fermionic non-critical string will persist in the next section.

5.3. Neveu-Schwarz 3-Point Function with (n,m) Screening Charges

As in the bosonic case it is in general necessary to include screening charges in the supermatter sector (n charges V_+ and m charges V_-) in order to study minimal models coupled to two-dimensional supergravity. The charges $V_\pm$ may be defined as vertex operators which do not need any dressing ($\beta_i = 0$):

$$V_\pm = e^{id_\pm \Phi_M} \quad , \tag{5.35}$$

with

$$\Delta(e^{id\Phi_M}) = \frac{1}{2}d(d-2\alpha_0) = \frac{1}{2} \quad , \tag{5.36}$$

$$d_\pm = \alpha_0 \pm \sqrt{\alpha_0^2+1} \quad . \tag{5.37}$$

In the Coulomb gas representation $V_\pm$ can be introduced directly in the action S_M as in (2.73). After integrating over the matter (X_0) and Liouville ϕ_0 zero mode we get

$$\left\langle \prod_{i=1}^{3} \int \Psi_{NS}(\tilde{\mathbf{z}}_i, k_i) \right\rangle_{S_M^\lambda + S_{SL}} = 2\pi\delta\left(\sum_{i=1}^{3} k_i + (n-1)d_+ + (m-1)d_-\right) \mathcal{A}_3^{nm}$$

$$\mathcal{A}_3^{nm}(k_1,k_2,k_3) = \frac{\lambda_+^n \lambda_-^m}{n!m!} \Gamma(-s)(i\overline{\mu})^s \left\langle \prod_{i=1}^{n} \int d^2\mathbf{t}_i e^{id_+\Phi_M(\mathbf{t}_i)} \prod_{i=1}^{m} \int d^2\mathbf{r}_i e^{id_-\Phi_M(\mathbf{r}_i)} \right.$$
$$\left. \times \int \prod_{i=1}^{3} d^2\tilde{\mathbf{z}}_i e^{ik_i\Phi_M(\tilde{\mathbf{z}}_i)+\beta_i\Phi_{SL}(\tilde{\mathbf{z}}_i)} \left(\int d^2\mathbf{z} e^{\alpha_+\Phi_{SL}(\mathbf{z})}\right)^s \right\rangle_{\mu=0} \quad . \tag{5.38}$$

Choosing the gauge (5.20) and integrating over the grassmannian variables we get:

$$\mathcal{A}_3^{nm}(k_1,k_2,k_3) = \Gamma(-s)\left(i\overline{\mu}\alpha_+^2\right)^s \frac{(-d_+^2)^n}{n!}\frac{(-d_-^2)^m}{m!}\lambda_+^n\lambda_-^m$$
$$\times \prod_{i=1}^{n}\int \mathrm{d}^2 t_i |t_i|^{2\tilde{\alpha}}|1-t_i|^{2\tilde{\beta}} \prod_{i<j=1}^{n} |t_i-t_j|^{4\tilde{\rho}}$$
$$\times \prod_{i=1}^{m}\int \mathrm{d}^2 r_i |r_i|^{2\tilde{\alpha}'}|1-r_i|^{2\tilde{\beta}'} \prod_{i<j=1}^{m} |r_i-r_j|^{4\tilde{\rho}'} \prod_{i=1}^{n}\prod_{j=1}^{m}|t_i-r_j|^{-2}$$
$$\times \prod_{i=1}^{s}\int \mathrm{d}^2 z_i |z_i|^{2\alpha}|1-z_i|^{2\beta} \prod_{i<j=1}^{s} |z_i-z_j|^{4\rho}$$
$$\times \left\langle \left(\beta_1^2\overline{\psi}\psi(0) - k_1^2\overline{\zeta}\zeta(0)\right)\prod_{i=1}^{n}\overline{\zeta}\zeta(t_i)\prod_{i=1}^{m}\overline{\xi}\xi(r_i)\prod_{i=1}^{s}\overline{\psi}\psi(z_i)\right\rangle_0 . \quad (5.39)$$

The above parameters are defined as follows,

$$\begin{aligned}
\tilde{\alpha} &= d_+k_1 \quad , \quad \tilde{\beta} = d_+k_2 \quad , \quad \tilde{\rho} = d_+^2 \quad , \\
\tilde{\alpha}' &= d_-k_1 \quad , \quad \tilde{\beta}' = d_-k_2 \quad , \quad \tilde{\rho}' = d_-^2 \quad , \\
\alpha &= -\alpha_+\beta_1 \quad , \quad \beta = -\alpha_+\beta_2 \quad , \quad \rho = -\frac{1}{2}\alpha_+^2 \quad .
\end{aligned} \quad (5.40)$$

Notice that now we have $|t_i - r_j|^{-2}$ instead of $|t_i - r_j|^{-4}$ because in the supersymmetric case $d_+d_- = -\alpha_+\alpha_- = -1$ instead of $d_+d_- = -2$. This is a consequence of the contribution 1 in the definition of Q in Eq. (5.8) instead of 2 from the bosonic case (see Eq. (2.3)).

Since the expectation value of an odd number of operators $\overline{\xi}\xi$ or $\overline{\psi}\psi$ vanishes, the above amplitude will not vanish only in two cases, namely $n+m$ even and s odd or $n+m$ odd and s even. From now on we concentrate ourselves on the first case. In fact the amplitude factorizes in a product of the previous term times the screening charges contribution as

$$\mathcal{A}_3^{nm}(k_1,k_2,k_3) = \Gamma(-s)\left(i\overline{\mu}\alpha_+^2\right)^s \frac{(-\lambda_+d_+^2)^n}{n!}\frac{(-\lambda_-d_-^2)^m}{m!}\beta_1^2 I_M^{nm} I_G^s \, , \quad (5.41)$$

where I_G^s corresponds to the integral (5.24a) from the last section, while I_M^{nm} is given by the expression

$$I_M^{nm} = \prod_{i=1}^{n}\int \mathrm{d}^2 t_i |t_i|^{2\tilde{\alpha}}|1-t_i|^{2\tilde{\beta}} \prod_{i<j}^{n} |t_i-t_j|^{2\tilde{\rho}}$$
$$\times \prod_{i=1}^{m}\int \mathrm{d}^2 r_i |r_i|^{2\tilde{\alpha}'}|1-r_i|^{2\tilde{\beta}'} \prod_{i<j}^{m} |r_i-r_j|^{2\tilde{\rho}'} \prod_{i=1}^{n}\prod_{j=1}^{m}|t_i-r_j|^{-2} \quad (5.42)$$
$$\times \left\langle \prod_{i=1}^{n}\overline{\zeta}\zeta(t_i)\prod_{i=1}^{m}\overline{\xi}\xi(r_i)\right\rangle_0 .$$

As in the case of I^s_G of the last section, it is very difficult to compute the integral I^{nm}_M, which corresponds to the supersymmetric generalization of the Eq. (B.10) of Dotsenko and Fatteev's work[30] (reproduced in (C.75)) and would be calculated later in refs. [81,82]. For this reason we use the asymptotic behavior and the symmetries of the integral as well as some relations with other integrals already calculated, to have some insight about the solution. This procedure is described in detail in Appendix C where we deduced that for n and m even (see formula (C.70))

$$
\begin{aligned}
I^{nm}_M(\tilde{\alpha},\tilde{\beta};\tilde{\rho}) &= (-)^{\frac{n+m}{2}}\frac{\pi^{n+m}}{2^{n+m}}n!m!\left(-\frac{\tilde{\rho}}{2}\right)^{-2nm}\left[\Delta\left(\frac{1}{2}-\frac{\tilde{\rho}}{2}\right)\right]^n\left[\Delta\left(\frac{1}{2}-\frac{\tilde{\rho}'}{2}\right)\right]^m \\
&\times\prod_1^{\frac{n}{2}}\Delta(i\tilde{\rho})\Delta\left(\frac{1}{2}+\tilde{\rho}\left(i-\frac{1}{2}\right)\right)\prod_1^{\frac{m}{2}}\Delta(i\tilde{\rho}'-\frac{n}{2})\Delta\left(\frac{1}{2}-\frac{n}{2}-\tilde{\rho}'\left(i-\frac{1}{2}\right)\right) \\
&\times\prod_{i=0}^{\frac{n}{2}-1}\Delta(1+\tilde{\alpha}+i\tilde{\rho})\Delta(1+\tilde{\beta}+i\tilde{\rho})\Delta(m-\tilde{\alpha}-\tilde{\beta}+\tilde{\rho}(i-n+1)) \\
&\times\prod_{i=1}^{\frac{n}{2}}\Delta(\frac{1}{2}+\tilde{\alpha}+(i-\frac{1}{2})\tilde{\rho})\Delta(\frac{1}{2}+\tilde{\beta}+(i-\frac{1}{2})\tilde{\rho})\Delta(-\frac{1}{2}-\tilde{\alpha}+m-\tilde{\beta}+\tilde{\rho}(i-n+\frac{1}{2})) \\
&\times\prod_{i=0}^{\frac{m}{2}-1}\Delta(1+\tilde{\alpha}'-\frac{n}{2}+i\tilde{\rho}')\Delta(1-\frac{n}{2}+\tilde{\beta}'+i\tilde{\rho}')\Delta(\frac{n}{2}-\tilde{\alpha}'-\tilde{\beta}'+\tilde{\rho}'(i-m+1)) \\
&\times\prod_{i=1}^{\frac{m}{2}}\Delta(\frac{1}{2}-\frac{n}{2}+\tilde{\alpha}'+(i-\frac{1}{2})\tilde{\rho}')\Delta(\frac{1}{2}-\frac{n}{2}+\tilde{\beta}'+(i-\frac{1}{2})\tilde{\rho}') \\
&\times\Delta(-\frac{1}{2}+\frac{n}{2}-\tilde{\alpha}'-\tilde{\beta}'+\tilde{\rho}'(i-m+\frac{1}{2}))\quad . \qquad (5.43)
\end{aligned}
$$

It is desirable to simplify the result for the amplitude using kinematic relations which permit us to eliminate some redundant variables. In the kinematic region $k_1, k_3 \geq \alpha_0$, $k_2 \leq \alpha_0 \leq 0$ we have the basic relations:

$$
\begin{aligned}
d_+ = -\alpha_+ \quad , \quad d_- = \alpha_- \quad ; \quad \alpha_+\alpha_- = 1 \\
Q = -(\alpha_+ + \alpha_-) \quad ; \quad \alpha_0 = \frac{\alpha_- - \alpha_+}{2} \quad .
\end{aligned}
\qquad (5.44a)
$$

Together with the energy-momentum conservation, taking the parameter s into account

$$
\begin{aligned}
k_1 + k_2 + k_3 &= (1-n)d_+ + (1-m)d_- = (n-1)\alpha_+ + (1-m)\alpha_- \\
k_1 - k_2 + k_3 &= -(s+1)\alpha_+ \qquad (5.44b) \\
\beta_1 = \alpha_+ + k_1 \quad &, \quad \beta_2 = \alpha_- - k_2 \quad , \quad \beta_3 = \alpha_+ + k_3 \quad .
\end{aligned}
$$

Using these relations we obtain for the parameters defined in (5.40) the results

$$(\beta,\tilde{\beta},\tilde{\beta}')=\left(-\frac{1}{2}-\frac{m}{2}-(n+s)\rho,-\beta-1,\frac{n+s}{2}+\frac{\rho^{-1}}{4}(m-1)\right)$$
$$(\tilde{\alpha},\tilde{\alpha}')=\left(\alpha-2\rho,-1+\frac{\rho^{-1}\alpha}{2}\right)\quad,\quad(\tilde{\rho},\tilde{\rho}')=\left(-2\rho,-\frac{\rho^{-1}}{2'}\right)\;. \tag{5.45}$$

Substituting back in Eq. (5.43) we find

$$
\begin{aligned}
&I_M^{mn}=\left(-\frac{\pi^2}{4}\right)^{(n+m)/2} n!m!\rho^{-2mn}\\
&\Delta^n\left(\frac{1}{2}+\rho\right)\Delta^m\left(\frac{1}{2}+\frac{\rho^{-1}}{4}\right)\prod_1^{n/2}\Delta(-2i\rho)\Delta\left(\frac{1}{2}+\rho(1-2i)\right)\\
&\prod_1^{m/2}\Delta\left(-\frac{n}{2}-\frac{i\rho^{-1}}{2}\right)\Delta\left(\frac{1}{2}-\frac{n}{2}+\left(\frac{1}{4}-\frac{i}{2}\right)\rho^{-1}\right)\\
&\prod_{i=0}^{n/2-1}\Delta(1+\alpha-2\rho(i+1))\Delta(\frac{1}{2}+\frac{m}{2}+\rho(n+2-2i))\Delta(\frac{1}{2}+\frac{m}{2}-\alpha-\rho(s-n+2i))\\
&\prod_{i=1}^{n/2}\Delta(\frac{1}{2}+\alpha-\rho(1+2i))\Delta(\frac{m}{2}+\rho(n+s+1-2i))\Delta(\frac{m}{2}-\alpha-\rho(2i-n-1+s))\\
&\prod_{i=0}^{m/2-1}\Delta(-\frac{n}{2}-\frac{\rho^{-1}}{2}(i-\alpha))\Delta(1+\frac{s}{2}-\frac{\rho^{-1}}{2}(i+\frac{1}{2}-\frac{m}{2}))\Delta(1-\frac{s}{2}-\frac{\rho^{-1}}{2}(i+\alpha+\frac{1}{2}-\frac{m}{2}))\\
&\prod_{i=1}^{m/2}\Delta(-\frac{1}{2}-\frac{n}{2}-\frac{\rho^{-1}}{2}(i-\frac{1}{2}-\alpha))\Delta(\frac{1}{2}+\frac{s}{2}-\frac{\rho^{-1}}{2}(i-\frac{m}{2}))\Delta(\frac{1}{2}-\frac{s}{2}-\frac{\rho^{-1}}{2}(i+\alpha-\frac{m}{2}))
\end{aligned}
\tag{5.46}$$

while for the super-Liouville tachyon contribution we have

$$
\begin{aligned}
&I_G^s=\pi^2\Delta(\frac{1}{2}-\rho)^s\prod_1^{\frac{s-1}{2}}\Delta(2i\rho)\prod_0^{\frac{s-1}{2}}\Delta(\frac{1}{2}+(2i+1)\rho)\\
&\prod_0^{\frac{s-1}{2}}\Delta(1+\alpha+2i\rho)\Delta(\frac{1}{2}-\frac{m}{2}+(2i-n-s)\rho)\Delta(-\alpha+(2i-s+2+n)\rho+\frac{1}{2}+\frac{m}{2})\\
&\prod_1^{\frac{s-1}{2}}\Delta(\frac{1}{2}+\alpha+(2i-1)\rho)\Delta(-\frac{m}{2}+(2i-1-n-s)\rho)\Delta(\frac{m}{2}-\alpha+(2i-s+1+n)\rho)
\end{aligned}
\tag{5.47}$$

Therefore we get for the product of the above terms, taking (5.39) into account, the rather long expression

$$
\begin{aligned}
&\mathcal{A}_3^{nm}(k_1,k_2,k_3) = \\
&= \Gamma(-s)\Gamma(s+1)\,(i\bar{\mu})^s\,\alpha_+^{2(s-1)}(2\rho)^{n-m}(-)^{\frac{n+m+s-1}{2}}\,\frac{\pi^{s+n+m}}{2^{m+n+s-1}}\rho^{-2mn} \\
&\times \left[\lambda_+\Delta(\frac{1}{2}+\rho)\right]^n \left[\lambda_-\Delta(\frac{1}{2}+\frac{\rho^{-1}}{4})\right]^m \\
&\times \prod_{i=1}^{\frac{n}{2}} \Delta(-2i\rho)\Delta(\frac{1}{2}+\rho(1-2i)) \prod_{i=1}^{\frac{m}{2}} \Delta(-\frac{n}{2}-\frac{i\rho^{-1}}{2})\Delta\left(\frac{1}{2}-\frac{n}{2}+\left(\frac{1}{4}-\frac{i}{2}\right)\rho^{-1}\right) \\
&\times \prod_{i=0}^{\frac{n}{2}-1} \Delta\left(\frac{1}{2}+\frac{m}{2}+\rho(n+s-2i)\right) \prod_{i=1}^{\frac{n}{2}} \Delta\left(\frac{m}{2}+\rho(n+s+1-2i)\right) \\
&\times \prod_{i=0}^{\frac{m}{2}-1} \Delta\left(1+\frac{s}{2}-\frac{\rho^{-1}}{2}(i+\frac{1}{2}-\frac{n}{2})\right) \prod_{i=1}^{\frac{m}{2}} \Delta\left(\frac{1}{2}+\frac{s}{2}-\frac{\rho^{-1}}{2}(i-\frac{m}{2})\right) \\
&\times \left[\Delta(\frac{1}{2}-\rho)\right]^s \prod_{i=1}^{\frac{s-1}{2}} \Delta(2i\rho) \prod_{i=0}^{\frac{s-1}{2}} \Delta\left(\frac{1}{2}+(2i+1)\rho\right) \\
&\times \prod_{i=0}^{\frac{s-1}{2}} \Delta(\frac{1}{2}-\frac{m}{2}+(2i-n-s)\rho) \prod_{i=1}^{\frac{s-1}{2}} \Delta(-\frac{m}{2}+(2i-1-n-s)\rho) \\
&\times \prod_{i=1}^{\frac{n}{2}} \overset{\bullet}{\Delta}(\frac{1}{2}+\alpha-\rho(2i+1))\,\overset{\bullet}{\Delta}(\frac{1}{2}+\frac{m}{2}-\alpha-\rho(s-n-2+2i)) \\
&\times \prod_{i=0}^{\frac{s-1}{2}} \overset{\bullet}{\Delta}(\frac{m}{2}+\frac{1}{2}-\alpha+(2i-s+n+2)\rho) \prod_{i=1}^{\frac{s-1}{2}} \overset{\bullet}{\Delta}(\frac{1}{2}+\alpha+(2i-1)\rho) \\
&\times \prod_{i=1}^{\frac{n}{2}} \overset{*}{\Delta}(\frac{m}{2}-\alpha-\rho(2i-n+s-1))\,\overset{*}{\Delta}(1+\alpha-2i\rho) \\
&\times \prod_{i=0}^{\frac{s-1}{2}} \overset{*}{\Delta}(1+\alpha+2i\rho) \prod_{i=1}^{\frac{s-1}{2}} \overset{*}{\Delta}(\frac{m}{2}-\alpha+(2i-s+n+1)\rho) \\
&\times \prod_{i=1}^{\frac{m}{2}} \overset{\star}{\Delta}(\frac{1}{2}-\frac{s}{2}-\frac{\rho^{-1}}{2}(i+\alpha-\frac{m}{2}))\,\overset{\star}{\Delta}(-\frac{n}{2}-\frac{\rho^{-1}}{2}(i-1-\alpha)) \\
&\times \prod_{i=1}^{\frac{m}{2}} \overset{\dagger}{\Delta}(-\frac{n}{2}-\frac{1}{2}-\frac{\rho^{-1}}{2}(i-\alpha-\frac{1}{2}))\,\overset{\dagger}{\Delta}(1-\frac{s}{2}-\frac{\rho^{-1}}{2}(i-\frac{(m+1)}{2}+\alpha)) \quad .
\end{aligned}
\tag{5.48}
$$

After a long but tedious calculation combining products of the matter and gravity sectors and using some properties of gamma functions we obtain a simple result. Indeed, let us first collect the terms with a • on the top; we find

$$(\bullet) = \prod_{i=2-s,\, i\,\mathrm{odd}}^{n+1} (-1)^{m/2} \left[\frac{\Gamma(\frac{1}{2} - \alpha + \frac{m}{2} + i\rho)}{\Gamma(\frac{1}{2} - \alpha + i\rho)}\right]^2 \Delta(\frac{1}{2} - \alpha + \rho) \quad . \tag{5.49}$$

For the terms having (*) on the top we find

$$(*) = \prod_{i=1-s,\, i\,\mathrm{even}}^{n} (-1)^{m/2} \left[\frac{\Gamma(-\alpha + \frac{m}{2} + i\rho)}{\Gamma(-\alpha + i\rho)}\right]^2 \Delta(1 - \frac{m}{2} + \alpha + (s - n - 1)\rho) \tag{5.50}$$

where in the above two expressions we just deal with the product and the definition of Δ. For the terms having the (†) on the top we expand the Δ's as a rational function of products of monomials, and rearrange terms to obtain the expression

$$(\dagger) = (2\rho)^{\frac{n(s+n+1)}{2}} \prod_{i=2-s,\, i\,\mathrm{odd}}^{n+1} (-)^{m/2} \left[\frac{\Gamma(-\alpha + \frac{1}{2} + i\rho)}{\Gamma(\frac{1}{2} + \frac{m}{2} - \alpha + i\rho)}\right]^2 \quad . \tag{5.51}$$

Finally, we use also

$$(\star) = (2\rho)^{\frac{m(s+n+1)}{2}} \prod_{i=1-s,\, i\,\mathrm{even}}^{n} (-)^{m/2} \left[\frac{\Gamma(-\alpha + i\rho)}{\Gamma(\frac{m}{2} - \alpha + i\rho)}\right]^2 \quad . \tag{5.52}$$

Thus we obtain for the product of the above terms,

$$(\star)(\bullet)(\dagger)(*) = (2\rho)^{m(n+s+1)} \Delta(\frac{1}{2} - \alpha + \rho)\Delta(1 - \frac{m}{2} + \alpha + \rho(s - n - 1)) \quad , \tag{5.53}$$

and the three point function (5.48) assumes the much simpler form

$$\begin{aligned}
&\mathcal{A}_3^{nm}(k_1, k_2, k_3) = \\
&\quad \left[\overline{\mu}\Delta(\frac{1}{2} - \rho)\right]^s \left[\lambda_- \Delta(\frac{1}{2} + \frac{\rho^{-1}}{4})\right]^m \left[\lambda_+ \Delta(\frac{1}{2} + \rho)\right]^n \\
&\quad \times \Delta\left(\rho - \alpha + \frac{1}{2}\right) \Delta\left(\frac{1}{2} - \frac{n+s}{2} - \frac{m\rho^{-1}}{4}\right) \Delta(1 - \frac{m}{2} + \alpha + (s - n - 1)\rho) \\
&\quad = \left[\overline{\mu}\Delta\left(\frac{1}{2} - \rho\right)\right]^s \left[\lambda_- \Delta\left(\frac{1}{2} + \frac{\rho^{-1}}{4}\right)\right]^m \left[\lambda_+ \Delta\left(\frac{1}{2} + \rho\right)\right]^n \\
&\quad \times \prod_{i=1}^{3} \Delta\left(\frac{1}{2} + \frac{1}{2}(\beta_i^2 - k_i^2)\right) \quad ,
\end{aligned} \tag{5.54}$$

where we used

$$\frac{1}{2}+\frac{1}{2}(\beta_1^2-k_1^2)=\frac{1}{2}+\frac{1}{2}(-\alpha_+^2+2\alpha_+\beta_1)$$

$$=\rho-\alpha+\frac{1}{2}\quad , \tag{5.55a}$$

$$\frac{1}{2}+\frac{1}{2}(\beta_2^2-k_2^2)=\frac{1}{2}+\frac{1}{2}(-\alpha_-^2+2\alpha_-\beta_2)$$

$$=\frac{1}{2}+\frac{1}{4}\rho^{-1}+\frac{\rho^{-1}}{2}\beta$$

$$=\frac{1}{2}-\frac{1}{4}\rho^{-1}+\frac{1}{2}\rho^{-1}(-\frac{1}{2}-\frac{m}{2}-(n+s)\rho)\quad ,$$

$$=\frac{1}{2}-\frac{n+s}{2}-\frac{1}{4}\rho^{-1}m\quad , \tag{5.55b}$$

and

$$\frac{1}{2}+\frac{1}{2}(\beta_3^2-k_3^2)=\frac{1}{2}+\frac{1}{2}(-\alpha_+^2+2\alpha_+\beta_3)$$

$$=\frac{1}{2}+\rho+\alpha_+\beta_2$$

$$=1-\frac{m}{2}+\alpha+(s-n-1)\rho\quad . \tag{5.55c}$$

Recalling that for each $V_+(V_-)$ one has $k_i = d_+(d_-) = -\alpha_+(\alpha_-)$ and a corresponding trivial dressing ($\beta_i = 0$) we see that the factors $\Delta(\frac{1}{2}+\rho)$ and $\Delta(\frac{1}{2}+\frac{\rho^{-1}}{4})$ in (5.54) have the form $\Delta(\frac{1}{2}+\frac{1}{2}(\beta_i^2-k_i^2))$ and again the result is similar to the $(s+n+m+3)$-point functions with special values of the momenta for the $s+n+m$ first vertices.

Integrating (5.54) with respect to $\overline{\mu}$ and setting $k_3 = 0$ we obtain the 2-point function:

$$\mathcal{A}_2^{\tilde{n}\tilde{m}}(k_1,k_2)=\frac{[\overline{\mu}\Delta(\frac{1}{2}-\rho)]^{-\alpha_-(|k_1-\alpha_0|+|k_2-\alpha_0|)}}{-\alpha_-(|k_1-\alpha_0|+|k_2-\alpha_0|)}$$

$$\times\left[\lambda_+\Delta(\frac{1}{2}+\rho)\right]^{\tilde{n}}\left[\lambda_-\Delta(\frac{1}{2}+\frac{\rho^{-1}}{4})\right]^{\tilde{m}}\prod_{i=1}^{2}\Delta\left(\frac{1}{2}+\frac{1}{2}(\beta_i^2-k_i^2)\right)\quad , \tag{5.56}$$

where $\tilde{n}$ and $\tilde{m}$ are such that the following conservation law (with $k_3 = 0$) is obeyed

$$k_1+k_2=(1-\tilde{n})d_++(1-\tilde{m})d_-\quad . \tag{5.57}$$

Integrating $\mathcal{A}_2^{\tilde{n}\tilde{m}}$ twice with respect to $\overline{\mu}$ and taking $k_1 = k_2 = 0$ (which implies $\tilde{n} = \tilde{m} = 1$) we obtain the partition function,

$$\mathcal{Z}=\frac{[\overline{\mu}\Delta(\frac{1}{2}-\rho)]^{1-\frac{\rho^{-1}}{2}}}{(2\rho-1)(2\rho+1)}(2\rho)^3\left[\lambda_+\Delta(\frac{1}{2}+\rho)\right]\left[\lambda_+\Delta(\frac{1}{2}+\frac{\rho^{-1}}{4})\right]\quad . \tag{5.58}$$

Calculating now the ratio (R) of the amplitudes, we obtain

$$R = \frac{(\mathcal{A}_3^{nm}(k_1,k_2,k_3))^2 \mathcal{Z}}{\mathcal{A}_2^{\tilde{n}_1\tilde{m}_1}(k_1,k_1)\mathcal{A}_2^{\tilde{n}_2\tilde{m}_2}(k_2,k_2)\mathcal{A}_2^{\tilde{n}_3\tilde{m}_3}(k_3,k_3)} = \frac{-8\alpha_-^3(\prod_{i=1}^3 |k_i - \alpha_0|)(2\rho)^3}{(2\rho-1)(2\rho+1)} \quad . \tag{5.59}$$

In the case of minimal models we have

$$k_i = \frac{(1-r_i')}{2}d_- + \frac{(1-r_i)}{2}d_+ \quad ; \tag{5.60}$$

using also that $\alpha_0 = \frac{(d_+ + d_-)}{2}$ we have for a generic kinematic region,

$$R = \frac{(2\alpha_+)^3 \prod_{i=1}^3 |r_i d_- + r_i' d_+|}{(2\rho-1)(2\rho+1)} \quad . \tag{5.61}$$

This result should be compared with the one obtained in ref. [82] for unitary minimal models[2)]. For diagonal operators ($r_i = r_i'$) we have:

$$R = \frac{(2\rho+1)^2}{(2\rho-1)} r_1 r_2 r_3 \quad . \tag{5.62}$$

This result agrees with ref. [84] taking into account that our definition of ρ corresponds to $-(2\rho)^{-1}$ of that work.

Although the analytical continuation of s used in refs. [77,79,82] are different the result for the amplitude does not depend on such details.

5.4. Neveu-Schwarz N-Point Function with (n,m) Screening Charges

In this section we generalize the previous sections for the case of N-point functions ($N \geq 4$) with screening charges in the matter sector[79]. This section is parallel to section 3.4 of the bosonic case.

After the integration over the zero modes of matter and Liouville sectors we get momentum conservation times the N-point amplitude, i.e.,

$$\left\langle \prod_{i=1}^N \Psi_{NS}(\tilde{\mathbf{z}}_i) \right\rangle_{S_{SL}+S_M^\lambda} = 2\pi\delta(\sum_{i=1}^N k_i + nd_+ + md_- - 2\alpha_0)\mathcal{A}_N^{nm} \quad , \tag{5.63}$$

where the amplitude $\mathcal{A}_N^{nm}$ is given by the correlator

$$\mathcal{A}_N^{nm}(k_1,\cdots,k_N) = \frac{(\lambda_+)^n}{n!}\frac{(\lambda_-)^m}{m!}\frac{\Gamma(-s)}{-\alpha_+}(i\overline{\mu})^s \left\langle \prod_{i=1}^N \int d^2\tilde{\mathbf{z}}_i e^{ik_i\Phi_M(\tilde{\mathbf{z}}_i)+\beta_i\Phi_{SL}(\tilde{\mathbf{z}}_i)} \right.$$
$$\left. \times \prod_{i=1}^n \int d^2\mathbf{t}_i e^{id_+\Phi_M(\tilde{\mathbf{t}}_i)} \prod_{j=1}^m \int d^2\mathbf{r}_j e^{id_-\Phi_M(\tilde{\mathbf{r}}_j)} \prod_{j=1}^s \int d^2\mathbf{z}_j e^{i\alpha_+\Phi_{SL}(\tilde{\mathbf{z}}_i)} \right\rangle_0 \quad , \tag{5.64}$$

---

2) The reader should notice that in ref. [82] $\alpha_0 = \frac{1}{\sqrt{q(q+2)}} > 0$ and therefore $d_+ = -\alpha_-$, $d_- = \alpha_+ = -\sqrt{\frac{q}{(q+2)}}$.

and s is defined as before by

$$s = -\frac{1}{\alpha_+}(\sum_{i=1}^{N} \beta_i + Q) \quad . \tag{5.65}$$

Using the free propagators (5.6) and choosing the gauge (5.20) we obtain, in components

$$\begin{aligned}
\mathcal{A}_N^{nm} &= \Gamma(-s)\left(-\frac{\pi}{2}\right)^3 \left(i\overline{\mu}\alpha_+^2\right)^s \frac{(-\lambda_+ d_+^2)^n}{n!} \frac{(-\lambda_- d_-^2)^m}{m!} \\
&\times \prod_{j=4}^{N} \int d^2\tilde{z}_j \prod_{i=1}^{n} \int d^2 t_i \prod_{i=1}^{m} \int d^2 r_i \prod_{i=1}^{s} d^2 z_i |z_i|^{2\alpha} |1-z_i|^{2\beta} \\
&\times \prod_{i<j=1}^{s} |z_i - z_j|^{4\rho} \prod_{i=1}^{s} \prod_{j=4}^{N} |z_i - \tilde{z}_j|^{2p_j} \\
&\times \prod_{j=4}^{N} |\tilde{z}_j|^{2\alpha_j} |1-\tilde{z}_j|^{2\beta_j} \prod_{j<l=4}^{N} |\tilde{z}_j - \tilde{z}_l|^{4\rho_{jl}} \\
&\times \prod_{i=1}^{n} |t_i|^{2\tilde{\alpha}} |1-t_i|^{2\tilde{\beta}} \prod_{i<j}^{n} |t_i - t_j|^{4\tilde{\rho}} \prod_{i=1}^{n} \prod_{j=1}^{m} |t_i - r_j|^{-2} \\
&\times \prod_{i=1}^{m} |r_i|^{2\tilde{\alpha}'} |1-r_i|^{2\tilde{\beta}'} \prod_{i<j}^{m} |r_i - r_j|^{4\tilde{\rho}'} \prod_{i=1}^{n} \prod_{j=4}^{N} |t_i - \tilde{z}_j|^{2\tilde{\alpha}_j} \prod_{i=1}^{m} \prod_{j=4}^{N} |r_i - \tilde{z}_j|^{2\tilde{\alpha}_j'} \\
&\times \Big\langle (\beta_1^2 \overline{\psi}\psi(0) - k_1^2 \overline{\xi}\xi(0)) \prod_{j=4}^{N} (\beta_j^2 \overline{\psi}\psi(\tilde{z}_j) - k_j^2 \overline{\xi}\xi(\tilde{z}_j)) \\
&\times \prod_{i=1}^{n} \overline{\xi}\xi(t_i) \prod_{i=1}^{m} \overline{\xi}\xi(r_i) \prod_{i=1}^{s} \overline{\psi}\psi(z_i) \Big\rangle_{\mu=0} \quad .
\end{aligned} \tag{5.66}$$

The parameters appearing in the above expression are defined as before (see formula (5.40)). There exist now several possibilities for N, s, n and m such that the above amplitude does not vanish. We could at this point simply obtain for each of these possibilities, the corresponding integral and sum over all the terms at the end. However we see that the role played by all such terms, in the expectation value is just to make the final result symmetric, with respect to interchanging of the momenta k_j and the screening charges. For this reason we found it more convenient to suggest an Ansatz for the final result instead of writing down an Ansatz for each case separately. As in the bosonic case we should use the kinematics to eliminate some parameters of (5.66) and write it as a function of $\alpha, \beta, \rho, s, p_j$ $(4 \le j \le N-1)$ before proposing a solution for the integral. After such elimination (see also [79]) we use the symmetries of the above amplitude under translation and inversion, and it is not difficult to find the relations

$$\mathcal{A}_N^{nm}(\alpha, \beta, \rho, p_j) = \mathcal{A}_N^{nm}(\beta, \alpha, \rho, p_j) \tag{5.67}$$

$$\mathcal{A}_N^{nm}(\alpha, \beta, \rho, p_j) = \mathcal{A}_N^{nm}(-\alpha - \beta + \frac{(m-1)}{2} - P + \rho(N+n-s-1), \beta, \rho, p_j)) \tag{5.68}$$

where $P = \sum_{j=4}^{N-1} p_j$. Notice that the referred elimination was possible only after the choice of a kinematic region, which by the way is the same of the bosonic case $k_1, k_2, \cdots k_{N-1} \geq \alpha_0, k_N \leq \alpha_0 \leq 0$.

The asymptotic behavior is given by

$$\mathcal{A}_N^{nm}(\alpha \to \infty) \sim \alpha^{1-m+2\beta+2\rho(s-N-n+3)+2P} \quad . \tag{5.69}$$

All the above conditions are satisfied by the Ansatz

$$\begin{aligned}\mathcal{A}_N^{nm} &= f_N^{nm}(\rho, p_j)\Delta(\frac{1}{2}+\rho-\alpha)\Delta(\frac{1}{2}+\rho-\beta)\Delta(1-\frac{m}{2}+P+\alpha+\beta+\rho(2+s-n-N)) \\ &= f_N^{nm}(\rho, p_j)\prod_{i=1}^{3}\Delta(\frac{1}{2}+\frac{1}{2}(\beta_i^2-k_i^2)) \quad .\end{aligned} \tag{5.70}$$

With the help of Eq. (2.98), which is also valid in the supersymmetric case we can determine f_N^{nm} and obtain

$$\begin{aligned}\mathcal{A}_N^{nm} = &\left[-\lambda_+\Delta(\frac{1}{2}+\rho)\right]^n \left[-\lambda_-\Delta(\frac{1}{2}+\frac{\rho^{-1}}{4})\right]^m \prod_{i=1}^{N}\Delta(\frac{1}{2}(1+\beta_i^2-k_i^2)) \\ &\times (s+N-3)(s+N-4)\cdots(s+1)\left[\overline{\mu}\Delta(\frac{1}{2}-\rho)\right]^s \quad .\end{aligned} \tag{5.71}$$

For $\hat{c} < 1$ we can redefine the constants λ_+ and λ_- similarly as we did for the cosmological constant before,

$$\lambda_+ = \frac{1}{\Delta\left(\frac{1}{2}+\rho\right)} \quad , \quad \lambda_- = \frac{1}{\Delta\left(\frac{1}{2}+\frac{\rho^{-1}}{4}\right)} \quad , \tag{5.72}$$

as well as renormalize $\overline{\mu}$ and the vertices Ψ_{NS} (see (5.31)) to obtain

$$\mathcal{A}_N^{nm} = \frac{\partial^{N-3}}{\partial\mu^{N-3}}(\mu)^{s+N-3} \quad . \tag{5.73}$$

For $\hat{c} = 1\,(\alpha_0 = 0)$ we are not allowed to redefine the vertices Ψ_{NS}, and we get

$$\mathcal{A}_N^{nm} = \frac{\partial^{N-3}}{\partial\mu}(\mu)^{s+N-3}\prod_{i=1}^{N}\frac{\Gamma(1-|k_i|)}{\Gamma(|k_i|)} \quad . \tag{5.74}$$

The results (5.73) and (5.74) are completely similar to the bosonic case (see (2.100a, b)). The similarity between the two cases for the renormalized amplitudes is quite surprising, since supersymmetry, in general, modify the ultra-violet and infra-red behaviours of the theory. However the simplicity of the bosonic and supersymmetric results is probably due to the low dimensionality of the embedding space which is a common feature of the two theories analysed here.

It is important to stress that the parallelism between the bosonic and the supersymmetric cases have already been noticed previously in the literature in the discrete approach[145], although the exact relation between results of ref. [145] and our results are still not clear and deserves a deeper understanding.

5.5. Ramond Sector

5.5.1. The Ramond Vertex

As we mentioned before, the non-critical superstring contains, beside the massless scalar field ψ_{NS}, also a massless spinor field represented by the Ramond vertex (V_R). In order to define V_R it is appropriate to represent the $SO(2)$ Lorentz generators $J^{\mu\nu}$ ($\mu, \nu = 0, 1$) in terms of fermion bilinears[146,147]

$$J^{\mu\nu} =: \psi^{\mu}(z)\psi^{\nu}(z): \quad , \tag{5.75}$$

where the covariant notation in the embedding space means $\psi^0 = \psi$, $\psi^1 = \zeta$. The Lorentz algebra can be checked from (5.75) with help of the free propagators (5.6), which can be written as

$$\langle \psi^{\mu}(z)\psi^{\nu}(w)\rangle = \delta^{\mu\nu}(z-w)^{-1} \tag{5.76}$$

and analogously for the anti-holomorphic part.

The spinor emission vertex V_R must behave as a two-dimensional spinor under $J^{\mu\nu}$. Let us consider a spin field , and denote it by S_α, with $\alpha = 0, 1$. Later we will relate V_R with S_α. Since $J^{\mu\nu}$ is a (1,1) conformal operator the above requirement amounts to the following operator product expansion

$$J^{\mu\nu}(z)S_\alpha(w) \sim \frac{1}{z-w}\left(\gamma^{[\mu,}\gamma^{\nu]}\right)_{\alpha}^{\ \delta} S_\delta(w) \quad , \tag{5.77}$$

where γ^μ are the usual two-dimensional gamma matrices which satisfy the Clifford algebra.

It can be shown[146,148] that expressions (5.75), (5.77) and the form of the energy momentum tensor for the fermions ψ^μ imply the following operator product expansions

$$\psi^{\mu}(z)S_\alpha(w) = \frac{1}{(z-w)^{\frac{1}{2}}}(\gamma^{\mu})_{\alpha}^{\ \delta} S_\delta(z) + \cdots \quad , \tag{5.78}$$

$$S^{\alpha}(z)S_\delta(w) = \frac{1}{(z-w)^{\frac{1}{4}}}\delta^{\alpha}_{\delta} + \cdots \quad . \tag{5.79}$$

Expression (5.78) shows that the spin field $S_\alpha(z)$ cannot be written as a local function of the fermions ψ^μ. At this point it is useful to bosonize the pair of Majorana fermions ψ and ζ into a bosonic massless field h following the usual procedure

$$\psi \pm i\zeta = \sqrt{2}e^{\mp ih} \quad , \tag{5.80}$$

where we have the propagator

$$\langle h(z)h(w)\rangle = -\ln(z-w) \quad , \tag{5.81a}$$

for the bosonized field h together with the superselection rule

$$\left\langle \sum_i e^{q_i h(z_i)} \right\rangle \neq 0 \quad \text{iff} \quad \sum_i q_i = 0 \quad . \tag{5.81b}$$

After the bosonization procedure, we can write the spin field $S_\alpha(z)$ in terms of $h(z)$, in such a way that (5.78) and (5.79) are satisfied, that is

$$S(z) = e^{\frac{i}{2}\epsilon h(z)} \quad , \tag{5.82}$$

where $\epsilon = \pm 1$ corresponds to the two spin components of the two-dimensional spinor field S_α, that is up($\uparrow$) and down ($\downarrow$). Notice that $\Delta(e^{ikh}) = k^2/2$, therefore $\Delta(S) = 1/8$, which is in agreement with (5.79). Furthermore we have to pay attention to the supersymmetry ghosts β, γ and to the relation (5.78), which tells us that the spin field $S_\alpha(w)$ introduces a cut at w which changes the boundary condition of the fermion field $\psi^\mu(z)$ (notice the factor $(z-w)^{-1/2}$ in (5.78)). The supersymmetry transformations of ψ^μ involve the ghost γ,

$$\delta_{susy}\psi^\mu = \lambda\gamma \, \partial\!\!\!/ X^\mu + \cdots \quad , \tag{5.83}$$

where $X^0 = i\phi$, $X^1 = X$ and λ is a Grassmannian constant such that $\lambda\gamma$ corresponds to the fermionic parameter $\theta = \lambda\gamma$ of the supersymmetry transformation. It is clear that due to supersymmetry[146,149] V_R must contain, besides the spin field S_α, also a ghost spinor field, which one calls Σ, changing also the boundary conditions of the ghost γ by introducing a cut. Once again bosonization can help us to have an explicit expression for Σ. Bosonizing the ghosts β and γ, we have to introduce, beside a massless bosonic field which is called $\sigma(z)$, a pair of Grassmannian fields $\eta(z)$ and $\xi(z)$, in order to ensure the correct statistics, once γ and β are bosonic. We have, explicitly

$$\gamma = e^{\sigma}\eta \quad ; \qquad \beta = e^{-\sigma}\partial\xi \quad . \tag{5.84}$$

The usual propagator $\langle\gamma(z)\beta(w)\rangle = (z-w)^{-1}$ can be recovered from

$$\begin{aligned} \langle \eta(z)\xi(w)\rangle &= (z-w)^{-1} \quad , \\ \langle \sigma(z)\sigma(w)\rangle &= -\ln(z-w) \quad . \end{aligned} \tag{5.85}$$

The energy momentum tensor $T_{\beta\gamma}$ given in (5.5) is replaced now by

$$T_{\beta\gamma} = T_\sigma + T_{\eta\xi} = -\frac{1}{2} :(\partial\sigma)^2: + \partial^2\sigma + :\partial\xi\eta: \quad . \tag{5.86}$$

As in the case of S_α we have now two solutions for Σ given by $\Sigma_{\pm 1/2} = e^{\pm\sigma/2}$. These two solutions have different conformal weights. From the coefficients of the operator product expansion of $e^{a\sigma}$ and the energy momentum tensor above, we have the conformal weight

$$\Delta\left(e^{a\sigma}\right) = -\frac{1}{2}a(a+2) \quad , \tag{5.87a}$$

where the 2 can be traced back to the ghost background charge. Therefore,

$$\Delta\left(e^{-\frac{1}{2}\sigma}\right) = \frac{3}{8} \quad , \tag{5.87b}$$

$$\Delta\left(e^{\frac{1}{2}\sigma}\right) = -\frac{5}{8} \quad . \tag{5.87c}$$

In order to choose the correct solution we have to check the conformal weight of the whole vertex V_R, which can be written as

$$V_R(k,\epsilon) = \int d^2 z e^{\frac{i}{2}\epsilon h(z)+ikX(z)+\beta\phi(z)}\Sigma(z) \quad . \tag{5.88}$$

Requiring conformal invariance we have to fix the conformal weight of the integrand to be unit, implying

$$\Delta\left(e^{\frac{i}{2}\epsilon h(z)+ikX(z)+\beta\phi(z)}\Sigma(z)\right) = \frac{1}{8} + \frac{1}{2}k(k-2\alpha_0) - \frac{1}{2}\beta(\beta+Q) + \Delta(\Sigma) = 1\, . \tag{5.89}$$

The vertex $V(k,\epsilon)$ should represent a massless particle, implying the on shell condition $E^2 - p^2 = 0$, which is equivalent to

$$\Delta\left(e^{+ikX(z)+\beta\phi(z)}\right) = \frac{1}{2}k(k-2\alpha_0) - \frac{1}{2}\beta(\beta+Q) = \frac{1}{2} \quad , \tag{5.90}$$

where we have used $E = \beta + \frac{Q}{2}$, and $p = k - \alpha_0$. Substituting (5.90) in (5.89) we end up with $\Delta(\Sigma) = 3/8$, which selects the solution $\Sigma = \Sigma_{-1/2} = e^{-\sigma/2}$, and therefore our spinor vertex can be written as

$$V_R = V_{-\frac{1}{2}}(k,\epsilon) = \int d^2 z e^{-\frac{1}{2}\sigma(z)+\frac{i}{2}\epsilon h(z)+ikX(z)+\beta\phi(z)} \quad , \tag{5.91}$$

with β satisfying (5.90). The fact that β satisfies (5.90) does not guarantee a priori that $V_{-1/2}$ is BRST invariant. This will be checked now. The whole BRST charge is given by[146]

$$Q_{BRST} = Q_0 + Q_1 + Q_2 \quad , \tag{5.92}$$

where Q_0 is the usual bosonic non-critical string charge[3)]

$$Q_0 = \oint \frac{dw}{2\pi i} c(T_{SL} + T_M + \frac{1}{2}T_{gh}) \quad . \tag{5.93}$$

Using (5.90) and the appropriate propagators one checks that $\left[Q_0, V_{-1/2}\right] = 0$, which is nothing but reparametrization invariance. The charge Q_1 is associated with supersymmetry,

$$Q_1 = \oint \frac{dw}{2\pi i}\eta e^{\sigma}\left[\psi\partial\phi + \zeta\partial X + Q\partial\psi - 2i\alpha_0\partial\zeta\right] \quad . \tag{5.94}$$

[3)] The factor 1/2 in front of T_{gh} is needed in order that the ghosts have the correct BRST transformation laws.

Since $V_{-1/2}$ is not explicitly supersymmetric we have to check supersymmetry in detail. We have for the holomorphic part of $V_{-1/2}$

$$\left[Q_1, V_{-\frac{1}{2}}\right] = \int \mathrm{d}z \oint \frac{\mathrm{d}w}{2\pi i}\eta(w)\mathrm{e}^{\sigma(w)}\mathrm{e}^{-\frac{1}{2}\sigma(z)}\times$$
$$\times\left[\psi\partial\phi + \zeta\partial X + Q\partial\psi - 2i\alpha_0\partial\zeta\right](w)\mathrm{e}^{\frac{i}{2}\epsilon h(z)+izX(z)+\beta\phi(z)} \quad .(5.95)$$

Using the bosonization formulas (5.80), we rewrite (5.95) as

$$\left[Q_1, V_{-\frac{1}{2}}\right] = \int \mathrm{d}z \oint \frac{\mathrm{d}w}{2\pi i}\left[(w-z)^{-1}\mathrm{e}^{\frac{1}{2}\sigma-\frac{i}{2}\epsilon h}\left(-\frac{Q}{2}+\alpha_0\epsilon-\beta-k\epsilon\right)+\text{finite}\right] \quad . \quad (5.96)$$

Therefore imposing that the commutator above vanishes we get the dispersion relation

$$E = \beta + \frac{Q}{2} = -\epsilon(k-\alpha_0) \quad . \quad (5.97)$$

Henceforth we assume that the dressing β of $V_{-1/2}$ in (5.88) satisfies (5.97), which assures supersymmetry and consequently BRST invariance of $V_{-1/2}$ since it is easy to check that $[Q_2, V_{-1/2}] = 0$ where Q_2 is needed for nilpotency of the BRST charge, being given by

$$Q_2 = \oint \frac{\mathrm{d}w}{2\pi i} b\gamma^2 \quad . \quad (5.98)$$

Collecting this information we arrive at

$$[Q_{BRST}, V_{-1/2}] = 0 \quad . \quad (5.99)$$

If we assume that $E > 0$, equation (5.97) shows that right moving ($k \geq \alpha_0$) and left-moving ($k \leq \alpha_0$) particles have the same helicity as usually, as a consequence of the Dirac equation in two-dimensions. Notice that $\beta(k)$ given in (5.97) satisfies also (5.90) (compare (5.97) with (2.50) as required by self consistency.

At this point we are able to compute correlation functions involving the spinor emission vertex. Before starting an important remark is in order. Notice that T_σ given in (5.86) is derived from the action

$$S_\sigma = \frac{1}{8\pi}\int \mathrm{d}^2 z\sqrt{\hat{g}}\left[\partial\sigma\overline{\partial}\sigma - \hat{R}\sigma\right] \quad , \quad (5.100)$$

therefore correlators involving the massless field $\sigma(z)$ satisfy the super selection rule determined by the zero mode of $\sigma(z)$,

$$\left\langle \sum_i \mathrm{e}^{q_i\sigma(z_i)}\right\rangle \neq 0 \quad \text{iff} \quad \sum q_i = -2 \quad . \quad (5.101)$$

This rule means that the only correlator involving only R-vertices which can be calculated by use of $V_{-1/2}$ is the four point function $\langle V_{-1/2}V_{-1/2}V_{-1/2}V_{-1/2}\rangle$. For the computation of higher amplitudes we need $V_{+1/2}$ which is another version of the R-vertex obtained from $V_{-1/2}$ via a *picture change* operation as follows. According to [146] any spinor vertex with σ-ghost charge $+1/2$ can only be found in the form $V_{+1/2} = [Q_{BRST}, V']$, where V' has σ-ghost charge $-1/2$. From the BRST point of view, $V_{+1/2}$ in this form is a trivial operator since when applied on physical states ($Q_{BRST}|phys\rangle = 0$) produces states orthogonal to any physical state (null state), i.e., $V_{+1/2}|phys\rangle = Q_{BRST}V'|phys\rangle \equiv 0$. We can get rid of this problem by going to the so-called "big algebra", by introducing the zero mode of the ghost $\xi(z)$ which was absent of our Hilbert space until now (see (5.84)). After such introduction one can show that[146] $Q_{BRST}\xi|0\rangle \neq 0$, and consequently $Q_{BRST}\xi|phys\rangle \neq 0$. This suggests to define $V_{+1/2} = \left[Q_{BRST}, \xi V_{-1/2}\right]$. Since ξ has zero conformal weight ($[Q_0, \xi] = 0 = [Q_0, V_{-1/2}]$) and satisfies $[Q_2, \xi] = 0 = [Q_2, V_{-1/2}]$ we have

$$\begin{aligned} V_{+1/2} &= \left[Q_{BRST}, \xi V_{-1/2}\right] = \left[Q_1, \xi V_{-1/2}\right] \\ &= \int \mathrm{d}z \oint \frac{\mathrm{d}w}{2\pi i}(w-z)^{-1} \mathrm{e}^{\frac{1}{2}\sigma}\Big[\mathrm{e}^{-\frac{i}{2}\epsilon h}\left(\partial\phi(w) - i\epsilon\partial X(w)\right) + \\ &+ \mathrm{e}^{\frac{3}{2}i\epsilon h}\left(\frac{Q}{2} + \alpha_0\epsilon - \beta + k\epsilon\right) + (w-z)^{-1}\mathrm{e}^{\frac{1}{2}i\epsilon h}\left(-\frac{Q}{2} + \alpha_0\epsilon - \beta - k\epsilon\right)\Big]\mathrm{e}^{ikX+\beta\phi} . \end{aligned} \tag{5.102}$$

Using (5.97) we get for the holomorphic part of $V_{+1/2}$

$$V_{+1/2}(k,\epsilon) = \int \mathrm{d}z \mathrm{e}^{\frac{1}{2}\sigma + ikX + \beta\phi}\left[\mathrm{e}^{\frac{3}{2}i\epsilon h}(Q + 2\epsilon k) + \mathrm{e}^{-\frac{1}{2}i\epsilon h}(\partial\phi - i\epsilon\partial X)\right] \quad . \tag{5.103}$$

An analogous formula holds for its anti-holomorphic part. Since $\xi(z)$ no longer appears in (5.102) it is clear that we can come back to the "small algebra" $(\eta, \partial\xi)$ and use (5.103) to calculate higher correlators, as the six-point amplitude

$$\langle V_{-1/2}V_{-1/2}V_{-1/2}V_{-1/2}V_{-1/2}V_{+1/2}\rangle \quad .$$

The BRST invariance of such correlators relies on the invariance of the small algebra vacuum ($Q_{BRST}|0\rangle = 0$) and BRST invariance[4)] of the vertices $\left[Q_{BRST}, V_{\pm 1/2}\right] = 0$.

At last, we leave for the reader to check the OPE $\langle V_{-1/2}V_{-1/2}\rangle \sim \psi_{NS}^{(-1)}$ where $\psi_{NS}^{(-1)}$ is another version of the NS-vertex which will be needed in the next section and is given by

$$\psi_{NS}^{(-1)} = \int \mathrm{d}^2 z \mathrm{e}^{-\sigma(z) + ikX(z) + \beta\phi(z)} \quad . \tag{5.104}$$

The above vertex is BRST invariant and can be "picture changed" into ψ_{NS}

$$\psi_{NS} = \left[Q_{BRST}, \xi\psi_{NS}^{(-1)}\right] \quad . \tag{5.105}$$

---

4) The BRST invariance of $V_{+1/2}$ can be proved by the Jacobi identity

5.6. Correlators Involving the Spinor Emission Vertex

After the preliminary vertex definitions of the last section we are now ready to compute correlators involving the spinor emission vertex. In general we have a mixed correlator

$$\mathcal{A}_N^{(n,N-n)}(k_1 \cdots k_N, \epsilon_1 \cdots \epsilon_n) = \left\langle \prod_{i=1}^{n} V_{-\frac{1}{2}}(k_i, \epsilon_i) \prod_{j=n+1}^{N} \psi_{NS}(k_j) \right\rangle_\mu \quad . \tag{5.106a}$$

After integrating over the matter (X_0) and Liouville (ϕ_0) bosonic zero modes we get

$$\mathcal{A}_N^{(n,N-n)} = \Gamma(-s)(i\mu)^s \left\langle \prod_{i=1}^{n} V_{-\frac{1}{2}}(k_i, \epsilon_i) \prod_{j=n+1}^{N} \psi_{NS}(k_j) \left(\int \mathrm{d}^2\theta \mathrm{d}^2 z e^{\alpha_+ \Phi} \right)^s \right\rangle_{\mu=0} , \tag{5.106b}$$

where the momentum conservation law and the definition of s are, respectively,

$$\sum_{i=1}^{N} k_i = 2\alpha_0 \quad , \tag{5.107a}$$

$$\sum_{i=1}^{N} \beta_i = -Q - \alpha_+ s \quad . \tag{5.107b}$$

We should remark that the expressions (5.106) indicate that the inclusion of the $V_{-1/2}$ vertices does not alter the scaling behavior of the partition function, that is $\mathcal{Z} \sim \mu^{-\frac{Q}{\alpha_+}}$, nor the scaling of the NS vertices $\sim \mu^{-\frac{\beta}{\alpha_+}}$. The vertices $V_{-1/2}(k,\epsilon)$ scale as $\mu^{-\frac{\beta}{\alpha_+}}$ with $\beta(k,\epsilon)$ given in (5.97), i.e.,

$$V_{-\frac{1}{2}}(k,\epsilon) \sim \mu^{-\frac{\beta(k,\epsilon)}{\alpha_+}} = \mu^{\frac{1}{\alpha_+}\left(\frac{Q}{2} + \epsilon(k-\alpha_0)\right)} \quad . \tag{5.108}$$

As in the case of the NS vertices, the strategy is to assume that s is a positive integer and to analytically continue the result for arbitrary real s. For integer s, it is enough to use free propagators to write (5.106b) in an integral representation. In the case of no spinor emission vertex ($n = 0$) these amplitudes have been calculated in section 5.4. for arbitrary values of N by successive derivatives of the three-point function with respect to the cosmological constant μ; the relevant integral for the $N = 3$ case is calculated in Appendix C. In the case where we include the spinor emission vertex $V_{-1/2}$ we have not been able to perform the integrals corresponding to the three point function[5)] $\langle V_{-1/2} V_{-1/2} \psi_{NS}^{(-1)} \rangle$ for non-vanishing integer s and therefore we cannot use the same strategy as before to obtain higher point amplitudes from the 3-point amplitude. Hence we restrict ourselves in this section to the so called bulk amplitudes ($s = 0$) where we take the finite part of

[5)] The only non-vanishing correlators are those with a net -2 charge for the σ-ghost which cancels the background +2 charge as we have seen in the previous section (see (5.101)).

$\Gamma(-s)\mu^s$ when $s \to 0$, namely the coefficient of $\ln\mu$. The simplest non-trivial bulk amplitude to be calculated (see ref. [75]) is the two spinor-two scalar scattering amplitude,

$$\mathcal{A}_4^{(2,2)} = \ln\mu \left\langle V_{-1/2}(k_1)V_{-1/2}(k_2)\psi_{NS}^{(-1)}(k_3)\psi_{NS}(k_4)\right\rangle \quad . \tag{5.109a}$$

It is convenient at this point to rewrite $\psi_{NS}(k_4)$ from (5.19) in the bosonized form, using (5.80), to obtain

$$\psi_{NS}(k_4) = \frac{1}{2}\int d^2z_4 e^{ik_4x+\beta_4\phi}\left(e^{ih(z_4)}(\beta_4-k_4)+e^{-ih(z_4)}(\beta_4+k_4)\right)\times \left(e^{ih(\bar z_4)}(\beta_4-k_4)+e^{-ih(\bar z_4)}(\beta_4+k_4)\right) \quad . \tag{5.109b}$$

From the selection rule derived from the integration over the zero mode associated with the massless field h (see (5.81)) we must have $\epsilon_1 = \epsilon_2$ in order to have $\mathcal{A}_4^{(2,2)} \neq 0$; i.e., the spinors must have the same spin polarization. We choose for instance $\epsilon_1 = \epsilon_2 = +1$. In this case if we follow Seiberg[21] and work only with positive energy particles ($E > 0$) we use (5.97) to obtain $k_1, k_2 \geq \alpha_0$. The remaining two momenta k_3 and k_4 are chosen such that $k_3 \geq \alpha_0$ and $k_4 \leq \alpha_0$, thus after fixing the residual[6)] $SL(2,\Re)$ invariance with $z_1 = 0, z_2 = 1, z_3 = \infty$ and calling $z_4 = z$ we have

$$\begin{aligned}\mathcal{A}_4^{(2,2)} &= \frac{(\beta_4+k_4)^2\ln\mu}{2}\int d^2z|z|^{2\theta_{14}-1}|1-z|^{2\theta_{24}-1}\\ &= \frac{\pi(\beta_4+k_4)^2\ln\mu}{2}\Delta(\theta_{14}+\frac{1}{2})\Delta(\theta_{24}+\frac{1}{2})\Delta(-\theta_{14}-\theta_{24}) \quad ,\end{aligned} \tag{5.110}$$

where $\theta_{ij} = k_ik_j - \beta_i\beta_j$ and $\Delta(x) = \Gamma(x)/\Gamma(1-x)$. Using (5.107) with $s = 0$ and the corresponding dispersion relations for the dressings $\beta(k)$, we can rewrite $\mathcal{A}_4^{(2,2)}$ in the given kinematic region as

$$\mathcal{A}_4^{(2,2)} = \frac{\pi\alpha_-^2\ln\mu}{2}\Delta(m_1)\Delta(m_2)\Delta(m_3+1/2) \quad . \tag{5.111}$$

where $m_i = \frac{1}{2}(\beta_i^2 - k_i^2)$.

This result was obtained in [75]. For $\hat c < 1$ the factors $\Delta(m_i)$ and $\Delta(m_i + \frac{1}{2})$ never diverge or vanish, thus we can renormalize the vertices according to

$$\begin{aligned}\Psi_{NS}(k) &= \frac{\widetilde\Psi_{NS}(k)}{\Delta(m+1/2)} \quad ,\\ V_{-\frac{1}{2}}(k,\epsilon) &= \frac{\widetilde V_{-\frac{1}{2}}(k,\epsilon)}{\Delta(m)} \quad .\end{aligned} \tag{5.112}$$

---

6) Notice that there is no fermionic residual symmetry anymore, since $\psi_{NS}^{(-1)}$ already corresponds to a $OSP(1,1)$ gauge fixed version of ψ_{NS}.

So we have the renormalized bulk amplitude (dropping the factor $\pi\alpha_-^2/2$)

$$\mathcal{A}_4^{(2,2)} = \ln\mu \quad . \tag{5.113}$$

In the case $\hat{c} = 1\,(\alpha_0 \to 0)$ we are not allowed to renormalize the vertices as in (5.112) since we have[7]

$$\mathcal{A}_4^{(2,2)} = \ln\mu \frac{\Gamma\left(\frac{1}{2} - |k_1|\right)\Gamma\left(\frac{1}{2} - |k_2|\right)\Gamma\left(1 - |k_1|\right)}{\Gamma\left(\frac{1}{2} + |k_1|\right)\Gamma\left(\frac{1}{2} + |k_2|\right)\Gamma\left(|k_3|\right)} \quad . \tag{5.114}$$

The above result shows that besides the discrete states in the NS-sector which appear for integer values of the momentum ($|k_3| = 1, 2, 3, \cdots$) we have also discrete states in the R-sector for half-integer values of the momenta ($|k_i| = \frac{1}{2}, \frac{3}{2}, \cdots, i = 1, 2$).

The next simplest amplitude to calculate is $\left\langle V_{-\frac{1}{2}}^{\epsilon_1} V_{-\frac{1}{2}}^{\epsilon_2} V_{-\frac{1}{2}}^{\epsilon_3} V_{-\frac{1}{2}}^{\epsilon_4} \right\rangle_{s=0}$ which cannot be calculated directly since the corresponding integral does not converge in the kinematic region $k_1, k_2 \geq \alpha_0$ $k_3, k_1 \leq \alpha_0$ which is required by selection rule imposed by the zero mode of h. Therefore, we treat instead the case of the 4- spinor, 1-scalar scattering ($\mathcal{A}_5^{(4,1)}$),

$$\mathcal{A}_5^{(4,1)} = \left\langle V_{-\frac{1}{2}}^{\epsilon_1}(k_1) \cdots V_{-\frac{1}{2}}^{\epsilon_4}(k_4) \psi_{NS}(k_5) \right\rangle \quad . \tag{5.115}$$

It can also be seen as corresponding to a four Ramond vertices amplitude with $s = 1$, for a suitable choice of k_5, in such a way that the result is still general. Due to the h zero mode selection rule we need three spinors with the same polarization and the last one with opposite polarization to those three; thus we choose $\epsilon_1 = +1 = -\epsilon_2 = -\epsilon_3 = -\epsilon_4$, and the kinematic region $k_i \geq \alpha_0$, ($i = 2, 3, 4, 5$), $k_1 \leq \alpha_0$. Fixing the gauge $z_2 = 1$, $z_4 = \infty$, $z_1 = 0$ and defining $z_5 = w$, $z_3 = z$ we have:

$$\begin{aligned}\mathcal{A}_5^{(4,1)} &= \frac{\alpha_+^2 \ln\mu}{2} \int \mathrm{d}^2 z \mathrm{d}^2 w |z|^{2(2m_3 - 1)} | \, |w|^{2(2m_5)} \times \\ &\times |1 - z|^{2(-m_2 - m_3)} |1 - w|^{2(-m_2 - m_5 - 1/2)} |z - w|^{2(-m_3 - m_5 - 1/2)} .\end{aligned} \tag{5.116}$$

In general, we do not know how to calculate such integrals, but in this specific case we find that after a convenient shift in m_5 ($m_5 \to m_5 - 1/2$), the integral above can be cast into the form of the bulk five-point correlator of scalars in the bosonic non-critical string and such an integral has been indirectly calculated in [74]–see a brief discussion after (5.124). We leave to the reader to verify that, using such a result we arrive, up to an overall constant at

$$\mathcal{A}_5^{(4,1)} = \ln\mu \Delta(m_5 + \frac{1}{2}) \prod_{i=1}^{4} \Delta(m_i) \quad . \tag{5.117}$$

[7] The reader should notice that, for $\hat{c} = 1$ in the infra-red limit $k \to 0$, the factor $\Delta(m(k))$ associated with the spinor particle is finite ($\Delta(m(k = 0)) = 1$) while the scalar factor $\Delta(m(k) + 1/2)$ vanishes indicating a divergent behavior which has been regularized to zero. The milder infra-red behavior of the massless spinor in comparison to the massless scalar particle is typical in local field theory.

The above result gives formula (5.113) for $\hat{c} < 1$ after vertices renormalization and provides an analogous expression to (5.114) for $\hat{c} = 1$, thus confirming the existence of discrete states in the R-sector for $\hat{c} = 1$ and the universality of the vertex renormalizations (5.112).

Now we come to the main computation of this section, namely the two spinors-three scalars scattering amplitude:

$$\mathcal{A}_5^{(2,3)} = \left\langle V^{\epsilon_1}_{-\frac{1}{2}} V^{\epsilon_2}_{-\frac{1}{2}} \psi_{NS}(k_3)^{(-1)}\psi_{NS}(k_4)\psi_{NS}(k_5)\right\rangle \quad . \tag{5.118}$$

Due to the h zero mode, the spinors must have opposite helicity, and we choose $\epsilon_1 = +1 = -\epsilon_2$, and the kinematic region $k_1 \le \alpha_0$, $k_i \ge \alpha_0$, $i = 2,3,4,5$. Fixing the gauge $z_1 = 0$, $z_2 = \infty$ and $z_3 = 1$, and defining $z_4 = z$, $z_5 = w$, it is easy to arrive at the following integral representation

$$\mathcal{A}_5^{(2,3)} = \int d^2w \int d^2z |z|^{2(2m_4-\frac{1}{2})}|w|^{2(2m_5-\frac{1}{2})}|1-z|^{2(-m_3-m_4)}|1-w|^{2(-m_3-m_5)} \times \vartheta \tag{5.119}$$

where the fermionic correlator (ϑ) is given by the expression

$$\begin{aligned}\vartheta &= \frac{1}{4}\left[\left\langle e^{\frac{i}{2}h(0)} e^{-\frac{i}{2}h(\infty)} \left(\alpha_+ e^{ih(z)} + 2\alpha_- m_4 e^{-ih(z)}\right)\left(\alpha_+ e^{ih(w)} + 2\alpha_- m_5 e^{-ih(w)}\right)\right\rangle \times \langle h.c.\rangle\right] \\ &= \frac{1}{2}\left[m_5(z-w)^{-1}z^{\frac{1}{2}}w^{-\frac{1}{2}} + m_4(z-w)^{-1}z^{-\frac{1}{2}}w^{\frac{1}{2}}\right] \times \langle h.c.\rangle \quad .\end{aligned} \tag{5.120}$$

Therefore we have,

$$\begin{aligned}\mathcal{A}_5^{(2,3)} &= \{[m_4(m_4+m_5)I_1(m_3,m_4,m_5) + \\ &\quad + m_5(m_4+m_5)I_1(m_3,m_5,m_4)] - m_4 m_5 I_2\} \quad ,\end{aligned} \tag{5.121}$$

where the first integral is given by the expression

$$I_1 = \int d^2z d^2w |z|^{2(2m_4-1)}|w|^{2(2m_5)}|1-z|^{2(-m_3-m_4)}|1-w|^{2(-m_3-m_5)}|z-w|^{2(-m_4-m_5-1)} \tag{5.122}$$

while the second one has already appeared in bosonic computations[74,75]:

$$I_2 = \int d^2z d^2w |z|^{2(2m_4-1)}|w|^{2(2m_5-1)}|1-z|^{2(-m_3-m_4)}|1-w|^{2(-m_3-m_5)}|z-w|^{2(-m_4-m_5)} \tag{5.123}$$

In general the amplitude $\mathcal{A}_5^{2,3}$ is a combination of two unknown integrals. However it is very fortunate that in the case where we place one of spinor vertices at the infinity we end up with the two integrals above where I_2 can be calculated (indirectly); Indeed, it has simple poles in m_4. We choose $m_4 = \epsilon$ and compute the residuum using $|z|^{2(2m_4-1)} \sim \frac{\pi}{4\epsilon}\delta^{(2)}(z)$ obtaining

$$\begin{aligned}I_2^{m_4=\epsilon} &= \frac{\pi}{4\epsilon}\int d^2w |w|^{2(m_5-1)}|1-w|^{2(-m_3-m_5)} \\ &= \frac{\pi^2}{\epsilon}\Delta(m_3)\Delta(m_5)\Delta(1-m_3-m_5) \quad .\end{aligned}$$

Since the integral is symmetric under interchange of m_4 and m_5, we arrive at the result:

$$\begin{aligned}I_2 &= \pi^2\Delta(m_3)\Delta(m_4)\Delta(m_5)\Delta(1-m_3-m_4-m_5)\\ &= \pi^2\Delta(m_3)\Delta(m_4)\Delta(m_5)\Delta(1/2+m_2) \quad ,\end{aligned} \tag{5.124}$$

where we have used the kinematic relation $m_2+m_3+m_4+m_5 = 1/2$. At this point it is quite surprising to have the result (5.124) contributing to $\mathcal{A}_5^{(2,3)}$ since the role of the factors $\Delta(m)$ and $\Delta(m+1/2)$ seem to be interchanged which might point to some non-universality of the vertices redefinition (5.112). We will convince the reader by calculating I_1 that this is actually not the case and the misplaced poles of I_2 are cancelled by corresponding poles of I_1. In order to obtain I_1 we follow a by now standard procedure (see appendix C). First of all it is convenient to rewrite I_1 using a change of variables after which we have:

$$\begin{aligned}I_1(m_3,m_4,m_5) = &\int d^2z d^2w \times\\ &\times|z|^{2(2m_3-1)}|w|^{2(-m_3-m_5)}|1-z|^{2(2m_4-1)}|1-w|^{2(-m_4-m_5-1)}|z-w|^{2(2m_5)} .\end{aligned} \tag{5.125}$$

Using a technique developed by Dotsenko and Fatteev[30] it is easy to derive from (5.122) and (5.125) the asymptotic behaviour:

$$\begin{aligned}&I_1(m_3,m_4\to\infty,m_5) \sim m_4^{-2-2(m_3+m_5)}\\ &I_1(m_3,m_4,m_5\to\infty) \sim m_5^{-2(m_3+m_4)}\\ &I_1(m_3\to\infty,m_4,m_5) \sim m_3^{-2(m_4+m_5)} \quad .\end{aligned} \tag{5.126}$$

Now we use the property that at the poles of intermediate states $\mathcal{A}_5$ factorizes in four point functions, for which we can use the Virasoro-Shapiro formula; in other words by using $|z|^{-2+\epsilon} = \frac{\pi}{\epsilon}\delta^{(2)}(z)$ we get from (5.122) and (5.125) the following constraints on I_1

$$\begin{aligned}I_1(m_4=\epsilon) &= \frac{\pi^2}{\epsilon}\Delta(m_5)\Delta(m_3)\Delta(1-m_3-m_5) \quad ,\\ I_1(m_3=\epsilon) &= \frac{\pi^2 m_5^2}{\epsilon(m_4+m_5)^2}\Delta(m_4)\Delta(m_5)\Delta(1-m_4-m_5) \quad ,\\ I_1(m_5=-\frac{1}{2}+\epsilon) &= \frac{-\pi^2}{\epsilon(m_4-1/2)^2}\Delta(1-m_3-m_4)\Delta(m_3+1/2)\Delta(m_4+1/2) \quad .\end{aligned} \tag{5.127}$$

Furthermore for $m_5 = 0$ the two integrals over z and w decouple, and we have (up to possible factors 1/2)

$$I_1(m_5=0) = \frac{-\pi^2}{m_4^2}\Delta(1/2-m_3-m_4)\Delta(m_3+1/2)\Delta(m_4+1/2) \quad . \tag{5.128}$$

Finally I_1 is also calculable at $m_4+m_5=\epsilon$ where it contains a double pole:

$$I_1(m_4+m_5=\epsilon) = -\frac{1}{\epsilon^2} \quad . \tag{5.129}$$

By taking formulae (5.126-129) and the result (5.124) for I_2 the reader can check that the poles at $m_3, m_4, m_5 \sim \epsilon$ and the double pole at $m_4+m_5 = \epsilon$ are cancelled and do not appear in the final expression for $\mathcal{A}_5^{(2,3)}$, only the pole at $m_5 = -1/2+\epsilon$ survives which is in agreement with our expectations. The reader may object at this point that we have only checked the calculation of the first poles of the gamma functions of I_2. However, using the factorization properties, i.e., $|z|^{-6+\epsilon} \sim \frac{\pi}{\epsilon}(\partial_z\partial_{\bar{z}})^2\delta^{(2)}(z)$ we have also obtained after a long algebra,

$$I_1(m_3 = -1+\epsilon) \approx -\frac{m_5^2}{6}(m_4+m_5-1)^2\Delta(m_4)\Delta(m_5)\Delta(-m_4-m_5) \quad . \tag{5.130}$$

Introducing the result above as well as (5.124) in (5.121) one checks that the first excited pole at $m_3 = -1$ of I_2 does not appear in $\mathcal{A}_5$ and this is a very striking cancellation which leads us to suggest the following result for I_1,

$$\begin{aligned} I_1 =& \frac{\pi^2 m_5}{(m_4+m_5)^2}\Delta(m_3)\Delta(m_4)\Delta(m_5)\Delta(1-m_3-m_4-m_5)+ \\ &- \frac{\pi^2}{(m_4+m_5)^2}\Delta(m_3+\frac{1}{2})\Delta(m_4+\frac{1}{2})\Delta(m_5+\frac{1}{2})\Delta(\frac{1}{2}-m_3-m_4-m_5) \quad . \end{aligned} \tag{5.131}$$

The previous result is in agreement with all[8)] formulae we have derived so far from I_1 including the asymptotic behaviors (5.126) (one uses Stirling formula to check it). Using (5.124) and (5.131) in (5.121) we have as expected[150]

$$\mathcal{A}_5^{(2,3)} = -\pi^2\Delta(m_2)\Delta(m_3+1/2)\Delta(m_4+1/2)\Delta(m_5+1/2) \quad . \tag{5.132}$$

From the result (5.132) we are reasonably safe to conclude that the correlation functions factorize, and each bosonic external leg contributes a factor $\Delta(m+\frac{1}{2})$, while a Ramond vertex gives a contribution $\Delta(m)$.

The factorizable result is actually highly non-trivial. At this moment, it is not clear at all how these results should be obtained from the recent super eigenvalue model techniques[84], which we present in Chapter 6, in such a way that a non-trivial check is also required. In some cases, the relevant spinor fields seem to contribute with a zero factor to several correlators (those involving more than two spinors) as opposite to our non-vanishing result (5.117), which involves four spinors. Nevertheless, there is a possibility of matching the results, disentangling the analytical contributions to the fermionic correlators in a similar way as it was done in the bosonic matrix models[31].

An important issue concerning the comparison with the matrix models results is the definition of the matter sector of the Ramond vertex, since the gravitational and matter fermions merge into $e^{\frac{i}{2}h}$ and a comparison of matter contribution with the Kac table results turn out to be difficult. We recall that in general for a Ramond vertex of the type (5.91) we are led to the condition

$$\frac{3}{8}+\frac{1}{8}+\frac{1}{2}k(k-\alpha_0)-\frac{1}{2}\beta(\beta+Q)=1 \quad , \tag{5.133}$$

8) Formula (5.131) does not agree with $I_1(m_4+m_5=\epsilon)$ and $I_2(m_5=-1/2+\epsilon)$ up to factors 1/2 which can presumably be traced back to symmetry factors.

where the 3/8 comes from the ghost contribution, and 1/8 from $e^{\frac{i}{2}\epsilon h}$. Therefore, since the latter is neither pure matter or pure gravity, we have to disentangle their contribution and find the relation between the pure matter conformal dimension $\Delta_{Kac}(\Delta_K)$ and the dressing.

In fact the gravitational dressing gives a contribution $-\frac{1}{2}\beta(\beta+Q)+\frac{1}{16}$ as argued in [50], the first being the usual contribution, while the 1/16 is a consequence of the fact that the fermionic component in the Neveu Schwarz sector is expanded in half integer components and displays no zero mode, while in the Ramond sector it is expanded in integer modes, and the zero mode contributes 1/16. Put in another way, in order to maintain the form of the Virasoro algebra we have to shift, in the Ramond sector L_0 into $L_0 + \frac{1}{16}$, where 1/16 represents the ground state energy of a single fermion. Therefore we have, for the dressing of the matter conformal dimension Δ_K the equation

$$\Delta_K + \frac{3}{8} + \frac{1}{16} - \frac{1}{2}\beta(\beta+Q) = 1 \quad , \tag{5.134}$$

or

$$\Delta_K - \frac{1}{2}\beta(\beta+Q) = \frac{9}{16} \quad , \tag{5.135}$$

which is equivalent to say that

$$\Delta_K = \Delta + \frac{1}{16} \quad , \tag{5.136}$$

where $\Delta = \frac{1}{2}k(k-\alpha_0)$.

5.7. Conclusion

The results presented in this Chapter are rather impressive. It was possible to compute all correlators involving either the NS or the R vertices in closed form for any value the parameters. The assumptions used were relatively mild.

In the next Chapter we present similar results from a discrete formulation, and some of the results obtained here can be checked in detail. We conclude that the methods used are very powerfull, and they open the possibility to obtain a complete solution to the problem on the sphere.

6. N=1 Super Eigenvalue Model

6.1. Introduction

Due to well known difficulties in realizing supersymmetry on the lattice, it is not a surprise that the discrete formulation of $N = 1$ superconformal theories coupled to two-dimensional supergravity still lags behind its purely bosonic version. A model in terms of generalized matrices has not yet been found: as shown in ref. [151], the most natural candidates, the supermatrices, do not define a theory of non-critical superstrings. In another attempt, the authors in ref. [145] proposed an approach to the double scaling limit of two-dimensional supergravity coupled to superconformal matter using a supersymmetric generalization of the KP hierarchy, inspired by the central role that the latter plays in the bosonic model.

The most successful description so far has been the discrete model introduced in ref. [84], whose double scaling limit generates the spectrum of anomalous dimensions of the dressed $(2, 4m)$ minimal superconformal models, both in the Neveu-Schwarz and Ramond sectors. That model was built up as a supersymmetric extension of the effective model obtained from random Hermitian matrices.

This Chapter is dedicated to the study of such super eigenvalue model. We concentrate on the planar limit of the theory in order to draw comparisons with the results found in the continuum approach. Higher genus[152] and non-perturbative results[153,154] are discussed in Chapter 8.

6.1.1. Eigenvalue Formulation of the Hermitian 1-Matrix Model

The Hermitian 1-matrix model was defined in Chapter 3 by its partition function

$$Z_B = \int \mathcal{D}\phi \exp\left\{-\frac{N}{\Lambda}\mathrm{tr}\, V(\phi)\right\} \quad , \tag{6.1}$$

where ϕ is an $N \times N$ Hermitian matrix, displaying a set of N^2 independent real variables (i.e., N real diagonal, and $N(N-1)/2$ complex off-diagonal independent components), whose integration measure is assumed to be

$$\mathcal{D}\phi = \prod_{i=1}^{N} \mathcal{D}\phi_{ii} \prod_{i<j}^{N} \mathcal{D}(\Re e\phi_{ij})\mathcal{D}(\Im m\phi_{ij}) \quad , \tag{6.2}$$

and we take a general polynomial potential,

$$V(\phi) = \sum_{k\geq 0} g_k \phi^k \quad . \tag{6.3}$$

The label B in Z_B was introduced in order to distinguish the purely bosonic partition function from the supersymmetric one Z_S, to be defined later. Any Hermitian matrix ϕ can be diagonalized by some unitary matrix U, i.e.,

$$\phi = U^\dagger \Lambda U \quad , \tag{6.4}$$

where Λ is the diagonal matrix of the eigenvalues $\{\lambda_i\}$ of ϕ,

$$\Lambda = \mathrm{diag}(\lambda_1, \cdots, \lambda_N) \quad . \tag{6.5}$$

Notice, however, that the unitary matrix U is not uniquely determined by (6.4): we can fix N parameters (corresponding to the phases of the N eigenvectors) among the original N^2 components of the unitary matrix leading to $N(N-1)$ parameters (which we will call "angular variables") that added to the N eigenvalues provide an alternative set for the degrees of freedom of the model.

Since the interaction term $\mathrm{tr}\, V(\phi) = \sum_{i=1}^{n} V(\lambda_i)$ does not depend on the angular variables, it is interesting to integrate out those degrees of freedom and find an effective theory for the eigenvalues.

The first step is to calculate the Jacobian factor for the change of variables $(\phi) \to (\Lambda, U)$, defined as

$$J \equiv \frac{\partial(\phi_{ii}, \phi_{ij})}{\partial(\Lambda_i, U_{ij})} \quad . \tag{6.6}$$

As a first approximation, let us expand the unitary matrix U around the identity 1:

$$U \sim 1 + \epsilon \quad , \quad \epsilon^\dagger = -\epsilon \quad , \tag{6.7}$$

where the anti-Hermitian matrix ϵ is infinitesimal. We can take its $N(N-1)$ off-diagonal components, $\{\Re e(\epsilon_{ij}), \Im m(\epsilon_{ij}) \; ; \; i < j\}$, as angular variables, and consider ϵ_{ii} fixed.

Up to first order in ϵ we have

$$\begin{aligned} \phi &\sim (1-\epsilon)\Lambda(1+\epsilon) = \Lambda + [\Lambda, \epsilon] + \mathcal{O}(\epsilon^2) \quad , \\ \phi_{ij} &= \lambda_i \delta_{ij} + (\lambda_i - \lambda_j)\epsilon_{ij} + \mathcal{O}(\epsilon^2) \quad , \end{aligned} \tag{6.8}$$

therefore

$$\begin{cases} \phi_{ii} &= \lambda_i + \mathcal{O}(\epsilon^2) \quad , \\ \phi_{ij} &= (\lambda_i - \lambda_j)\epsilon_{ij} + \mathcal{O}(\epsilon^2) \quad , \quad i \neq j \end{cases} \tag{6.9}$$

from where we construct the Jacobian matrix

$$\begin{pmatrix} \frac{\partial \phi_{ii}}{\partial \lambda_a} & \frac{\partial(\Re e\phi_{ij})}{\partial \lambda_a} & \frac{\partial(\Im m\phi_{ij})}{\partial \lambda_a} \\ \frac{\partial \phi_{ii}}{\partial(\Re e\epsilon_{ab})} & \frac{\partial(\Re e\phi_{ij})}{\partial(\Re e\epsilon_{ab})} & \frac{\partial(\Im m\phi_{ij})}{\partial(\Re e\epsilon_{ab})} \\ \frac{\partial \phi_{ii}}{\partial(\Im m\epsilon_{ab})} & \frac{\partial(\Re e\phi_{ij})}{\partial(\Im m\epsilon_{ab})} & \frac{\partial(\Im m\phi_{ij})}{\partial(\Im m\epsilon_{ab})} \end{pmatrix} \quad , \tag{6.10}$$

which is diagonal at lowest order in ϵ, namely

$$\begin{pmatrix} 1 & 0 & 0 \\ 0 & (\lambda_i - \lambda_j) & 0 \\ 0 & 0 & (\lambda_i - \lambda_j) \end{pmatrix} \quad , \tag{6.11}$$

and whose determinant gives the leading term of the Jacobian

$$J = \prod_{i<j}(\lambda_i - \lambda_j)^2 + \mathcal{O}(\epsilon) \quad . \tag{6.12}$$

Comparing to the exact result[126],

$$J = f(U)\prod_{i<j}(\lambda_i - \lambda_j)^2 \quad , \tag{6.13}$$

where $f(U)$ is a volume factor depending on the angular variables, we see that the leading term in (6.12) has already the factorization property exhibited by (6.13). There are some different methods to demonstrate (6.13): in the book by M.L. Mehta[126] one introduces a set of generic angular variables $\{p_\mu\}$ encoded in the matrix U. The Jacobian is carefully constructed in terms of the matrices $U^{-1}\frac{\partial\phi}{\partial\lambda_i}U$ and $U^{-1}\frac{\partial\phi}{\partial p_\mu}U$ (the technique is applied to symmetric, quaternionic and Hermitian matrix ensembles). One can also apply the method of Diffusion Equation[126] to integrate matrix variables in multimatrix models; or derive (6.13) via the usual Fadeev-Popov method, where the reasons for the success of the approximation (6.12) become more transparent. We can use (6.13) to integrate (6.1) over the angular variables:

$$\begin{aligned} Z_B &= \int \mathcal{D}U f(U) \int \prod_{a=1}^{N} \mathrm{d}\lambda_a \prod_{i<j}^{N}(\lambda_i - \lambda_j)^2 \exp\left\{-\frac{N}{\Lambda}\sum_i V(\lambda_i)\right\} \\ &= C(N)\int \prod_{a=1}^{N} \mathrm{d}\lambda_a \mathrm{e}^{-\frac{N}{\Lambda}V(\lambda_a)} \prod_{i<j}^{N}(\lambda_i - \lambda_j)^2 \quad , \end{aligned} \tag{6.14}$$

where

$$C(N) \equiv \int \mathcal{D}U f(U) \quad , \tag{6.15}$$

although exactly calculable[126], is just a non-interesting multiplicative (N-dependent) constant which can be absorbed in the normalization of the partition function. Therefore we define the eigenvalue model simply by:

$$Z_B = \int \prod_{i=1}^{N} \mathrm{d}\lambda_i \mathrm{e}^{-\frac{N}{\Lambda}V(\lambda_i)} \Delta(\lambda) \quad , \tag{6.16}$$

where

$$\Delta(\lambda)^2 = \prod_{i<j}^{N}(\lambda_i - \lambda_j)^2 \tag{6.17}$$

is the square of the so called Vandermonde determinant, $\Delta(\lambda) = \prod_{i<j}(\lambda_i - \lambda_j)$.

The expansion (6.16) can be interpreted[70] as the partition function of a one-dimensional Coulomb gas at temperature Λ/N. The Vandermonde determinant corresponds to a logarithmic interaction, whose remarkable properties allow one to solve the statistical mechanical problem exactly. We shall discuss this solution in Chapter 8 using the powerful method of orthogonal polynomials.

6.1.2. Virasoro Constraints

In this section we rederive the Virasoro constraints, satisfied by the 1-matrix model partition function, using invariance properties of the effective eigenvalue model, thus preparing grounds for the construction of its supersymmetric extension.

Consider the following infinitesimal change of integration variables in (6.16),

$$\delta_n \lambda_i = \epsilon_n \lambda_i^{n+1} \quad , \quad n \geq -1 \quad , \quad i = 1, \cdots, N \tag{6.18}$$

under which the partition function is obviously invariant. This means that the contributions arising from the transformed measure, potential and Vandermonde terms must cancel each other.

The measure contributes with a Jacobian

$$\begin{aligned} j_n &= \left| \det \frac{\partial(\lambda_i + \delta_n \lambda_i)}{\partial \lambda_i} \right| = |\det(1 + (n+1)\epsilon_n \mathrm{diag}(\lambda_i^n))| \\ &= 1 + \epsilon_n (n+1) \sum_{i=1}^{N} \lambda_i^n \ , \end{aligned} \tag{6.19}$$

which we write concisely as

$$\delta_n \mathcal{D}\lambda = \mathcal{D}\lambda\, \epsilon_n (n+1) \sum_{i=1}^{N} \lambda_i^n \quad . \tag{6.20}$$

On the other hand, the potential and Vandermonde terms, being simply analytical functions of the eigenvalues, transform according to the action of the differential operators

$$l_n = \sum_{i=1}^{N} \lambda_i^{n+1} \frac{\partial}{\partial \lambda_i} \quad , \tag{6.21}$$

as suggested by (6.18). We find

$$\begin{aligned} \delta_n \left(\mathrm{e}^{-\frac{N}{\Lambda} \sum_i V(\lambda_i)} \right) &= \epsilon_n l_n \mathrm{e}^{-\frac{N}{\Lambda} \sum_i V(\lambda_i)} \\ &= \mathrm{e}^{-\frac{N}{\Lambda} \sum_i V(\lambda_i)} \left(-\frac{N}{\Lambda} \right) \epsilon_n \sum_{i=1}^{N} l_n V(\lambda_i) \\ &= \mathrm{e}^{-\frac{N}{\Lambda} \sum_i V(\lambda_i)} \left(-\frac{N}{\Lambda} \right) \epsilon_n \sum_{i=1}^{N} \sum_{k \geq 0} k g_k \lambda_i^{k+n} \quad , \end{aligned} \tag{6.22}$$

and

$$\begin{aligned} \delta_n \Delta(\lambda)^2 &= \epsilon_n l_n \prod_{i<j} (\lambda_i - \lambda_j)^2 \\ &= 2\epsilon_n \prod_{i<j} (\lambda_i - \lambda_j)^2 \sum_{k<l} \frac{(\lambda_k^{n+1} - \lambda_l^{n+1})}{\lambda_k - \lambda_l} \\ &= \Delta(\lambda)^2 \epsilon_n \sum_{k \neq l} \frac{(\lambda_k^{n+1} - \lambda_l^{n+1})}{\lambda_k - \lambda_l} \\ &= \Delta(\lambda)^2 \epsilon_n \sum_{k \neq l} \sum_{a=0}^{n} \lambda_k^a \lambda_l^{n-a} \quad . \end{aligned} \tag{6.23}$$

Summing up all contributions we get the following Ward identities

$$\int \mathcal{D}\lambda\, e^{-\frac{N}{\Lambda}\sum_i V(\lambda_i)} \Delta(\lambda)^2 \left\{ \sum_{k\geq 0} k g_k \left(-\frac{N}{\Lambda} \sum_{i=1}^{N} \lambda_i^{n+k} \right) + \right.$$
$$\left. +(n+1)\sum_{i=1}^{N} \lambda_i^n + \sum_{i\neq j} \sum_{a=0}^{n} \lambda_i^a \lambda_j^{n-a} \right\} = 0 \quad . \tag{6.24}$$

The last two terms merge as

$$(n+1)\sum_{i=1}^{N} \lambda_i^n + \sum_{a=0}^{n} \sum_{i\neq j}^{N} \lambda_i^a \lambda_j^{n-a} = \sum_{a=0}^{n} \left(\sum_{i=1}^{N} \lambda_i^a \right) \left(\sum_{j=1}^{N} \lambda_j^{n-a} \right) \quad . \tag{6.25}$$

We also notice the eminent presence of trace-like terms, $\sum_i \lambda_i^a$, which can be written in terms of the potential as

$$\sum_{i=1}^{N} \lambda_i^a = -\frac{\Lambda}{N} \frac{\partial}{\partial g_a} \left(-\frac{N}{\Lambda} \sum_i V(\lambda_i) \right) \quad . \tag{6.26}$$

This allows one to rewrite the identities (6.24) as

$$\int \mathcal{D}\lambda \Delta(\lambda)^2 \left\{ \sum_{k\geq 0} k g_k \frac{\partial}{\partial g_{k+n}} + \frac{\Lambda^2}{N^2} \sum_{k=0}^{n} \frac{\partial^2}{\partial g_k \partial g_{n-k}} \right\} e^{-\frac{N}{\Lambda}\sum_i V(\lambda_i)} = 0 \quad , \tag{6.27}$$

or simply

$$L_n Z_B = 0 \quad , \quad n \geq -1 \quad , \tag{6.28a}$$

where

$$L_n = \sum_{k\geq 0} k g_k \frac{\partial}{\partial g_{k+n}} + \frac{\Lambda^2}{N^2} \sum_{k=0}^{n} \frac{\partial^2}{\partial g_k \partial g_{n-k}} \quad , \tag{6.28b}$$

i.e., the same set of Virasoro constraints found in Chapter 3 (recall (3.27)).

The operators (6.21) and (6.28b) can be regarded as two representations of the same invariance generators, the first ones acting on the space of "fields" (eigenvalues), and the latter on the space of "sources" (coupling constants). This duality will be used in the construction of the supersymmetrization of the model.

6.1.3. N=1 Supersymmetric Extension of the Eigenvalue Model

We shall introduce the discrete model proposed in ref. [84], where the basic assumption was a set of super Virasoro constraints imposed on the partition function of the system.

Let us consider the Virasoro constraints (6.28) as a fundamental property satisfied by the partition function of the model, in the following sense: given the self-interaction (i.e., the potential term, which depends on the coupling constants), and defined a measure for the set of "fields", the Virasoro constraints determine the partition function (i.e., the remaining Jacobian factor).

We can take that as a guiding property to build up a supersymmetric partition function Z_s. Take as super Virasoro generators the following set:

$$G_{n+1/2} = \sum_{k\geq 0} k g_k \frac{\partial}{\partial \xi_{k+n+1/2}} + \sum_{k\geq 0} \xi_{k+1/2} \frac{\partial}{\partial g_{k+n+1}} + \frac{\Lambda^2}{N^2} \sum_{k=0}^{n} \frac{\partial^2}{\partial g_k \partial g_{n-k}} \, , n \geq -1 \tag{6.29}$$

$$\begin{aligned} L_n = \sum_{k\geq 0} k g_k \frac{\partial}{\partial g_{k+n}} + \frac{\Lambda^2}{2N^2} \sum_{k=0}^{n} \frac{\partial^2}{\partial g_k \partial g_{n-k}} + \frac{1}{2} \sum_{k\geq 0} (n+2k+1) \xi_{k+1/2} \frac{\partial}{\partial \xi_{k+n+1/2}} + \\ + \frac{\Lambda^2}{4N^2} \sum_{k=0}^{n-1} (n-2k-1) \frac{\partial^2}{\partial \xi_{k+1/2} \partial \xi_{n-k-1/2}} \, , n \geq -1 , \end{aligned} \tag{6.30}$$

with $\{g_k\}$ $(\{\xi_{k+1/2}\})$ a set of (anti-) commuting coupling constants. The generators above satisfy

$$\begin{aligned} [L_n, L_m] &= (n-m) L_{n+m} \quad , \\ [L_n, G_{m+1/2}] &= \left(\frac{n-1-2m}{2} \right) G_{n+m+1/2} \quad , \\ \{G_{n+1/2}, G_{m+1/2}\} &= 2 L_{n+m+1} \quad , \end{aligned} \tag{6.31}$$

which constitutes a closed subalgebra (remember that $n, m \geq -1$) of the $N=1$ superconformal algebra in the Neveu-Schwarz sector. Now we assume the partition function to be annihilated by $\{L_n, G_n; n \geq -1\}$. However, the commutators in (6.31) show that it is enough to impose

$$G_{n+1/2} Z_s = 0 \quad , \quad n \geq -1 \quad , \tag{6.32}$$

since the constraints $L_n Z_s = 0$ come out as consistency conditions of (6.32).

In analogy with the purely bosonic model, suppose that the dependence of Z_s with respect to the coupling constants (g, ξ), over which the $G_{n+1/2}$ generators act, comes from a potential V_s. We shall also introduce a set of anti-commuting variables $\{\theta_i\}$, each θ_i playing the role of the supersymmetric partner of the eigenvalue λ_i.

The most general polynomial potential for the self-interaction (i.e., without interaction between different pairs (λ_i,θ_i)) would contain terms like λ_i^k and $\theta_i\lambda_i^k$ at most. So we take:

$$V_s(\lambda_i,\theta_i)=\sum_{k=0}^{\infty}\left(g_k\lambda_i^k+\xi_{k+1/2}\theta_i\lambda_i^k\right) \quad . \tag{6.33}$$

As regards the integration measure, we simply assume

$$\mathcal{D}\lambda\mathcal{D}\theta=\prod_{i=1}^{N}\mathrm{d}\lambda_i\,\mathrm{d}\theta_i \quad . \tag{6.34}$$

Writing the supersymmetric partition function as

$$Z_s=\int\mathcal{D}\lambda\mathcal{D}\theta\,\mathrm{e}^{-\frac{N}{\Lambda}\sum_i V_s(\lambda_i)}\Delta(\lambda,\theta) \quad , \tag{6.35}$$

we can determine the leftover factor, the supersymmetric extension of the Vandermonde determinant $\Delta(\lambda,\theta)$, using the constraints (6.32). From the explicit form of the generators (6.29) and the potential (6.33) we obtain

$$\begin{aligned}
&G_{n+1/2}Z_s=\\
&=\int\mathcal{D}\lambda\mathcal{D}\theta\,\mathrm{e}^{-\frac{N}{\Lambda}\sum_i V_s(\lambda_i)}\Delta(\lambda,\theta)\left[\sum_{k\geq 0}kg_k\left(-\frac{N}{\Lambda}\sum_{i=1}^{N}\theta_i\lambda_i^{k+n}\right)+\right.\\
&\left.+\sum_{k\geq 0}\xi_{k+1/2}\left(-\frac{N}{\Lambda}\sum_{i=1}^{N}\lambda_i^{k+n+1}\right)+\frac{\Lambda^2}{N^2}\sum_{k=0}^{n}\left(-\frac{N}{\Lambda}\sum_{i=1}^{N}\theta_i\lambda_i^k\right)\left(-\frac{N}{\Lambda}\sum_{j=1}^{N}\lambda_j^{n-k}\right)\right]\\
&=0 \quad .
\end{aligned} \tag{6.36}$$

The first two terms between brackets in the above equation, which depend explicitly on the coupling constants, can otherwise be seen as coming from the variation of the potential under the action of a differential operator over the (λ,θ) –variables, as follows

$$\begin{aligned}
&\left\{\sum_{k\geq 0}kg_k\left(-\frac{N}{\Lambda}\sum_{i=1}^{N}\theta_i\lambda_i^{k+n}\right)+\sum_{k\geq 0}\xi_{k+1/2}\left(-\frac{N}{\Lambda}\sum_{i=1}^{N}\lambda_i^{k+n+1}\right)\right\}\mathrm{e}^{-\frac{N}{\Lambda}\sum_i^N V_s(\lambda_i,\theta_i)}\\
&=\left\{\sum_{j=1}^{N}\left(\theta_j\lambda_j^{n+1}\frac{\partial}{\partial\lambda_j}-\lambda_j^{n+1}\frac{\partial}{\partial\theta_j}\right)\right\}\mathrm{e}^{-\frac{N}{\Lambda}\sum_i V_s(\lambda_i,\theta_i)} \quad ,
\end{aligned} \tag{6.37}$$

which represents the supersymmetric analog of the operator in (6.21); we therefore define

$$\mathcal{G}_{n+1/2}\equiv\sum_{i=1}^{N}\left(\theta_i\lambda_i^{n+1}\frac{\partial}{\partial\lambda_i}-\lambda_i^{n+1}\frac{\partial}{\partial\theta_i}\right) \quad , \quad n\geq -1 \quad . \tag{6.38}$$

Thus we rewrite the constraints (6.36) as

$$\int \mathcal{D}\lambda\mathcal{D}\theta\Delta(\lambda,\theta)\left\{\sum_{j=1}^{N}\left(\theta_j\lambda_j^{n+1}\frac{\partial}{\partial\lambda_j}-\lambda_j^{n+1}\frac{\partial}{\partial\theta_j}\right)\right.$$
$$\left.+\sum_{k=0}^{n}\left(\sum_{i=1}^{N}\theta_i\lambda_i^k\right)\left(\sum_{j=1}^{N}\lambda_j^{n-k}\right)\right\}\mathrm{e}^{-\frac{N}{\Lambda}\sum_i V_s(\lambda_i,\theta_i)}=0\,. \quad (6.39)$$

After some simple integrations by parts, the above equation is translated into a set of differential equations obeyed by $\Delta(\lambda,\theta)$:

$$\left\{\sum_{i=1}^{N}\left(\frac{\partial}{\partial\theta_i}\lambda_i^{n+1}-\frac{\partial}{\partial\lambda_i}\theta_i\lambda_i^{n+1}\right)+\sum_{k=0}^{n}\left(\sum_{i=1}^{N}\theta_i\lambda_i^k\right)\left(\sum_{j=1}^{N}\lambda_j^{n-k}\right)\right\}\Delta(\lambda,\theta)=$$
$$=\left\{\sum_{i=1}^{N}\left(\lambda_i^{n+1}\frac{\partial}{\partial\theta_i}-\theta_i\lambda_i^{n+1}\frac{\partial}{\partial\lambda_i}\right)+\sum_{i\neq j}^{N}\theta_i\sum_{k=0}^{n}\lambda_i^k\lambda_j^{n-k}\right\}\Delta(\lambda,\theta)=0 \quad . \quad (6.40)$$

Notice the resurgence of the $\mathcal{G}_{n+1/2}$ operator in Eq. (6.40). Moreover, notice that

$$\begin{aligned}\sum_{i\neq j}^{N}\theta_i\sum_{k=0}^{n}\lambda_i^k\lambda_j^{n-k} &= \sum_{i\neq j}^{N}\theta_i\frac{(\lambda_i^{n+1}-\lambda_j^{n+1})}{(\lambda_i-\lambda_j)}=\\ &=\frac{1}{2}\sum_{i\neq j}^{N}(\theta_i+\theta_j)\frac{(\lambda_i^{n+1}-\lambda_j^{n+1})}{(\lambda_i-\lambda_j)}=\\ &=\sum_{i<j}^{N}\frac{(\theta_i\lambda_i^{n+1}-\theta_j\lambda_j^{n+1}+\theta_j\lambda_i^{n+1}-\theta_i\lambda_j^{n+1})}{(\lambda_i-\lambda_j-\theta_i\theta_j)}\\ &=\mathcal{G}_{n+1/2}\sum_{i<j}^{N}\ln(\lambda_i-\lambda_j-\theta_i\theta_j)=\\ &=\mathcal{G}_{n+1/2}\ln\left(\prod_{i<j}(\lambda_i-\lambda_j-\theta_i\theta_j)\right) \quad ,\end{aligned} \quad (6.41)$$

which allows us to rewrite Eq. (6.40) as

$$\frac{1}{\Delta}\mathcal{G}_{n+1/2}\Delta=\frac{1}{\prod_{i<j}^{N}(\lambda_i-\lambda_j-\theta_i\theta_j)}\mathcal{G}_{n+1/2}\prod_{i<j}^{N}(\lambda_i-\lambda_j-\theta_i\theta_j) \quad , \quad (6.42)$$

implying that the supersymmetric extension of the Vandermonde determinant is given by

$$\Delta(\lambda,\theta)=\mathrm{const.}\prod_{i<j}^{N}(\lambda_i-\lambda_j-\theta_i\theta_j) \quad . \quad (6.43)$$

The constant factor in (6.43) just leads to a unimportant multiplicative constant for the partition function, thus we can abandon it. The resulting supersymmetric partition function is therefore

$$Z_s = \int \prod_{n=1}^{N} \mathrm{d}\lambda_n \, \mathrm{d}\theta_n e^{-\frac{N}{\Lambda}\sum_i V_s(\lambda_i,\theta_i)} \prod_{i<j}(\lambda_i - \lambda_j - \theta_i\theta_j) \quad . \tag{6.44}$$

We close this introductory section emphasizing the role of the operators (6.38): as in the bosonic case, the Ward identities (6.36) can be related to a set of infinitesimal (supersymmetric) transformations of "fields", namely

$$\begin{cases} s_n\lambda_i & = \varepsilon_n \mathcal{G}_{n+1/2}\lambda_i = \varepsilon_n \theta_i \lambda_i^{n+1} \\ s_n\theta_i & = \varepsilon_n \mathcal{G}_{n+1/2}\theta_i = -\varepsilon_n \lambda_i^{n+1} \end{cases} \quad , \quad n \geq -1 \tag{6.45}$$

where ε_n are constant anti-commuting infinitesimal parameters. The bosonic sector of the transformations, related to the constraints $L_n Z_s = 0$, come from the commutation of two independent super transformations (6.45):

$$\begin{aligned} \frac{1}{2}[s_n, s_m]\lambda_i &= -(\varepsilon_m\varepsilon_n)\lambda_i^{n+m+2} \quad , \\ \frac{1}{2}[s_n, s_m]\theta_i &= -(\varepsilon_m\varepsilon_n)\left(\frac{m+n+2}{2}\right)\theta_i\lambda_i^{n+m+1} \quad , \end{aligned} \tag{6.46}$$

from which we define

$$\begin{aligned} \delta_n\lambda_i &= \epsilon_n \, \lambda_i^{n+1} \quad , \\ \delta_n\theta_i &= \epsilon_n \left(\frac{n+1}{2}\right)\theta_i\lambda_i^n \quad , \end{aligned} \qquad n \geq -1 \tag{6.47}$$

generated by the action of the differential operators

$$l_n \equiv \sum_{i=1}^{N}\left(\lambda_i^{n+1}\frac{\partial}{\partial\lambda_i} + \frac{(n+1)}{2}\theta_i\lambda_i^n\frac{\partial}{\partial\theta_i}\right) \quad , \quad n \geq -1 \, , \tag{6.48}$$

with ϵ_n infinitesimal commuting parameters. The set of l_n's and $\mathcal{G}_{n+1/2}$'s realize the same closed superconformal subalgebra (6.31).

To conclude, we shall take (6.44) as the definition of a super eigenvalue model, with the additional condition $N \in 2\mathbb{N}$ (to assure a Grassmanian even partition function).

6.2. Planar Superloop Equations

In analogy with the purely bosonic model, we shall derive the planar limit ($N \to \infty$) of the constraints satisfied by the free energy of the supersymmetric model. These constraints will be the microscopic superloops, constituting the starting point of the planar solution of the theory.

Let us consider a permutation of the N pairs of fields (λ_i, θ_i): the potential term is obviously invariant; the measure has odd parity against an elementary permutation $(\lambda_i, \theta_i) \to (\lambda_j, \theta_j)$, as well as the Jacobian term; therefore, the measure together with the integrand is even against any permutation, which guarantees that the partition function is not identically zero[1]. Such Z_N-symmetry is reminiscent of the $U(N)$-symmetry of the matrix model (remember that the microscopic loops and scaling operators were $U(N)$ invariant). This indicates that, in the eigenvalue formulation, we should study correlation functions of Z_N-invariant operators, built up from the trace-like operators: $\hat{\omega}^{(n)} \sim \sum_{i=1}^N \lambda_i^n$ and $\hat{\nu}^{(n)} \sim \sum_{i=1}^N \theta_i \lambda_i^n$.

Indeed, the super Virasoro constraints, as (6.36), correspond to Ward identities obeyed by such operators. We shall use them to calculate 1- point functions.

First we define the connected 1- point functions of microscopic superloop operators as

$$\omega^{(k)} = -\frac{\Lambda^2}{N^2}\frac{1}{Z_s}\frac{\partial}{\partial g_k} Z_s = \left\langle \frac{\Lambda}{N}\sum_i \lambda_i^k \right\rangle \quad , \quad \omega^{(0)} = \Lambda \, , \tag{6.49}$$

$$\nu^{(k)} = -\frac{\Lambda^2}{N^2}\frac{1}{Z_s}\frac{\partial}{\partial \xi_{k+1/2}} Z_s = \left\langle \frac{\Lambda}{N}\sum_i \theta_i \lambda_i^k \right\rangle \quad , \tag{6.50}$$

which depend basically on the free energy, which is defined in analogy to the bosonic theory (recall (3.30)), as

$$F_s \equiv \frac{1}{N^2} \ln Z_s \quad . \tag{6.51}$$

We rewrite the constraint (6.36) in terms of the free energy as:

$$\begin{aligned} 0 = {} & \frac{1}{N^2}\frac{1}{Z_s} G_{n+1/2} Z_s = \\ = {} & \left\{ \sum_{k\geq 0} k g_k \left(\frac{\partial F_s}{\partial \xi_{k+n+1/2}}\right) + \sum_{k\geq 0} \xi_{k+1/2}\left(\frac{\partial F_s}{\partial g_{k+n+1}}\right) + \Lambda^2 \sum_{k=0}^{n}\left(\frac{\partial F_s}{\partial \xi_{k+1/2}}\right)\left(\frac{\partial F_s}{\partial g_{n-k}}\right)\right\} \\ & + \frac{1}{N^2}\left\{ \Lambda^2 \sum_{n=0}^{n} \frac{\partial^2}{\partial \xi_{k+1/2}\partial g_{n-k}} F_s \right\} \quad . \end{aligned} \tag{6.52}$$

Notice the presence of a 2-point function, the second-order derivative term in (6.52). But this term disappears in the $N \to \infty$ limit (the planar approximation),

[1] This is not a trivial check up. We could have a null partition function and yet non-zero correlators for some set of operators.

and we are left with simple algebraic (non-linear) equations between 1- point functions:

$$0=\lim_{N\to\infty}\frac{1}{N^2}\frac{1}{Z_s}G_{n+1/2}Z_s$$
$$=\sum_{k\geq 0}kg_k\left(\frac{\partial F_s}{\partial\xi_{k+n+1/2}}\right)+\sum_{k\geq 0}\xi_{k+1/2}\left(\frac{\partial F_s}{\partial g_{k+n+1}}\right)+\Lambda^2\sum_{k=0}^{n}\left(\frac{\partial F_s}{\partial\xi_{k+1/2}}\right)\left(\frac{\partial F_s}{\partial g_{n-k}}\right). \tag{6.53}$$

These planar equations can be written in terms of the loops (6.49,50) as

$$\sum_{k\geq 0}kg_k\nu^{(k+n)}+\sum_{k\geq 0}\xi_{k+1/2}\omega^{(k+n+1)}=\sum_{k=0}^{n}\omega^{(n-k)}\nu^{(k)}\quad,\quad n\geq -1\ . \tag{6.54}$$

The same planar limit can be taken for the bosonic constraints (6.30), allowing another set of non-linear equations obeyed by 1- point functions, as follows:

$$0=\lim_{N\to\infty}\frac{1}{N^2}\frac{1}{Z_s}L_nZ_s=$$
$$=\sum_{k\geq 0}kg_k\frac{\partial F_s}{\partial g_{k+n}}+\sum_{k\geq 0}\frac{(n+2k+1)}{2}\xi_{k+1/2}\frac{\partial F_s}{\partial\xi_{k+n+1/2}}+$$
$$+\frac{\Lambda^2}{2}\sum_{k=0}^{n}\left(\frac{\partial F_s}{\partial g_k}\right)\left(\frac{\partial F_s}{\partial g_{n-k}}\right)+\frac{\Lambda^2}{2}\sum_{k=0}^{n-1}\frac{(n-2k-1)}{2}\left(\frac{\partial F_s}{\partial\xi_{k+1/2}}\right)\left(\frac{\partial F_s}{\partial\xi_{n-k-1/2}}\right) \tag{6.55}$$

implying other equations for the microscopic superloops:

$$\sum_{k\geq 0}kg_k\omega^{(k+n)}+\sum_{k\geq 0}\frac{(n+2k+1)}{2}\xi_{k+1/2}\nu^{(k+n)}=$$
$$=\frac{1}{2}\sum_{k=0}^{n}\omega^{(k)}\omega^{(n-k)}+\frac{1}{2}\sum_{k=0}^{n-1}\frac{(n-2k-1)}{2}\nu^{(k)}\nu^{(n-k-1)}\quad,n\geq -1. \tag{6.56}$$

The complete set of equations (6.54) and (6.56), together with the "initial condition"

$$\omega^{(0)}=\Lambda\quad, \tag{6.57}$$

compose the planar superloop equations of the model.

As in the purely bosonic model, these equations do not determine all loops: indeed it is shown in Appendix D that, if the bosonic piece of the potential, $V(\lambda)=\sum_{k=0}^{a}g_a\lambda^a$, is a polynomial of degree $\underline{a}$, the general solution can be given in terms of a finite set of undetermined loops, namely $\omega^{(1)}\cdots\omega^{(a-2)}$ and $\nu^{(1)}\cdots\nu^{(a-2)}$.

We proceed on the analogy to the bosonic case, introducing the bosonic and fermionic loop generators, respectively given by

$$\omega(p)=\sum_{n\geq 0}\frac{\omega^{(n)}}{p^{n+1}}\quad, \tag{6.58}$$

$$\nu(p)=\sum_{n\geq 0}\frac{\nu^{(n)}}{p^{n+1}}\quad, \tag{6.59}$$

which are to be compared with the definition (3.38). After some simple algebraic manipulations, one can show that equations (6.54) imply

$$0=\sum_{k\geq 0} kg_k p^{k-1}\frac{\nu^{(k-1)}}{p^k}+\sum_{k\geq 0}\xi_{k+1/2}p^k\frac{\omega^{(k)}}{p^{k+1}}+$$
$$+\sum_{n\geq 0}\left(\sum_{k\geq 0} kg_k p^{k-1}\frac{\nu^{(k+n)}}{p^{k+n+1}}+\sum_{k\geq 0}\xi_{k+1/2}p^k\frac{\omega^{(k+n+1)}}{p^{k+n+2}}-\sum_{k=0}^{n}\frac{\omega^{(n-k)}}{p^{n-k+1}}\frac{\nu^{(k)}}{p^{k+1}}\right) \quad (6.60)$$

which can be rewritten in terms of the expansion (6.58) and (6.59) as

$$0=-\left(\sum_{m\geq 0}\frac{\omega^{(m)}}{p^{m+1}}\right)\left(\sum_{n\geq 0}\frac{\nu^{(n)}}{p^{n+1}}\right)+\left(\sum_{k\geq 0} kg_k p^{k-1}\right)\left(\sum_{m\geq 0}\frac{\nu^{(m)}}{p^{m+1}}\right)+$$
$$+\left(\sum_{k\geq 0}\xi_{k+1/2}p^k\right)\left(\sum_{m\geq 0}\frac{\omega^{(m)}}{p^{m+1}}\right)-\sum_{k\geq 2} kg_k\sum_{m=2}^{k}\nu^{(k-m)}p^{m-2}+$$
$$-\sum_{k\geq 1}\xi_{k+1/2}\sum_{m=2}^{k+1}\omega^{(k-m+1)}p^{m-2} \quad . \quad (6.61)$$

The above equation suggests the definitions of the following analytical functions

$$V'(p)=\sum_{k=0}^{a} kg_k p^{k-1} \quad , \quad (6.62)$$

$$\xi(p)=\sum_{k\geq 0}\xi_{k+1/2}p^k \quad , \quad (6.63)$$

$$Q_1(p)=-\sum_{k\geq 2} kg_k\sum_{m=2}^{k}\nu^{(k-m)}p^{m-2}-\sum_{k\geq 1}\xi_{k+1/2}\sum_{m=2}^{k+1}\omega^{(k-m+1)}p^{m-2} \quad . (6.64)$$

Notice that the function $Q_1(p)$ contains the undetermined coefficients $\{w^{^{(k)}}, k = 1,\cdots,a-2;\, \nu^{^{(l)}}, l=0,\cdots,a-2\}$. Now we can rewrite Eq. (6.61) shortly as

$$\omega(p)\nu(p)-V'(p)\nu(p)-\xi(p)\omega(p)=-Q_1(p) \quad , \quad (6.65)$$

or rather

$$(\omega(p)-V'(p))(\nu(p)-\xi(p))=\Delta_1(p) \quad , \quad (6.66)$$

where

$$\Delta_1(p)=V'(p)\xi(p)-Q_1(p) \quad (6.67)$$

is also an analytical function of p. This is the odd part of the superloop equations we are looking for.

Equations (6.56) also lead to a loop equation: indeed they imply

$$
\begin{aligned}
0 = & \sum_{n\geq -1}\sum_{k\geq 0} k g_k p^{k-1}\frac{\omega^{(k+n)}}{p^{k+n+1}} + \sum_{n\geq -1}\sum_{k\geq 0}\frac{(n+2k+1)}{2}\xi_{k+1/2}\frac{\nu^{(k+n)}}{p^{n+2}} + \\
& -\frac{1}{2}\sum_{n\geq 0}\sum_{k=0}^{n}\frac{\omega^{(k)}}{p^{k+1}}\frac{\omega^{(n-k)}}{p^{n-k+1}} - \frac{1}{2}\sum_{n\geq 1}\sum_{k=0}^{n-1}\frac{(n-2k-1)}{2}\frac{\nu^{(k)}\nu^{(n-k-1)}}{p^{n+2}} = \\
= & -\frac{1}{2}\omega(p)^2 + V'(p)\omega(p) + \frac{1}{2}\xi'(p)\nu(p) - \frac{1}{2}\xi(p)\nu'(p) + \\
& + \frac{1}{2}\nu(p)\nu'(p) - \sum_{k\geq 2} k g_k \sum_{n=2}^{k}\omega^{(k-n)}p^{n-2} - \frac{1}{2}\sum_{k\geq 2}\xi_{k+1/2}\sum_{n=2}^{k}(2k+1-n)\nu^{(k-n)}p^{n-2}
\end{aligned}
\tag{6.68}
$$

which can be written as

$$
\omega^2(p) - 2V'(p)\omega(p) + \nu'(p)\nu(p) - \xi'(p)\nu(p) + \xi(p)\nu'(p) = -Q_0(p) \quad . \tag{6.69}
$$

The analytical function $Q_0(p)$ is given by the expression

$$
Q_0(p) = 2\sum_{k\geq 2} k g_k \sum_{n=2}^{k}\omega^{(k-n)}p^{n-2} - \sum_{k\geq 2}\xi_{k+1/2}\sum_{n=2}^{k}(2k+1-n)\nu^{(k-n)}p^{n-2} \quad . \tag{6.70}
$$

Alternatively, (6.69) can be rewritten as

$$
\left(\omega(p) - V'(p)\right)^2 + \left(\nu(p) - \xi(p)\right)'\left(\nu(p) - \xi(p)\right) = \Delta_0(p) \quad , \tag{6.71}
$$

where

$$
\Delta_0(p) = V'(p)^2 + \xi(p)'\xi(p) - Q_0(p) \tag{6.72}
$$

is also an analytical function of p (it represents the even partner of the function $\Delta_1(p)$ defined in (6.67)).

The Grassmanian even function $Q_0(p)$ (resp. $\Delta_0(p)$) is the "bosonic" partner of $Q_1(p)$ (resp. $\Delta_1(p)$), and also involves the unknown coefficients $w^{(1)}\cdots w^{(a-2)}$, $\nu^{(0)}\cdots\nu^{(a-2)}$.

The reader can find in ref. [84] two alternative derivations of the superloop equations (6.66) and (6.71). They will be taken as the supersymmetric version of Eq. (3.39). Now we can proceed to the following steps: to solve the superloop equations, take the scaling limit and find the critical points and critical dimensions.

6.2.1. Solving the Superloop Equations: The One-Cut Solution

In this section we explain the one-cut solution of the superloop equations proposed by the authors in ref. [84]. We will make the simplifying assumption that the bosonic part of the potential is even. A detailed discussion of the one-cut Ansatz can be found in Appendix D.

The fermionic component $\nu(p)$ can be solved in terms of the bosonic one $\omega(p)$, using equation (6.66):

$$\nu(p) = \xi(p) + \frac{\Delta_1(p)}{\omega(p) - V'(p)} \quad , \tag{6.73}$$

which is then substituted in (6.71), to give a quartic equation for the loop $w(p)$,

$$(\omega(p) - V'(p))^4 - (\omega(p) - V'(p))^2 \, \Delta_0(p) + \Delta_1' \Delta_1(p) = 0 \quad . \tag{6.74}$$

The above equation can be easily solved: in particular, using the fact that Δ_1 is Grassmannian-odd ($\Delta_1^2 = 0$) we find

$$\omega(p) = V'(p) + \sqrt{\Delta_0} - \frac{1}{2} \frac{\Delta_1' \Delta_1}{\Delta_0^{3/2}} \quad . \tag{6.75}$$

Finally replacing the solution in Eq. (6.75), and using that $\Delta_1^2 = 0$ once again we get

$$\nu(p) = \xi(p) + \frac{\Delta_1}{\sqrt{\Delta_0}} \quad . \tag{6.76}$$

In principle Δ_0 , Δ_1 are respectively Grassmanian-even and odd series in the fermionic couplings $\{\xi_{k+1/2}\}$, and a similar conclusion applies to the components $w(p)$ and $\nu(p)$. It is straightforward to check that the partition function (6.44) is non-trivial when one turns off the fermionic couplings, $\xi_{k+1/2} \to 0$, and therefore, among the non-trivial solutions in (6.75) we must choose the one which has a contribution of zeroth order in $\xi_{k+1/2}$.

6.2.2. Even Bosonic Potential

One can proceed using analogies, extending the solution to the planar bosonic case to the present situation. In this spirit, a single cut solution of the form

$$\begin{aligned} w(p) &= V'(p) - M(p)\sqrt{(p-A)(p-B)} + \frac{C(p)}{[(p-A)(p-B)]^{3/2}} \quad , \\ \nu(p) &= \xi(p) - \frac{N(p)}{\sqrt{(p-A)(p-B)}} \quad , \end{aligned} \tag{6.77}$$

was presented in refs. [84,152]. Above, the cut $[A, B]$ is a function of the couplings and the cosmological constant Λ and is given by Eqs. (D.43) and (D.44). The functions $M(p), N(p)$ and $C(p)$ are analytical in p, and respectively of order zero, one and two in the fermionic couplings. It is rather surprising[84] that the solution proposed is at most bilinear in fermions, and one might be tempted to add higher order terms in the ξ-couplings. In Appendix D we show that the Ansatz (6.77) corresponds to a consistent choice of the undetermined coefficients. Actually, it was shown in ref. [154] that the partition function (6.44) is bilinear in the ξ-couplings, and therefore this is not a peculiar property of the planar approximation. We discuss the non-perturbative aspects of the model in Chapter 8.

As in the purely bosonic model, we simplify the problem assuming that the bosonic part of the potential is an even polynomial of degree $a = 2b$,

$$V(p) = \sum_{k=0}^{b} g_{2k} p^{2k} \quad , \tag{6.78}$$

More general potentials are considered in ref. [152]. When $V(p) = V(-p)$, equation $(D.43)$ tells us that the cut $[A, B]$ is actually symmetric

$$A = -B \equiv \sqrt{R} \quad , \quad (p-A)(p-B) = p^2 - R \quad , \tag{6.79}$$

where the cut function R is introduced to stress the similarities with the bosonic theory. On the other hand, the initial condition $\omega^{(0)} = \Lambda$ leads to the string equation

$$\Lambda = \oint \frac{dp}{2\pi i} \frac{pV'(p)}{\sqrt{p^2 - R}} = \sum_{k=1}^{b} 2k g_{2k} \frac{R^k}{4^k} \binom{2k}{k} \quad , \tag{6.80}$$

as follows from $(D.44)$ which implicitly provides the solution $R = R(g_{2k}, \Lambda)$. Compare Eq. (6.80) with (3.48): they only differ by a factor of 2.

Concerning the bosonic loop, we split it in two pieces: $\omega(p) = \omega_0(p) + \omega_2(p)$, where ω_0 (resp. ω_2) is the zeroth (resp. second-order) term with respect to the ξ-couplings. Its zeroth-order piece is given by:

$$\frac{\partial \omega_0(p)}{\partial \Lambda} = \frac{1}{\sqrt{p^2 - R}} \quad \text{or} \quad \frac{\partial \omega_0^{(2k)}(p)}{\partial \Lambda} = \binom{2k}{k} \frac{R^k}{4^k} \quad , \tag{6.81}$$

$$M(R) = \partial_R \Lambda(R) = 1/\partial_\Lambda R \quad ,$$

exactly the same solution found in the purely bosonic model (recall (3.53)), while the 2nd-order microscopic loops are taken from $(D.61)$, as follows:

$$\begin{aligned} \omega_2(p) &= \frac{C_0 + C_1 p}{(p^2 - R)^{3/2}} = (C_0 + C_1 p) \sum_{k \geq 0} \binom{2k}{k} \frac{R^k}{4^k} \frac{(2k+1)}{p^{2k+3}} \\ &= \sum_{k \geq 0} \left[C_1 (2k+1) \frac{R^k}{4^k} \binom{2k}{k} \right] \frac{1}{p^{(2k+1)+1}} + \sum_{k \geq 0} \left[\frac{C_0}{R} 2k \frac{R^k}{4^k} \binom{2k}{k} \right] \frac{1}{p^{2k+1}} \quad , \end{aligned} \tag{6.82}$$

which means

$$\omega_2^{(2k)} = 2k \frac{R^k}{4^k} \binom{2k}{k} \frac{C_0}{R} \quad , \tag{6.83a}$$

$$\omega_2^{(2k+1)} = (2k+1) \frac{R^k}{4^k} \binom{2k}{k} C_1 \quad , \quad k \geq 0 \quad . \tag{6.83b}$$

Notice that $\omega_2^{(0)} = 0$ as expected[2]. The coefficients C_0 and C_1 must be taken from the function $N(p)$, which brings us to the fermionic loop $\nu(p)$: we begin with splitting the series $N(p)$ into even and odd pieces,

[2] The initial condition $\omega^{(0)} = \Lambda$ holds to all orders in the ξ-couplings.

$$N(p) \equiv \widehat{N}_-(p^2) + pN_+(p^2) \quad , \tag{6.84}$$

where $\widehat{N}_-(p^2)$ contains the undetermined coefficient n_0 as shown in Appendix D. Using (6.79) we write

$$N(A) = \widehat{N}_-(R) + \sqrt{R}N_+(R) \quad , \tag{6.85a}$$

$$N(B) = \widehat{N}_-(R) - \sqrt{R}N_+(R) \quad , \tag{6.85b}$$

and since $\widehat{N}_-^2 = 0$, we find

$$N(A)N(B) = 2\sqrt{R}N_+(R)\widehat{N}_-(R) \quad . \tag{6.86}$$

Now we can impose the condition $(D.70)$: we choose n_0 such that

$$\widehat{N}_-(R) = 0 \quad ; \tag{6.87}$$

as a consequence, the l.h.s. of the expression (6.86) vanishes. The above choice suggests the definition of the function $N_-(p^2)$ as follows,

$$\widehat{N}_-(p^2) \equiv (p^2 - R)N_-(p^2) \quad . \tag{6.88}$$

Substituting (6.84) and (6.88) back into (6.77), the fermionic loop becomes

$$\begin{aligned} \nu(p) &= \xi(p) - \frac{(p^2 - R)N_-(p^2) + pN_+(p^2)}{\sqrt{p^2 - R}} \\ &= \xi(p) - N_-(p^2)\sqrt{p^2 - R} - \frac{pN_+(p^2)}{\sqrt{p^2 - R}} \quad . \end{aligned} \tag{6.89}$$

This is the solution proposed in ref. [84]. In order to find $N_+(p^2)$ and $N_-(p^2)$ we split (6.89) in two equations of opposite parity with respect to p: first we define

$$\nu_-(p) = \sum_{k \geq 0} \frac{\nu^{(2k)}}{p^{2k+1}} \quad , \quad \nu_+(p^2) = \sum_{k \geq 0} \frac{\nu^{(2k+1)}}{p^{2k+2}} \quad , \tag{6.90}$$

$$\xi_-(p) = \sum_{k \geq 0} \xi_{2k+3/2} p^{2k+1} \quad , \quad \xi_+(p^2) = \sum_{k \geq 0} \xi_{2k+1/2} p^{2k} \quad , \tag{6.91}$$

so that the component ν_- obeys

$$\nu_-(p) = \xi_-(p) - N_-(p^2)\sqrt{p^2 - R} \quad , \tag{6.92}$$

which is formally the same equation satisfied by $\omega_0(p)$ (see $(D.47)$), while the ν_+ piece satisfies

$$\nu_+(p) = \xi_+(p) - \frac{pN_+(p^2)}{\sqrt{p^2 - R}} \quad . \tag{6.93}$$

We can solve the above equations following the same "analyticity strategy" used in ref. [130] to solve the bosonic model: multiply Eq. (6.92) by p, and define

$$t = p^{-2} \quad , \tag{6.94}$$

$$\nu_-(t) = p\nu_-(p) = \sum_{k\geq 0} \nu^{(2k)} t^k \tag{6.95}$$

$$\xi_-(t^{-1}) = p\xi_-(p) = \sum_{k\geq 0} \xi_{2k+3/2} t^{-k-1} = \sum_{k\geq 1} \xi_{2k+1/2} t^{-k} \quad , \tag{6.96}$$

$$N_-(t^{-1}) = N_-(p^2) = \sum_{k\geq 0} n_k^{(-)} t^{-k} \quad , \tag{6.97}$$

which obey

$$\nu_-(t) = \xi_-(t^{-1}) - N_-(t^{-1}) t^{-1} \sqrt{1 - Rt} \quad . \tag{6.98}$$

Since $\nu_-(t)/\sqrt{1-Rt}$ is analytical around $t = 0$, we have

$$\oint_0 \mathrm{d}t\, t^n \frac{\xi_-(t^{-1})}{\sqrt{1 - Rt}} = \oint_0 \mathrm{d}t\, t^{n-1} N_-(t^{-1}) \quad , \tag{6.99}$$

from which we determine the coefficients of N_- in terms of the couplings and the cut R,

$$\begin{gathered}\sum_{m\geq 1} \xi_{2m+1/2} \sum_{k\geq 0} \binom{2k}{k} \frac{R^k}{4^k} \oint_0 \mathrm{d}t\, t^{n+k-m} = \sum_{k\geq 0} n_k^{(-)} \oint_0 \mathrm{d}t\, t^{n-k-1} \quad , \\ \sum_{k\geq 0} \xi_{2(n+k)+3/2} \frac{R^k}{4^k} \binom{2k}{k} = n_n^{(-)} \quad .\end{gathered} \tag{6.100}$$

We eliminate the potential term ξ_- from (6.98) calculating

$$\frac{\partial}{\partial R} \nu_-(t) = \frac{1}{\sqrt{1 - Rt}} \left(\frac{1}{2} N_-(t^{-1}) - \frac{\partial N_-(t^{-1})}{\partial R} (t^{-1} - R) \right) \quad , \tag{6.101}$$

and since $\sqrt{1 - Rt}\; \partial\nu_-/\partial R$ is also analytical about $t = 0$, we conclude that the term in parenthesis is t-independent. We can evaluate it at $t^{-1} = 0$,

$$\begin{aligned}\left[\frac{1}{2} N_-(t^{-1}) - \frac{\partial N_-(t^{-1})}{\partial R}(t^{-1} - R)\right]_{t^{-1}=0} &= \frac{1}{2} N_-(0) + R \frac{\partial}{\partial R} N_-(0) \\ &= \frac{1}{2} \sum_{k\geq 0} \xi_{2k+3/2} \frac{R^k}{4^k} \binom{2k}{k} + \sum_{k\geq 0} \xi_{2k+3/2} \frac{R^k}{4^k} k \binom{2k}{k} \\ &= \frac{1}{2} \sum_{k\geq 0} (2k+1) \xi_{2k+3/2} \frac{R^k}{4^k} \binom{2k}{k} \quad ,\end{aligned} \tag{6.102}$$

as well as at $t^{-1} = R$,

$$\left[\frac{1}{2} N_-(t^{-1}) - \frac{\partial N_-(t^{-1})}{\partial R}(t^{-1} - R)\right]_{t^{-1}=0} = \frac{1}{2} N_-(R) \quad . \tag{6.103}$$

The right hand sides of (6.102) and (6.103) must be equal. Therefore the solution for the component ν_- is given by

$$\frac{\partial}{\partial R}\nu_-(t) = \frac{1}{2}\frac{N_-(R)}{\sqrt{1-Rt}} \quad , \tag{6.104a}$$

or else

$$\frac{\partial}{\partial R}\nu_-(p) = \frac{1}{2}\frac{N_-(R)}{\sqrt{p^2-R}} \quad , \quad \frac{\partial}{\partial R}\nu^{(2k)} = \frac{R^k}{4^k}\binom{2k}{k}\frac{N_-(R)}{2} \quad , \tag{6.104b}$$

where

$$N_-(R) = \sum_{k\geq 0}(2k+1)\xi_{2k+3/2}\frac{R^k}{4^k}\binom{2k}{k} \quad . \tag{6.105}$$

Now we apply the same procedure to calculate $\nu_+(p)$: starting from Eq. (6.93) multiplied by p^2 and the definitions

$$\nu_+(t) = p^2\nu_+(p) = \sum_{k\geq 0}\nu^{(2k+1)}t^k \quad , \tag{6.106}$$

$$\xi_+(t^{-1}) = p^2\xi_+(p) = \sum_{k\geq 0}\xi_{2k+1/2}t^{-k-1} \quad , \tag{6.107}$$

$$N_+(t^{-1}) = N_+(p^2) = \sum_{k\geq 0}n_k^{(+)}t^{-k} \quad , \tag{6.108}$$

Eq. (6.93) implies

$$\nu_+(t) = \xi_+(t^{-1}) - \frac{N_+(t^{-1})t^{-1}}{\sqrt{1-Rt}} \quad . \tag{6.109}$$

From the analyticity of $\sqrt{1-Rt}\,\nu_+(t)$ around $t=0$ we conclude that

$$\oint_0 \mathrm{d}t\, t^n \xi_+(t^{-1})\sqrt{1-Rt} = \oint_0 \mathrm{d}t\, t^{n-1}N_+(t^{-1}) \quad , \tag{6.110}$$

which determines the coefficients of the series $N_+(p^2)$:

$$\sum_{k\geq 0}\frac{\xi_{2(n+k)+1/2}}{(1-2k)}\frac{R^k}{4^k}\binom{2k}{k} = n_n^{(+)} \quad . \tag{6.111}$$

Deriving Eq. (6.93) with respect to the cut R, we get

$$\frac{\partial}{\partial R}\nu_+(t^{-1}) = -\frac{1}{(1-Rt)^{3/2}}\left(\frac{1}{2}N_+(t^{-1}) + \frac{\partial N_+(t^{-1})}{\partial R}(t^{-1}-R)\right) \quad , \tag{6.112}$$

where analyticity implies that the factor between parenthesis must be t-independent and can be calculated at $t^{-1}=0$ or $t^{-1}=R$. We find as a result

$$N_+(R) = \sum_{k\geq 0}\xi_{2k+1/2}\frac{R^k}{4^k}\binom{2k}{k} \quad , \tag{6.113}$$

and finally the solution for the odd fermionic component ν_+ is

$$\frac{\partial}{\partial R}\nu_+(p) = -\frac{1}{2}\frac{pN_+(R)}{(p^2-R)^{3/2}} \ ,$$
$$\frac{\partial}{\partial R}\nu^{(2k+1)} = (2k+1)\frac{R^k}{4^k}\binom{2k}{k}\left(-\frac{N_+(R)}{2}\right) \ . \tag{6.114}$$

Now we can calculate the coefficients C_0 and C_1 to determine the second order bosonic loop $\omega_2(p)$. From Eq. $(D.64)$ we learn that we need to calculate $M(p^2)$ and the product $N'N(p)$ at the points $p = A = \sqrt{R}$ and $p = B = -\sqrt{R}$. Using the results (6.81) and (6.84-88), and remembering that $N_+^2 = 0 = N_-^2$, we find

$$N'(A)N(A) = -2RN_+(R)N_-(R) - 2R\sqrt{R}N_+(R)N_+'(R) \ ,$$
$$N'(B)N(B) = -2RN_+(R)N_-(R) + 2R\sqrt{R}N_+(R)N_+'(R) \ , \tag{6.115}$$

$$\frac{1}{M(A)} = \frac{1}{M(B)} = \frac{1}{\partial_R\Lambda} = \frac{\partial}{\partial\Lambda}R \ , \tag{6.116}$$

whose replacement in Eq. $(D.64)$ gives

$$C_0 = \frac{1}{2}\frac{\sqrt{R}}{2\sqrt{R}}\frac{\partial R}{\partial\Lambda}(-4RN_+(R)N_-(R)) = -N_+(R)N_-(R)R\frac{\partial}{\partial\Lambda}R \ , \tag{6.117}$$

$$C_1 = \frac{1}{2}\frac{1}{2\sqrt{R}}\frac{\partial R}{\partial\Lambda}\left(-4R\sqrt{R}N_+(R)N_+'(R)\right) = -N_+(R)N_+'(R)R\frac{\partial}{\partial\Lambda}R \,. \tag{6.118}$$

The solution (6.83) becomes

$$\omega_2^{(2k)} = -2k\frac{R^k}{4^k}\binom{2k}{k}N_+(R)N_-(R)\frac{\partial}{\partial\Lambda}R \ , \tag{6.119a}$$

$$\omega_2^{(2k+1)} = -(2k+1)\frac{R^k}{4^k}\binom{2k}{k}N_+(R)N_+'(R)R\frac{\partial}{\partial\Lambda}R \ . \tag{6.119b}$$

This completes the planar solution of the superloop equations: the loop components can be expressed in terms of given functions of the cut R, which in turn is related to the coupling constants through the string equation. For the sake of clarity we summarize the final results below

$$\frac{\partial\omega_0^{(2k)}}{\partial\Lambda} = \binom{2k}{k}\frac{R^k}{4^k} \ , \tag{6.120a}$$

$$\omega_2^{(2k)} = -2k\frac{R^k}{4^k}\binom{2k}{k}N_+(R)N_-(R)\frac{\partial R}{\partial\Lambda} \ , \tag{6.120b}$$

$$\omega_2^{(2k+1)} = -(2k+1)\frac{R^k}{4^k}\binom{2k}{k}N_+(R)N_+'(R)R\frac{\partial R}{\partial\Lambda} \ , \tag{6.120c}$$

$$\frac{\partial\nu^{(2k)}}{\partial\Lambda} = \binom{2k}{k}\frac{R^k}{4^k}\frac{N_-(R)}{2}\frac{\partial R}{\partial\Lambda} \ , \tag{6.120d}$$

$$\frac{\partial\nu^{(2k+1)}}{\partial\Lambda} = -(2k+1)\binom{2k}{k}\frac{R^k}{4^k}\frac{N_+(R)}{2}\frac{\partial R}{\partial\Lambda} \ . \tag{6.120e}$$

6.3. Double Scaling Limit

In this section we take the scaling limit of the previous expressions. Many calculations are similar to those of the purely bosonic theory, therefore we translate most of the results from Chapters 3 and 4. In particular we define bosonic and fermionic scaling operators and recover the Neveu-Schwarz and Ramond sectors of the theory of $(2,4m)$ minimal superconformal models coupled to supergravity.

We begin with observing that the 1-cut solution obtained in the previous section depends on three basic functions of the cut variable R, gathered below

$$\Lambda(R) = \sum_{k\geq 0} 2k g_{2k} \frac{R^k}{4^k}\binom{2k}{k} = R\frac{\partial}{\partial R}\left[\sum_{k\geq 0} 2g_{2k}\frac{R^k}{4^k}\binom{2k}{k}\right] \quad , \tag{6.121}$$

$$N_+(R) = \sum_{k\geq 0} \xi_{2k+1/2}\frac{R^k}{4^k}\binom{2k}{k} \quad , \tag{6.122}$$

$$N_-(R) = \sum_{k\geq 0} (2k+1)\xi_{2k+3/2}\frac{R^k}{4^k}\binom{2k}{k} \quad . \tag{6.123}$$

They are series in R whose coefficients depend on three independent sets of coupling constants, namely $\{g_{2k}\}$, $\{\xi_{2k+1/2}\}$ and $\{\xi_{(2k+1)+1/2}\}$. As in the bosonic model, we are interested in the results close to criticality, where $\Lambda \to \Lambda_c$ and $R \to R_c$. Therefore we expand the above series around the critical point R_c: the string equation (6.121) is, up to a factor of 2, the same one given in Eq. (3.57), and we can borrow the result (3.60-62) with the replacement $g_{2k} \to 2g_{2k}$, thus obtaining

$$\Lambda = \sum_{n\geq 0} t_n^B \left[(R_c - R)^n - R_c^{-1}(R_c - R)^{n+1}\right] \quad , \tag{6.124}$$

where

$$t_n^B \equiv -\sum_{k\geq n+1}\left[\frac{(n+1)}{4^k}\binom{2k}{k}\binom{k}{n+1}(-1)^{n+1}R_c^{k-n}\right] 2g_{2k} \quad , \tag{6.125a}$$

$$t_0^B = \Lambda_c \quad , \tag{6.125b}$$

$$g_{2k} = \frac{4^k}{2\binom{2k}{k}}\sum_{n\geq k}\binom{n}{k}(-1)^k\left(-\frac{t_{n-1}^B}{n}R_c^{n-k-1}\right) \quad ,$$

$$= \sum_{n+1\geq k}\left[-\frac{(-4)^k}{2(n+1)}\frac{\binom{n+1}{k}}{\binom{2k}{k}}R_c^{n-k}\right] t_n^B \quad . \tag{6.126}$$

Using the general transformation formula (3.58) we rewrite the functions N_+ and N_- as

$$N_+ = \sum_{n\geq 0} \tau_n^{B+}(R_c - R)^n \quad , \tag{6.127}$$

$$\tau_n^{B+} = \sum_{k\geq n} \binom{k}{n} (-1)^n \xi_{2k+1/2} \frac{R_c^{k-n}}{4^k} \binom{2k}{k} \quad , \tag{6.128}$$

$$\xi_{2k+1/2} = \sum_{n\geq k} (-4)^k \frac{\binom{n}{k}}{\binom{2k}{k}} \tau_n^{B+} R_c^{n-k} \quad , \tag{6.129}$$

and

$$N_- = \sum_{n\geq 0} \tau_n^{B-}(R_c - R)^n \quad , \tag{6.130}$$

$$\tau_n^{B-} = \sum_{k\geq n} \binom{k}{n} (-1)^n (2k+1) \xi_{2k+3/2} \frac{R_c^{k-n}}{4^k} \binom{2k}{k} \quad , \tag{6.131}$$

$$\xi_{2k+3/2} = \sum_{n\geq k} \frac{(-4)^k}{(2k+1)} \frac{\binom{n}{k}}{\binom{2k}{k}} \tau_n^{B-} R_c^{n-k} \quad . \tag{6.132}$$

The string equation (6.124) coincides with Eq. (3.60) obtained in the bosonic model: this means that one can tune the coupling constants as in (3.63) to induce the same family of scaling regimes labelled by the string susceptibility $\gamma = -1/m\,,\, m = 1, 2, 3 \cdots$ as in (3.64). One can further proceed with the analogy, introducing a scaling parameter to control the scaling limit as in (3.67), defining renormalized couplings and their corresponding scaling operators and therefore establishing the continuum limit of the model.

We follow ref. [84] and introduce a set of bare scaling operators, as an intermediate step in the double scaling limit. The bare scaling operators are defined as the ones coupled to the bare coupling constants: the bosonic ones are

$$\begin{aligned} \langle \sigma_n^B \rangle \equiv N^2 \Lambda^2 \frac{\partial}{\partial t_n^B} F &= -N^2 \sum_{k\geq 0} \left(\frac{\partial g_{2k}}{\partial t_n^B} \right) \left\langle \frac{\Lambda}{N} \sum_i \lambda_i^{2k} \right\rangle \\ &= -N^2 \sum_{k=0}^{n+1} -\frac{1}{2} \frac{(-4)^k}{(n+1)} \frac{\binom{n+1}{k}}{\binom{2k}{k}} R_c^{n-k} w^{(2k)} \quad . \end{aligned} \tag{6.133}$$

The renormalization factor $N^2\Lambda^2$ is introduced for later convenience. Using the solution (6.120) we obtain

$$\frac{\partial}{\partial\Lambda}\langle\sigma_n^B\rangle_0 = \frac{N^2}{2R_c}\frac{(R_c - R)^{n+1}}{(n+1)} \quad , \tag{6.134}$$

$$\langle\sigma_n^B\rangle_2 = -\frac{N^2}{2}\left[(R_c - R)^n - R_c^{-1}(R_c - R)^{n+1}\right] N_+ N_- \frac{\partial}{\partial\Lambda}(R_c - R) \quad , \tag{6.135}$$

where the indices 0 and 2 indicate the 0th and 2nd order in fermionic couplings respectively. The fermionic scaling operators are defined analogously as

$$\langle\nu_n^{B+}\rangle \equiv N^2\Lambda^2 \frac{\partial}{\partial\tau_n^{B+}} F = -N^2 \sum_{k=0}^{n} (-4)^k \frac{\binom{n}{k}}{\binom{2k}{k}} R_c^{n-k} \nu^{(2k)} \quad , \tag{6.136}$$

$$\langle\nu_n^{B-}\rangle \equiv N^2\Lambda^2 \frac{\partial}{\partial\tau_n^{B-}} F = -N^2 \sum_{k=0}^{n} \frac{(-4)^k}{(2k+1)} \frac{\binom{n}{k}}{\binom{2k}{k}} R_c^{n-k} \nu^{(2k+1)} \quad , \tag{6.137}$$

and the planar solution (6.120) implies

$$\frac{\partial}{\partial\Lambda}\langle\nu_n^{B+}\rangle = +\frac{N^2}{2}\left[N_-(R_c - R)^n \frac{\partial}{\partial\Lambda}(R_c - R)\right] \quad , \tag{6.138}$$

hence

$$\frac{\partial}{\partial\Lambda}\langle\nu_n^{B-}\rangle = -\frac{N^2}{2}\left[N_+(R_c - R)^n \frac{\partial}{\partial\Lambda}(R_c - R)\right] \quad . \tag{6.139}$$

We introduce a parameter a to control the critical limit, defining

$$\Lambda \sim \Lambda_c(1 - a^2 t) \quad , \tag{6.140}$$

$$R \sim R_c(1 - a^{2/m} u) \quad , \tag{6.141}$$

near the critical point, which is therefore given by the limit $a \to 0$; t is the renormalized "cosmological constant" and u is the scaling variable corresponding to R. We also define the renormalized bosonic couplings as in Chapter 3

$$t_n^B = \begin{cases} \Lambda_c R_c^{-n}(1 + a^{2(1-n/m)} t_n) & , \quad n < m \\ \Lambda_c R_c^{-n} a^{2(1-n/m)} t_n & , \quad n \geq m \end{cases} \tag{6.142}$$

leading (6.124) to the same continuum string equation as in the bosonic model, (recall (3.70))

$$t = u^m - \sum_{n\geq 0} t_n u^n \quad , \tag{6.143}$$

the exact m-th critical point being given by $t_n = 0\,,\ u^m = t$.

Now we define the renormalized bosonic scaling operators coupled to t_n,

$$\langle\sigma_n\rangle \equiv N^2\frac{\Lambda^2}{\Lambda_c^2}\frac{\partial}{\partial t_n}F = \frac{1}{\Lambda_c^2}\frac{\partial t_n^B}{\partial t_n}\langle\sigma_n^B\rangle = \Lambda_c^{-1}R_c^{-n}a^{2(1-n/m)}\langle\sigma_n^B\rangle \quad . \tag{6.144}$$

From (6.134) we obtain its zeroth order 1-point function,

$$-a^2\Lambda_c\frac{\partial}{\partial\Lambda}\langle\sigma_n\rangle_0 = -\frac{(Na^{2+1/m})^2}{2(n+1)}\left[a^{-2/m}\left(1-\frac{R}{R_c}\right)\right]^{n+1} \quad . \tag{6.145}$$

The above expression is suitably written for the scaling limit. In particular we notice the presence of the special combination $(Na^{2+1/m})$ that characterized the double scaling limit in the purely bosonic model. As before we shall bind the large N limit to the scaling limit keeping fixed the constant.

$$Na^{2+1/m} \equiv \kappa^{-1} \quad . \tag{6.146}$$

We call κ the string coupling constant. Therefore the double scaling limit of (6.145) is

$$\frac{\partial}{\partial t}\langle\sigma_n\rangle_0 = -\frac{1}{2\kappa^2}\frac{u^{n+1}}{(n+1)} \quad . \tag{6.147}$$

Up to a factor 1/2, this zeroth order 1-point function coincides with the bosonic result (3.92). Notice also that, in the double limit, Eq. (6.144) reads

$$\langle\sigma_n\rangle = \frac{\partial}{\partial t_n}\ln\, Z_S(N,\Lambda_c) \quad , \tag{6.148}$$

which justifies, a posteriori, the renormalization in (6.133).

For the 2nd-order bosonic scaling operator, equation (6.144) implies

$$\begin{aligned}\langle\sigma_n\rangle_2 &= \Lambda_c^{-1}R_c^{-n}a^{2(1-n/m)}\langle\sigma_n^B\rangle_2\\ &= \Lambda_c^{-1}R_c^{-n}a^{2(1-n/m)}\frac{N^2}{2}\left((R_c-R)^n - R_c^{-1}(R_c-R)^{n+1}\right)N_+N_-\frac{\partial}{\partial\Lambda}R\\ &= \Lambda_c^{-2}R_c\frac{N^2}{2}a^{2/m}(N_+(R)N_-(R))[u^n - a^{2/m}u^{n+1}]\frac{\partial}{\partial t}u\\ &\Longrightarrow \frac{(Na^{2+1/m})^2}{2}\left(\Lambda_c^{-2}R_c a^{-4}N_+(R)N_-(R)\right)u^n\frac{\partial}{\partial t}u \quad ,\end{aligned} \tag{6.149}$$

which becomes finite in the double scaling limit if N_+ and N_- behave like

$$\begin{aligned}N_+(R) &\to a^{2+2\beta}\Lambda_c/\sqrt{R_c}\tau_+(u) \quad ,\\ N_-(R) &\to a^{2-2\beta}\Lambda_c/\sqrt{R_c}\tau_-(u) \quad ,\end{aligned} \tag{6.150}$$

near the critical point, so that $\Lambda_c^{-2}R_c a^{-4}N_+(R)N_-(R) \to \tau_+(u)\tau_-(u)$, and therefore the double scaling limit of (6.149) reads

$$\langle\sigma_n\rangle_2 = \frac{1}{2\kappa^2}\tau_+(u)\tau_-(u)u^n\frac{\partial}{\partial t}u \quad . \tag{6.151}$$

Notice that the exponent β in (6.150) is invisible to the 1-point function above. Nevertheless it shows up in the scaling of the auxiliary function $N_\pm$: indeed, from Eqs. (6.127, 130) and the definitions (6.150) we have

$$\begin{aligned} N_\pm(R) &= \sum_{n\geq 0} \tau_n^{B\pm}(R-R_c)^n \to \sum_{n\geq 0} a^{2n/m} \tau_n^{B\pm} R_c^n u^n \\ &= a^{2\pm 2\beta} \sum_{n\geq 0} \left[a^{-2(1\pm\beta-n/m)} R_c^n \tau_n^{B\pm} \right] u^n \quad , \end{aligned} \tag{6.152}$$

thus fixing the renormalized fermionic coupling constants $\{\tau_n^\pm\}$ as

$$\tau_n^\pm = \Lambda_c^{-1} R_c^{n+1/2} a^{-2(1\pm\beta-n/m)} \tau_n^{B\pm} \quad , \tag{6.153}$$

in terms of which the scaling functions $\tau_\pm(u)$ read

$$\tau_\pm(u) = \sum_{n\geq 0} \tau_n^\pm u^n \quad . \tag{6.154}$$

Coupled to the renormalized fermionic couplings we find a set of renormalized scaling operators $\{\nu_n^\pm\}$ satisfying

$$\begin{aligned} \frac{\partial}{\partial\Lambda}\langle\nu_n^\pm\rangle &\equiv \frac{\partial}{\partial\Lambda}\left(N^2 \frac{\Lambda^2}{\Lambda_c^2} \frac{\partial}{\tau_n^\pm} F \right) = \frac{1}{\Lambda_c^2} \frac{\partial \tau_n^{B\pm}}{\partial t_n^\pm} \frac{\partial}{\partial\Lambda} \langle \nu_n^{B\pm} \rangle \\ &= \frac{1}{\Lambda_c^2}\left(\Lambda_c R_c^{-n-1/2} a^{2(1\pm\beta-n/m)}\right)\left(\pm\frac{N^2}{2} N_\mp(R)(R_c-R)^n \frac{\partial}{\partial\Lambda}(R_c-R)\right) \\ &\to \pm\frac{(Na^{2+1/m})^2}{2} \tau_\mp(u) u^n \frac{\partial}{\partial\Lambda} u \quad , \end{aligned} \tag{6.155}$$

which tend, in the double scaling limit, to

$$\frac{\partial}{\partial t}\langle\nu_n^\pm\rangle = \pm\frac{1}{2\kappa^2} \tau_\mp(u) u^n \frac{\partial}{\partial t} u \quad . \tag{6.156}$$

Notice that the scaling exponent β has not been determined so far. It could be calculated from the scaling limit of an operator whose 1-point function depended on some product of the $N_\pm$ functions other than $N_+ \times N_-$. The odd bosonic loops $w_2^{(2k+1)}$ depend on $N_+ N'_+$, as stated in (6.119), and seem to be interesting candidates, but they have not a well defined scaling behavior; some odd bosonic scaling operators, built up from $w_2^{(2k+1)}$, ought to be defined. Since we assumed an even potential, i.e. $g_{2k+1} = 0$, there is no odd analog to the string equation $\Lambda = \Lambda(g_{2k}; R)$ to guide us in the construction of such odd scaling operators and respective coupling constants. In fact this shortage can be satisfactorily overcome by considering just small odd perturbations of the bosonic potential, i.e. it will be enough to find the solution of the loop equations up to 1st. order in the odd bosonic couplings $\{g_{2k+1}\}$.

Since $w_0(p)$ satisfies the same loop equation (up to a factor 2) of the purely bosonic theory, in which the perturbative odd solution is given in Appendix D, we just rewrite the final result

$$w^{(2k+1)} = -\Lambda^2 \frac{\partial}{\partial g_{2k+1}} F = \langle \frac{\Lambda}{N} \sum_i \lambda_i^{2k+1} \rangle \quad , \tag{6.157}$$

$$\frac{\partial}{\partial \Lambda} w_0^{2k+1} = (2k+1) \frac{R^k}{4^k} \binom{2k}{k} W_-(R) \frac{\partial R}{\partial \Lambda} \quad , \tag{6.158}$$

$$W_-(R) = -\frac{1}{2} \sum_{k \geq 0} g_{2k+1} (2k+1) \frac{R^k}{4^k} \binom{2k}{k} \quad . \tag{6.159}$$

In analogy to $N_-(R)$ we expand the auxiliary function $W_-(R)$ around the critical point, thus preparing it to the scaling limit

$$W_-(R) = -\frac{1}{2} \sum_{n \geq 0} t_n^{B-} (R_c - R)^n \quad , \tag{6.160}$$

$$t_n^{B-} = \sum_{k \geq n} \binom{k}{n} (-1)^n (2k+1) g_{2k+1} \frac{R_c^{k-n}}{4^k} \binom{2k}{k} \quad . \tag{6.161}$$

$$g_{2k+1} = \sum_{n \geq k} \frac{(-4)^k}{(2k+1)} \frac{\binom{n}{k}}{\binom{2k}{k}} t_n^{B-} R_c^{n-k} \quad . \tag{6.162}$$

Coupled to the bare constants $\{t_n^{B-}\}$ we find the following odd bare bosonic scaling operators

$$\begin{aligned} \langle \sigma_n^{B-} \rangle \equiv N^2 \Lambda^2 \frac{\partial}{\partial t_n^{B-}} F &= -N^2 \sum_{k \geq 0} \left(\frac{\partial g_{2k+1}}{\partial t_n^{B-}} \right) w^{(2k+1)} \\ &= -N^2 \sum_{k=0}^{n} \frac{(-4)^k}{(2k+1)} \frac{\binom{n}{k}}{\binom{2k}{k}} R_c^{n-k} w^{(2k+1)} \quad . \end{aligned} \tag{6.163}$$

Equation (6.158) gives the zeroth order (in fermionic couplings) 1-point functions,

$$\begin{aligned} \frac{\partial}{\partial \Lambda} \langle \sigma_n^{B-} \rangle_0 &= -N^2 \sum_{k=0}^{n} \binom{n}{k} R_c^{n-k} (-R)^k W_-(R) \frac{\partial R}{\partial \Lambda} \\ &= -\frac{N^2}{2} \sum_{k \geq 0} t_k^{B-} (R_c - R)^{k+n} \frac{\partial}{\partial \Lambda} (R_c - R) \\ &= \frac{\partial}{\partial \Lambda} \left(-\frac{N^2}{2} \sum_{k \geq 0} t_k^{B-} \frac{(R_c - R)^{k+n+1}}{(k+n+1)} \right) \quad , \end{aligned} \tag{6.164}$$

which can be integrated to give

$$\langle\sigma_n^{B-}\rangle_0 = -\frac{N^2}{2}\sum_{k\geq 0} t_k^{B-}\frac{(R_c - R)^{k+n+1}}{(k+n+1)} \quad . \tag{6.165}$$

This 1-point function tends to zero when $t_k^{B-} \to 0$, as expected, and therefore fixes no scaling at all. But the following 2-point functions

$$\begin{aligned}\langle\sigma_i^{B-}\sigma_j^{B-}\rangle_0 &= \Lambda^2\frac{\partial}{\partial t_j^{B-}}\langle\sigma_i^{B-}\rangle = -\frac{\Lambda^2 N^2}{2}\frac{(R_c - R)^{i+j+1}}{(i+j+1)} \quad , \\ &= -\frac{1}{2}\Lambda_c^2 R_c^{i+j+1} a^{-4+2(i+j)/m}(Na^{2+1/m})^2\frac{u^{i+j+1}}{(i+j+1)} \quad , \end{aligned} \tag{6.166}$$

tell us to renormalize the odd operators as

$$\sigma_n^- = \Lambda^{-1} R_c^{-n} a^{2(1-n/m)}\sigma_n^{B-} \quad , \tag{6.167}$$

so that the 2-point function of odd bosonic operators has a non-vanishing double scaling limit

$$\langle\sigma_i^-\sigma_j^-\rangle_0 = -\frac{R_c}{2\kappa^2}\frac{u^{i+j+1}}{(i+j+1)} \quad . \tag{6.168}$$

Back to the 1-point function (6.165) renormalization (6.167) gives

$$\begin{aligned}\langle\sigma_n^-\rangle_0 &= -\frac{1}{2}\Lambda_c^{-1} R_c^{-n} a^{2(1-n/m)} N^2\sum_{k\geq 0} t_k^{B-}\frac{(R_c - R)^{k+n+1}}{(k+n+1)} \\ &= -\frac{R_c}{2}(Na^{2+1/m})^2\sum_{k\geq 0}(\Lambda_c^{-1} R_c^k a^{-2(1-k/m)} t_k^{B-})\frac{u^{k+n+1}}{(k+n+1)} \\ &= -\frac{R_c}{2\kappa^2}\sum_{k\geq 0} t_k^-\frac{u^{k+n+1}}{(k+n+1)} \quad , \end{aligned} \tag{6.169}$$

where the renormalized odd bosonic couplings are defined as

$$t_n^- = \Lambda_c^{-1} R_c^k a^{-2(1-k/m)} t_n^{B-} \quad . \tag{6.170}$$

Now that we have a well defined odd bosonic scaling operator we can return to (6.118) and calculate its 1-point function at 2nd-order in the fermionic couplings:

$$\begin{aligned}\langle\sigma_n^-\rangle_2 &= \Lambda_c^{-1} R_c^{-n} a^{2(1-n/m)}\langle\sigma_n^{B-}\rangle_2 \\ &= -N^2\Lambda_c^{-1} R_c^{-n} a^{2(1-n/m)}\sum_{k=0}^{n}\frac{(-4)^k}{(2k+1)}\frac{\binom{n}{k}}{\binom{2k}{k}} R_c^{n-k} w_2^{(2k+1)} \\ &= N^2\Lambda_c^{-1} R_c^{-n} a^{2(1-n/m)}(R_c - R)^n N_+(R) N_+'(R) R\frac{\partial}{\partial\Lambda} R \\ &= -a^{4(\beta-1/2m)}(Na^{2+1/m})^2\tau_+(u)\tau_+'(u) u^n\frac{\partial}{\partial t} u \quad , \end{aligned} \tag{6.171}$$

which, in the double scaling limit, is finite and reads

$$\langle \sigma_n^- \rangle_2 = -\frac{1}{\kappa^2} \tau_+(u) \tau_+'(u) u^n \frac{\partial}{\partial t} u \quad , \tag{6.172}$$

as long as the exponent β is taken to be $\beta = \frac{1}{2m}$.

Therefore we have fixed the scaling of all couplings, renormalized operators and equations, whose double scaling limit is taken as the definition of the continuum limit of the original discrete model.

6.3.1. Free Energy and Critical Exponents

As in the bosonic model, we shall take the double scaled 1-point functions from the previous section to define a renormalized free energy for the system. From the scaling behavior of correlation functions with respect to the cosmological constant, we shall also define the string susceptibility and the scaling dimensions associated to bosonic and fermionic operators.

The 1-point functions calculated along the previous section depend on the bosonic couplings t_n's implicitly through the function $u(t; t_n)$, while the fermionic ones $\{\tau_n^\pm\}$ are concentrated in the auxiliary functions $\tau_\pm(u; \tau_n^\pm)$. Since the insertion of the scaling operators $\sigma_n(\nu_n^\pm)$ correspond to the action of the derivatives $\frac{\partial}{\partial t_n} (\frac{\partial}{\partial \tau_n^\pm})$ on correlation functions, it is important to know how these operators act on u and $\tau_\pm$ – in other words, the flows of the functions u and $\tau_\pm$.

From Eq. (6.143) we obtain exactly the same t-flows found in Chapter 3 (recall Eq. (3.71)), namely

$$\frac{\partial u}{\partial t_n} = u^n \frac{\partial u}{\partial t} \quad . \tag{6.173}$$

On the other hand, by construction the function u does not involve the odd couplings, and therefore

$$\frac{\partial u}{\partial \tau_n^\pm} = 0 \quad . \tag{6.174}$$

Concerning the fermionic functions $\tau_\pm$, the definition (6.154) tells us that

$$\frac{\partial \tau_\pm}{\partial \tau_n^\pm} = u^n \quad , \quad \frac{\partial \tau_\pm}{\partial \tau_n^\mp} = 0 \quad , \tag{6.175}$$

while its even flows are easily calculated using (6.173),

$$\begin{aligned} \frac{\partial \tau_\pm}{\partial t_n} &= \sum_{a \geq 0} a \tau_a^\pm u^{a-1} \frac{\partial u}{\partial t_n} = \sum_{a \geq 0} a \tau_a^\pm u^{a-1+n} \frac{\partial u}{\partial t} \\ &= u^n \frac{\partial}{\partial t} \sum_{a \geq 0} \tau_a^\pm u^a = u^n \frac{\partial \tau_\pm}{\partial t} \quad . \end{aligned} \tag{6.176}$$

Assuming that the double-scaled correlation functions can be directly calculated from a double-scaled (continuum) free energy $\mathcal{F}_S$ we can read it from the results (6.147) and (6.151),

$$\begin{aligned}\langle\sigma_n\rangle &\equiv \frac{\partial}{\partial t_n}\mathcal{F}_S = \langle\sigma_n\rangle_0 + \langle\sigma_n\rangle_2 \\ &= -\frac{1}{2\kappa^2}\partial_t^{-1}\left(\frac{u^{n+1}}{(n+1)}\right) + \frac{1}{2\kappa^2}\tau_+(u)\tau_-(u)u^n\partial_t u \\ &= -\frac{1}{2\kappa^2}\left[\partial_t^{-2}(u^n\partial_t u) - \partial_t^{-1}((u^n\partial_t(\tau_+\tau_-))\partial_t u + \tau_+\tau_-(u^n\partial_t u))\right] \\ &= \frac{\partial}{\partial t_n}\left[-\frac{1}{2\kappa^2}\left(\partial_t^{-2}u - \partial_t^{-1}(\tau_+\tau_-\partial_t u)\right)\right] \quad, \end{aligned} \tag{6.177}$$

as well as from the fermionic 1-point functions (6.156) which we rewrite as

$$\begin{aligned}\langle\nu_n^\pm\rangle &\equiv \frac{\partial}{\partial\tau_n^\pm}\mathcal{F}_S = \pm\frac{1}{2\kappa^2}\partial_t^{-1}\left[\tau_\mp(u)u^n\partial_t u\right] \\ &= \frac{1}{2\kappa^2}\partial_t^{-1}\left[\frac{\partial}{\partial\tau_n^\pm}(\tau_+\tau_-\partial_t u)\right] \\ &= \frac{\partial}{\partial\tau_n^\pm}\left[-\frac{1}{2\kappa^2}\left(\partial_t^{-2}u - \partial_t^{-1}(\tau_+\tau_-\partial_t u)\right)\right] \quad . \end{aligned} \tag{6.178}$$

The above expressions are compatible with the definition

$$\mathcal{F}_S = -\frac{1}{2\kappa^2}\left(\partial_t^{-2}u - \partial_t^{-1}(\tau_+\tau_-\partial_t u)\right) \quad . \tag{6.179}$$

Such free energy can now be taken to calculate higher order correlation functions and therefore measure scaling exponents. In the zero coupling (zc) regime, when $t_n \to 0$ and $\tau_n^\pm \to 0$, the scaling of the free energy defines the string susceptibility γ,

$$\left(\partial_t^2\mathcal{F}_S\right)_{zc} \sim t^{-\gamma} \quad , \tag{6.180}$$

while the scaling dimension d of some scaling operator $\mathcal{O}$ is read from

$$\partial_t^2\langle\prod_i \mathcal{O}_i\rangle_{zc} \sim t^{-\gamma+\sum_i(d_i-1)} \quad . \tag{6.181}$$

For the model defined by the string equation (6.143), the functions (6.154) and the free energy (6.179), we find

$$\left(\partial_t^2\mathcal{F}_S\right)_{zc} \sim (u)_{zc} = t^{1/m} \quad , \tag{6.182}$$

generating the same family of susceptibilities $\gamma = -\frac{1}{m}$, $m = 1, 2, 3\cdots$ detected in the purely bosonic theory.

For the even bosonic operators we find

$$\partial_t^2 \langle \sigma_n \rangle_{zc} \sim \left(\frac{\partial}{\partial t_n} u \right)_{zc} = (u^n \partial_t u)_{zc}$$
$$\sim t^{n/m+1/m-1} = t^{-\gamma+(n/m-1)} \quad , \tag{6.183}$$

thus giving the same spectrum of dimensions found in Chapter 3,

$$d_{\sigma_n} = \frac{n}{m} \quad . \tag{6.184}$$

Equation (6.168) allows us to calculate also the dimensions of the odd bosonic operators:

$$\partial_t^2 \langle \sigma_i^- \sigma_j^- \rangle_{zc} \sim \partial_t^2 (t^{i/m+j/m+1/m}) \sim t^{1/m+i/m+j/m-2}$$
$$= t^{-\gamma+(i/m-1)+(j/m-1)} \quad , \tag{6.185}$$

thus duplicating the even spectrum:

$$d_{\sigma_n^-} = \frac{n}{m} \quad , \tag{6.186}$$

a phenomenon present in the purely bosonic theory too.

In order to measure the dimension of a fermionic operators unambiguously, we start from the 1-point function (6.172) and calculate the following 3-point function

$$\langle \sigma_n^- \nu_i^+ \nu_j^+ \rangle \sim \frac{\partial^2}{\partial \tau_i^+ \partial \tau_j^+} (\tau_+ \tau_+') u^n \partial_t u$$
$$= \frac{\partial^2}{\partial \tau_i^+ \partial \tau_j^+} \left(\sum_{k,l \geq 0} \tau_k^+ l \tau_l^+ u^{k+l-1} \right) u^n \partial_t u$$
$$= \frac{\partial}{\partial \tau_i^+} \left(\sum_{l \geq 0} l \tau_l^+ u^{j+l-1} - \sum_{k \geq 0} \tau_k^+ j u^{k+j-1} \right) u^n \partial_t u$$
$$= (i-j) u^{i+j+n-1} \partial_t u \sim \partial_t u^{i+j+n} \quad , \tag{6.187}$$

which, in the zero-coupling regime, gives the scaling

$$\partial_t^2 \langle \sigma_n^- \nu_i^+ \nu_j^+ \rangle_{zc} \sim t^{i/m+j/m+n/m-3}$$
$$= t^{1/m+(n/m-1)+(i/m-1/2m-1)+(j/m-1/2m-1)} \quad . \tag{6.188}$$

Since we already know the string susceptibility and the scaling of σ_n^-, we conclude that the dimensions of the even fermionic operators are

$$d_{\nu_n^+} = \frac{n}{m} - \frac{1}{2m} \quad . \tag{6.189}$$

Finally we take the 1-point function (6.156) and calculate

$$\partial_t \langle \nu_i^+ \nu_j^- \rangle = -\frac{1}{2\kappa^2} u^{i+j} \partial_t u \sim \partial_t (u^{i+j+1}) \quad , \tag{6.190}$$

whose zero-coupling limit,

$$\begin{aligned}\partial_t^2 \langle \nu_i^+ \nu_j^- \rangle_{zc} &\sim \partial_t^2 (t^{i/m+j/m+1/m}) \sim t^{i/m+j/m+1/m-2} \\ &= t^{1/m+(i/m-1/2m-1)+(j/m+1/2m-1)} \quad ,\end{aligned} \tag{6.191}$$

provides the dimension of the remaining odd fermionic operators

$$d_{\nu_n^-} = \frac{n}{m} + \frac{1}{2m} \quad . \tag{6.192}$$

Comparing the spectra (6.189) and (6.192), we see that, except for the operator ν_0^+, there is a duplication of dimensions as in the bosonic sector[3)].

Notice that the scaling dimensions above are consistent with the renormalization of the operators: indeed, equations (6.144), (6.153) and (6.167) can be written as

$$\sigma_n^\pm = \Lambda_c^{-1} R_c^{-n} [a^{-2}]^{(d_{\sigma_n^\pm} - 1)} \sigma_n^{B\pm} \quad , \tag{6.193}$$
$$\nu_n^\pm = \Lambda_c^{-1} R_c^{-n-1/2} [a^{-2}]^{(d_{\nu_n^\pm} - 1)} \nu_n^\pm \quad . \tag{6.194}$$

With the free energy at hands, we can proceed in the calculation of higher order correlation functions.

First we rewrite Eq. (6.179) as follows

$$\begin{aligned}-2\kappa^2 \partial_t^2 \mathcal{F}_S &= u - \partial_t (\tau_+ \tau_- \partial_t u) \\ &= (1 - \partial_t \tau_+ \tau_- \partial_t) u \quad ,\end{aligned} \tag{6.195}$$

and the insertion of an operator σ_n corresponds to

$$\begin{aligned}\frac{\partial}{\partial t_n} (-2\kappa^2 \partial_t^2 \mathcal{F}_S) &= (1 - \partial_t \tau_+ \tau_- \partial_t) \frac{\partial}{\partial t_n} u - (\partial_t \left[\frac{\partial}{\partial t_n} (\tau_+ \tau_-) \right] \partial_t) u \\ &= (1 - \partial_t \tau_+ \tau_- \partial_t) u^n \partial_t u - \partial_t (\partial_t (\tau_+ \tau_-)) u^n \partial_t u \\ &= \partial_t (1 - \partial_t \tau_+ \tau_- \partial_t) \left(\frac{u^{n+1}}{n+1} \right) \quad ,\end{aligned} \tag{6.196}$$

suggesting the following definition of "generalized polynomials"

$$\rho_n[u] \equiv (1 - \partial_t \tau_+ \tau_- \partial_t) \frac{u^n}{n} \quad . \tag{6.197}$$

---

3) Spectrum duplication is related to the assumption of even potentials. In ref. [152] the authors consider a general polynomial potential and show that this degeneracy is removed.

In particular the free energy can be written as

$$\mathcal{F}_S = -\frac{1}{2\kappa^2}\partial_t^{-2}\rho_1[u] \quad . \tag{6.198}$$

The polynomials thus defined generalize the monomials $R_n[u]$ introduced in Chapter 3 to the supersymmetric case.

We calculate their even and odd flows as they will be useful in the evaluation of N-point functions; we have

$$\begin{aligned}
\frac{\partial}{\partial t_n}\rho_m[u] &= (1-\partial_t\tau_+\tau_-\partial_t)\frac{\partial}{\partial t_n}\frac{u^m}{m} - \partial_t\left(\frac{\partial}{\partial t_n}(\tau_+\tau_-)\right)\partial_t\frac{u^m}{m} \\
&= (1-\partial_t\tau_+\tau_-\partial_t)u^{n+m-1}\partial_t u - \partial_t(\partial_t(\tau_+\tau_-))u^n\partial_t\frac{u^m}{m} \\
&= \partial_t(1-\partial_t\tau_+\tau_-\partial_t)\frac{u^{n+m}}{(n+m)} \Longrightarrow
\end{aligned}$$

$$\Longrightarrow \frac{\partial}{\partial t_n}\rho_m[u] = \frac{\partial}{\partial t}\rho_{n+m}[u] \quad , \tag{6.199}$$

and

$$\begin{aligned}
\frac{\partial}{\partial \tau_n^{\pm}}\rho_m[u] &= (\mp\partial_t\tau_\mp u^n\partial_t)\frac{u^m}{m} \\
&= (\mp\partial_t\tau_\mp\partial_t)\frac{u^{n+m}}{(n+m)} \\
&= \frac{\partial}{\partial\tau_\pm}(1-\partial_t\tau_+\tau_-\partial_t)\frac{u^{n+m}}{(n+m)} \Longrightarrow
\end{aligned}$$

$$\Longrightarrow \frac{\partial}{\partial \tau_n^{\pm}}\rho_m[u] = \frac{\partial}{\partial\tau_\pm}\rho_{n+m}[u] \quad . \tag{6.200}$$

The expressions above should be compared to their purely bosonic counterpart, Eq. (3.73). Now we can calculate all correlators using (6.198) and the flows (6.199, 200): the non-vanishing multi-point functions are

$$\langle\prod_{i=1}^n \sigma_{a_i}\rangle = -\frac{1}{2\kappa^2}\partial_t^{n-2}\rho_{a+1}[u] \quad , \quad a = \sum_i a_i \quad , \tag{6.201}$$

$$\langle\nu_k^{\pm}\prod_{i=1}^n \sigma_{a_i}\rangle = -\frac{1}{2\kappa^2}\partial_t^{n-2}\frac{\partial}{\partial\tau_\pm}\rho_{k+a+1}[u] \quad , \tag{6.202}$$

$$\langle\nu_k^{+}\nu_l^{-}\prod_{i=1}^n \sigma_{a_i}\rangle = -\frac{1}{2\kappa^2}\partial_t^{n-2}\frac{\partial}{\partial\tau_+}\frac{\partial}{\partial\tau_-}\rho_{k+l+a+1}[u] \quad . \tag{6.203}$$

In particular, in the zero-coupling regime, we find

$$
\begin{aligned}
\langle\prod_{i=1}^{a}\sigma_{a_i}\rangle_{zc} &= -\frac{1}{2\kappa^2}\partial_t^{n-2}\frac{(t^{1/m})^{a+1}}{a+1} = -\frac{1}{2m\kappa^2}\partial_t^{n-3}t^{-1+1/m+a/m} \\
&= -\frac{1}{2m\kappa^2}\frac{\Gamma(1/m+a/m)}{\Gamma(3-n+1/m+a/m)}\,t^{2+1/m+a/m-n} \\
&= \frac{\gamma}{2\kappa^2}\frac{\Gamma\left(n-\gamma+\sum(d_{\sigma_{a_i}}-1)\right)}{\Gamma\left(3-\gamma+\sum(d_{\sigma_{a_i}}-1)\right)}\,t^{2-\gamma+\sum(d_{\sigma_{a_i}}-1)}
\end{aligned}
\tag{6.204}
$$

$$
\langle\nu_k^{\pm}\prod_{i=1}^{n}\sigma_{a_i}\rangle_{zc} = 0 \quad , \tag{6.205}
$$

$$
\begin{aligned}
\langle\nu_k^{+}\nu_l^{-}\prod_{i=1}^{n}\sigma_{a_i}\rangle_{zc} &= -\frac{1}{2\kappa^2}\partial_t^{n-2}\partial_t^2\frac{t^{(k+l+a+1)/m}}{k+l+a+1} \\
&= -\frac{1}{2m\kappa^2}\partial_t^{(2+n)-1}\,t^{(l+k+a+1)/m-1} \\
&= \frac{\gamma}{2\kappa^2}\frac{\Gamma\left((2+n)-\gamma+(d_{\nu_k^+}-1)+(d_{\nu_l^-}-1)+\sum(d_{\sigma_{a_i}}-1)\right)}{\Gamma\left(3-\gamma+(d_{\nu_k^+}-1)+(d_{\nu_l^-}-1)+\sum(d_{\sigma_{a_i}}-1)\right)}\times \\
&\qquad\times t^{2-\gamma+(d_{\nu_k^+}-1)+(d_{\nu_l^-}-1)+\sum(d_{\sigma_{a_i}}-1)} \quad .
\end{aligned}
\tag{6.206}
$$

The reader is invited to compare the above results with the correlation functions (3.106). Notice that non-vanishing correlators involve at most two fermionic operators. This is a consequence of the bilinear dependence on the $\tau^{\pm}$-couplings exhibited by the free energy (6.195).

We shall finish this section observing that the generalized polynomials (6.197) can be related to their bosonic counterparts $R_n[u]$ as follows:

$$
\begin{aligned}
\rho_n[u] &= (1-\partial_t\tau_+\tau_-\partial_t)\frac{u^n}{n} \\
&= \frac{u^n}{n}-\sum_{i,j}\tau_i^+\tau_j^-\partial_t\left[u^{i+j}\partial_t\frac{u^n}{n}\right] \\
&= \frac{u^n}{n}-\sum_{i,j}\tau_i^+\tau_j^-\partial_t^2\left[\frac{u^{i+j+n}}{(i+j+n)}\right] \\
&= \frac{u^n}{n}-\sum_{i,j}\tau_i^+\tau_j^-\frac{\partial^2}{\partial t_i\partial t_j}\left[\frac{u^n}{n}\right]\Longrightarrow \\
\Longrightarrow\rho_n[u] &= \left(1-\sum_{i,j}\tau_i^+\tau_j^-\frac{\partial^2}{\partial t_i\partial t_j}\right)R_n[u]
\end{aligned}
\tag{6.207}
$$

thus establishing a simple connection between the free energy of the purely bosonic (recall (3.94)) and supersymmetric models,

$$
\mathcal{F}_S = \frac{1}{2}\left(1-\sum_{i,j}\tau_i^+\tau_j^-\frac{\partial^2}{\partial t_i\partial t_j}\right)\mathcal{F}_B \quad . \tag{6.208}
$$

In ref. [154] a similar relation was proved to hold for general potentials and beyond the planar approximation. The arguments are based upon an exact calculation of the integral (6.44), showing that the at most bilinear dependence of $\mathcal{F}_S$ with respect to the fermionic couplings, which was found[84] originally in the planar solution of the superloop equations, is valid for all genera. We shall comment on the non-perturbative results of [154] in Chapter 8.

6.3.2. Macroscopic Superloops

Before comparing results with the continuum theory, let us introduce[84] the macroscopic superloop operators, characterized by finite super lengths.

As in the bosonic theory, we associate macroscopic loops to the insertion of operators $w^{(2k)}$ and $\nu^{(k)}$ in the large k limit. The bosonic part of a macroscopic loop is therefore built up considering the limit

$$\lim_{k\to\infty}\langle\sum_i \lambda_i^{2k}\rangle = \frac{N}{\Lambda}\lim_{k\to\infty} w^{(2k)} \quad , \tag{6.209}$$

whose zeroth order contribution comes from Eq. (6.120a)

$$\begin{aligned}\lim_{k\to\infty} w_0^{(2k)} &= \partial_\Lambda^{-1}\lim_{k\to\infty}\partial_\Lambda w_0^{2k} = \partial_\Lambda^{-1}\lim_{k\to\infty}\binom{2k}{k}\frac{R^k}{4^k}\\ &= R_c^k\partial_\Lambda^{-1}\lim_{k\to\infty}\binom{2k}{k}\frac{(R/R_c)^k}{4^k}\\ &\to \frac{R_c^k}{\sqrt{\pi k}}\partial_\Lambda^{-1}\,\mathrm{e}^{k\ln(R/R_c)} \quad ,\end{aligned} \tag{6.210}$$

while the 2nd order piece follows from the 1-point function (6.120b),

$$\begin{aligned}\lim_{k\to\infty} w_2^{(2k)} &= \lim_{k\to\infty}(-k)\frac{R^k}{4^k}\binom{2k}{k}N_+(R)N_-(R)\partial_\Lambda R\\ &= -N_+(R)N_-(R)\partial_\Lambda\lim_{k\to\infty}k\binom{2k}{k}\frac{R^k}{4^k}\\ &\longrightarrow -N_+(R)N_-(R)\partial_\Lambda R\sqrt{\frac{k}{\pi}}R_c^k\mathrm{e}^{k\ln(R/R_c)} \quad .\end{aligned} \tag{6.211}$$

Indeed, we are interested in the DSLim (double scaling limit, $a\to 0$ and $N\to a^{-2-1/m}\kappa^{-1}$) of Eq. (6.209), given by

$$\begin{aligned}\underset{k\to\infty}{\mathrm{DSlim}}\langle\sum_i\lambda_i^{2k}\rangle &\to -R_c^kNa^2\sqrt{\frac{k}{\pi}}\left(\partial_t^{-1}\frac{\mathrm{e}^{-ka^{2/m}u}}{k}+a^{2/m}\tau_+\tau_-\mathrm{e}^{-ka^{2/m}u}\partial_t u\right)\\ &= R_c^k(Na^{2+1/m})\sqrt{\frac{ka^{2/m}}{\pi}}\left(-\partial_t^{-1}\frac{\mathrm{e}^{-ka^{2/m}u}}{ka^{2/m}}-\tau_+\tau_-\mathrm{e}^{-ka^{2/m}u}\partial_t u\right)\end{aligned} \tag{6.212}$$

thus suggesting the definition of the renormalized loop length

$$l \equiv ka^{2/m} \quad , \tag{6.213}$$

as in the bosonic theory. Therefore we define a macroscopic bosonic loop of length l as

$$\begin{aligned}\langle U(l)\rangle &\equiv \underset{k\to la^{-2/m}}{\text{DSlim}} \quad R_c^{-k}\langle\sum_i \lambda_i^{2k}\rangle = \\ &= -\frac{1}{\kappa}\sqrt{\frac{l}{\pi}}\left(\partial_t^{-1}\frac{e^{-lu}}{l} + \tau_+\tau_- e^{-lu}\partial_t u\right) \quad ,\end{aligned} \tag{6.214}$$

as well as the associated rescaled loop

$$\begin{aligned}\langle \tilde{U}(l)\rangle \equiv \sqrt{\frac{\pi}{l}}\langle U(l)\rangle &= -\frac{1}{\kappa l}\partial_t^{-1}e^{-lu} - \frac{1}{\kappa}\tau_+\tau_- e^{-lu}\partial_t u \\ &= \frac{1}{\kappa l}\left(-\partial_t^{-1} + \tau_+\tau_-\partial_t\right)e^{-lu} \quad ,\end{aligned} \tag{6.215}$$

introduced (as in the bosonic model) to simplify the expression of the multiloop correlation functions. By analogy, we start the construction of fermionic loops by studying the following large k limit: the even fermionic operators

$$\begin{aligned}\lim_{k\to\infty}\langle\sum_i \theta_i\lambda_i^{2k}\rangle &= \frac{N}{\Lambda}\lim_{k\to\infty}\nu^{(2k)} = \frac{N}{\Lambda}\partial_\Lambda^{-1}\lim_{k\to\infty}\partial_\Lambda\nu^{(2k)} \\ &= \Lambda^{-1}N\partial_\Lambda^{-1}\left(\frac{1}{2}N_-(R)\partial_\Lambda R\lim_{k\to\infty}\frac{R^k}{4^k}\binom{2k}{k}\right) \\ &\to \Lambda^{-1}R_c^k N\partial_\Lambda^{-1}\left(\frac{1}{2}N_-(R)\partial_\Lambda R\frac{1}{\sqrt{\pi R}}e^{k\ln(R/R_c)}\right)\end{aligned} \tag{6.216}$$

and their double scaling limit

$$\begin{aligned}&\underset{k\to\infty}{\text{DSlim}}\langle\sum_i \theta_i\lambda_i^{2k}\rangle = \\ &= \frac{\Lambda_c^{-1}R_c^k N}{2\sqrt{\pi R}}(-a^2\Lambda_c\partial_t^{-1})\left(a^{2-1/m}\Lambda_c R_c^{-1/2}\tau_- a^{2/m-2}\Lambda_c^{-1}R_c\partial_t u e^{-ka^{2/m}u}\right) \\ &= -\frac{a^{1/m}R_c^{k+1/2}}{2}\frac{(Na^{2+1/m})}{\sqrt{\pi k a^{2/m}}}\partial_t^{-1}\left(\tau_-(u)e^{-ka^{2/m}u}\partial_t u\right) \\ &= -a^{1/m}\frac{R_c^{k+1/2}}{2}\frac{1}{\kappa\sqrt{\pi l}}\partial_t^{-1}\left(\tau_-(u)e^{-lu}\partial_t u\right) \quad ,\end{aligned} \tag{6.217}$$

where $ka^{2/m}$ is once again taken as a length l. We also consider the odd fermionic operators

$$\begin{aligned}\lim_{k\to\infty}\langle\sum_i \theta_i\lambda_i^{2k+1}\rangle &= \frac{n}{\Lambda}\lim_{k\to\infty}\nu^{(2k+1)} = \Lambda^{-1}N\partial_\Lambda^{-1}\lim_{k\to\infty}\partial_\Lambda\nu^{(2k+1)} \\ &= \Lambda^{-1}N\partial_\Lambda^{-1}\left(-\frac{1}{2}N_+(R)\partial_\Lambda R\lim_{k\to\infty}(2k+1)\frac{R^k}{4^k}\binom{2k}{k}\right) \quad ,\end{aligned} \tag{6.218}$$

whose double scaling limit reads

$$
\begin{aligned}
\underset{k\to\infty}{\mathrm{DSlim}}\,\langle\sum_i \theta_i\lambda_i^{2k+1}\rangle = \\
&= R_c^{k+1/2}\sqrt{\frac{k}{\pi}}(Na^{2+1/m})a^{2/m}\partial_t^{-1}\left(\tau_+(u)\mathrm{e}^{-ka^{2/m}u}\partial_t u\right) \\
&= a^{1/m}R_c^{k+1/2}(Na^{a+1/2})\sqrt{\frac{ka^{2/m}}{\pi}}\partial_t^{-1}\left(\tau_+(u)\mathrm{e}^{-ka^{2/m}u}\partial_t u\right) \\
&= a^{1/m}R_c^{k+1/2}\sqrt{\frac{l}{\pi}\frac{1}{\kappa}}\partial_t^{-1}(\tau_+(u)\mathrm{e}^{-lu}\partial_t u) \quad . \qquad (6.219)
\end{aligned}
$$

The results (6.217) and (6.219) induce the following definitions of macroscopic fermionic loops

$$
\begin{aligned}
\langle V_+(l)\rangle &\equiv \underset{k\to\infty}{\mathrm{DSlim}}\ 2R_c^{-k-1/2}\langle\sum_i \theta_i\lambda_i^{2k}\rangle \\
&= -a^{1/m}\frac{1}{\kappa\sqrt{\pi l}}\partial_t^{-1}(\tau_-(u)\mathrm{e}^{-lu}\partial_t u) \quad ; \qquad (6.220)
\end{aligned}
$$

$$
\begin{aligned}
\langle V_-(l)\rangle &\equiv \underset{k\to\infty}{\mathrm{DSlim}}\ R_c^{-k-1/2}\langle\sum_i \theta_i\lambda_i^{2k+1}\rangle \\
&= +a^{1/m}\frac{1}{\kappa}\sqrt{\frac{l}{\pi}}\partial_t^{-1}(\tau_+(u)\mathrm{e}^{-lu}\partial_t u) \quad ; \qquad (6.221)
\end{aligned}
$$

and the respective rescaled loops,

$$
\langle\tilde{V}_\pm(l)\rangle \equiv \sqrt{\pi}l^{\pm 1/2}\langle V_\pm(l)\rangle = \mp\frac{a^{1/m}}{\kappa}\partial_t^{-1}(\tau_\mp(u)\mathrm{e}^{-lu}\partial_t u) \quad . \qquad (6.222)
$$

The presence of the $a^{1/m}$ factor above is not really a trouble: in fact, if we now define [84] a superloop with length l and (bare) super length $\theta_\pm^B$ as

$$
\tilde{W}_\pm(l,\theta_\pm) \equiv \tilde{U}(l) + \theta_\pm\tilde{V}(l) \quad , \qquad (6.223)
$$

with

$$
\theta_\pm \equiv a^{1/m}\theta_\pm^B \qquad (6.224)
$$

as the corresponding renormalized super length, its 1-point function follows from Eqs. (6.215) and (6.222),

$$
\begin{aligned}
\langle\tilde{W}_\pm(l,\theta)\rangle &= \frac{1}{\kappa l}(-\partial_t^{-1} + \tau_+\tau_-\partial_t)\mathrm{e}^{-lu} \mp \theta_\pm\partial_t^{-1}(\tau_\mp(u)\mathrm{e}^{-lu}\partial_t u) \\
&= \frac{1}{\kappa l}(-\partial_t^{-1} + \tau_+\tau_-\partial_t \pm \theta_\pm\partial_t^{-1}\tau_\mp\partial_t)\mathrm{e}^{-lu} \quad . \qquad (6.225)
\end{aligned}
$$

The renormalization (6.224) gives to the superlength $\theta_\pm$ a well defined dimension, as discussed in [84].

Before facing the multiloop correlators, it is useful to note that the fermionic loop functions can be related to the bosonic ones as follows

$$\begin{aligned}\langle \tilde{V}_{\pm}(l)\rangle &= a^{1/m}\partial_t^{-1}\left(\pm\frac{1}{\kappa l}\tau_{\mp}(u)\partial_t \mathrm{e}^{-lu}\right)\\ &= a^{1/m}\partial_t^{-1}\frac{\partial}{\partial\tau_{\pm}}\left[\frac{1}{\kappa l}(-\partial_t^{-1}+\tau_+\tau_-\partial_t)\mathrm{e}^{-lu}\right]\\ &= a^{1/m}\partial_t^{-1}\frac{\partial}{\partial\tau_{\pm}}\langle\tilde{U}(l)\rangle \quad ,\end{aligned} \tag{6.226}$$

and therefore a superloop can be constructed from its bosonic part as given below

$$\langle \tilde{W}_{\pm}(l,\theta_{\pm})\rangle = \left(1+\theta_{\pm}\partial_t^{-1}\frac{\partial}{\partial\tau_{\pm}}\right)\langle\tilde{U}(l)\rangle \quad . \tag{6.227}$$

This relation can now be used to write a multi-superloop correlation function as

$$\begin{aligned}&\langle \tilde{W}_+(l_{+1},\theta_{+1})\cdots\tilde{W}_+(l_{+a},\theta_{+a})\tilde{W}_-(l_{-1},\theta_{-1})\cdots\tilde{W}_-(l_{-b},\theta_{-b})\rangle =\\ &= \prod_{i=1}^{a}\left(1+\theta_{+i}\partial_t^{-1}\frac{\partial}{\partial\tau_+}\right)\prod_{j=1}^{b}\left(1+\theta_{-j}\partial_t^{-1}\frac{\partial}{\partial\tau_-}\right)\times\\ &\qquad\times\langle\tilde{U}(l_{+1})\cdots\tilde{U}(l_{+a})\tilde{U}(l_{-1})\cdots\tilde{U}(l_{-b})\rangle\\ &= \left[1+\left(\sum_{i=1}^{a}\theta_{+i}\right)\partial_t^{-1}\frac{\partial}{\partial\tau_+}\right]\left[1+\left(\sum_{j=1}^{b}\theta_{-j}\right)\partial_t^{-1}\frac{\partial}{\partial\tau_-}\right]\times\\ &\qquad\times\langle\tilde{U}(l_{+1})\cdots\tilde{U}(l_{+a})\tilde{U}(l_{-1})\cdots\tilde{U}(l_{-b})\rangle\end{aligned} \tag{6.228}$$

i.e., one can easily evaluate multi-loop correlators starting from the simpler bosonic multi-loop functions, which we calculate as in the bosonic model: from the definitions (6.210,214), we construct the following 2-loop function,

$$\begin{aligned}\langle\tilde{U}(l_1)\tilde{U}(l_2)\rangle &= \sqrt{\frac{\pi}{l_1}}\mathrm{DSlim}_{k_1\to\infty}\ R_c^{-k_1}\left(-\frac{\Lambda}{N}\frac{\partial}{\partial g_{2k_1}}\right)\langle\tilde{U}(l_2)\rangle\\ &= \sqrt{\frac{\pi}{l_1}}\mathrm{DSlim}_{k_1\to\infty}\left(-\Lambda_c R_c^{-k_1}N^{-1}\frac{\partial}{\partial g_{2k_1}}\right)(-\partial_t^{-1}+\tau_+\tau_-\partial_t)\frac{\mathrm{e}^{-l_2 u}}{\kappa l_2}\\ &= \sqrt{\frac{\pi}{l_1}}(\partial_t^{-1}-\tau_+\tau_-\partial_t)\left[\mathrm{e}^{-l_2 u}\mathrm{DSlim}_{k_1\to\infty}\frac{(-\Lambda_c R_c^{-k_1}N^{-1})}{\kappa}\frac{\partial u}{\partial g_{2k_1}}\right]+\\ &\quad -\sqrt{\frac{\pi}{l_1}}\left[\mathrm{DSlim}_{k_1\to\infty}\frac{(-\Lambda_c R_c^{-k_1}N^{-1})}{\kappa}\frac{\partial(\tau_+\tau_-)}{\partial g_{2k_1}}\right]\mathrm{e}^{-l_2 u}\partial_t u\end{aligned} \tag{6.229}$$

which requires the evaluation of the following two limits

$$\begin{aligned}
\operatorname*{DSlim}_{k\to\infty} \frac{(-\Lambda_c R_c^{-k} N^{-1})}{\kappa} \frac{\partial u}{\partial g_{2k}} &= \operatorname*{DSlim}_{k\to\infty} \frac{(-\Lambda_c R_c^{-k} N^{-1})}{\kappa} \sum_n \left(\frac{\partial t_n}{\partial g_{2k}}\right) \frac{\partial u}{\partial t_n} \\
&= \operatorname*{DSlim}_{k\to\infty} -2\frac{N^{-1}a^{-2}}{\kappa} \frac{1}{4^k} \binom{2k}{k} k \sum_{n=0}^{k-1} \binom{k-1}{n} (-1)^n a^{2n/m} u^n \frac{\partial u}{\partial t} \\
&\to -2a^{1/m} \sqrt{\frac{k}{\pi}} (1 - a^{2/m} u)^{k-1} \frac{\partial u}{\partial t} \\
&\to -2\sqrt{\frac{a^{2/m}k}{\pi}} \mathrm{e}^{-(k-1)a^{2/m}u} \frac{\partial u}{\partial t} \to -2\sqrt{\frac{l}{\pi}} \mathrm{e}^{-lu} \partial_t u
\end{aligned} \tag{6.230}$$

$$\begin{aligned}
\operatorname*{DSlim}_{k\to\infty} \frac{(-\Lambda_c R_c^{-k} N^{-1})}{\kappa} \frac{\partial \tau_\pm}{\partial g_{2k}} &= \sum_n \tau_n^\pm n u^{n-1} \operatorname*{DSlim}_{k\to\infty} \frac{(-\Lambda_c R_c^{-k} N^{-1})}{\kappa} \frac{\partial u}{\partial g_{2k}} \\
&\to \sum_n \tau_n^\pm n u^{n-1} (-2) \sqrt{\frac{l}{\pi}} \mathrm{e}^{-lu} \partial_t u \\
&= -2\sqrt{\frac{l}{\pi}} \mathrm{e}^{-lu} \partial_t \tau_\pm \quad .
\end{aligned} \tag{6.231}$$

Therefore we find

$$\begin{aligned}
\langle \tilde{U}(l_1) \tilde{U}(l_2) \rangle &= -2(\partial_t^{-1} - \tau_+\tau_-\partial_t) \mathrm{e}^{-(l_1+l_2)u} \partial_t u + 2\partial_t(\tau_+\tau_-) \mathrm{e}^{-(l_1+l_2)u} \partial_t u \\
&= 2\partial_t(\partial_t^{-1} - \tau_+\tau_-)\partial_t \left(\frac{\mathrm{e}^{-(l_1+l_2)u}}{(l_1+l_2)} \right) \\
&= (-2\kappa\partial_t)(-\partial_t^{-1} + \tau_+\tau_-\partial_t) \left(\frac{\mathrm{e}^{-(l_1+l_2)u}}{\kappa(l_1+l_2)} \right) \quad ,
\end{aligned} \tag{6.232}$$

which, compared with (6.215), implies

$$\langle \tilde{U}(l_1) \tilde{U}(l_2) \rangle = (-2\kappa\partial_t) \langle \tilde{U}(l_1 + l_2) \rangle \quad . \tag{6.233}$$

Up to factor 2, this is precisely the same result (see Eq. (3.83)) found in the bosonic theory, in spite of a 2nd-order (in fermionic couplings) term present in the bosonic loop $\widetilde{U}(l)$. As in Chapter 3 one also proves, by induction that a multi-loop function reads

$$\langle \tilde{U}(l_1) \cdots \tilde{U}(l_n) \rangle = (-2\kappa\partial_t)^{n-1} \left\langle \tilde{U}(\sum_{i=1}^n l_i) \right\rangle \quad . \tag{6.234}$$

If we finally substitute the above correlator back into (6.228) we will obtain the multi-superloop functions

$$\langle \tilde{W}_+(l_{+1},\theta_{+1})\cdots \tilde{W}_+(l_{+a},\theta_{+a})\tilde{W}_-(l_{-1},\theta_{-1})\cdots \tilde{W}_-(l_{-b},\theta_{-b})\rangle =$$

$$= \left[1+\left(\sum_{i=1}^{a}\theta_{+i}\right)\partial_t^{-1}\frac{\partial}{\partial\tau_+}\right]\left[1+\left(\sum_{j=1}^{b}\theta_{-j}\right)\partial_t^{-1}\frac{\partial}{\partial\tau_-}\right]\times$$

$$\times(-2\kappa\partial_t)^{a+b-1}\langle\tilde{U}(\sum_{i=1}^{a}l_{+i}+\sum_{j=1}^{b}l_{j-})\rangle$$

$$= (-2\kappa)^{a+b-1}\left[\partial_t^a+\left(\sum_{i=1}^{a}\theta_{+i}\right)\partial_t^{a-1}\frac{\partial}{\partial\tau_+}\right]\left[\partial_t^b+\left(\sum_{j=1}^{b}\theta_{-j}\right)\partial_t^{b-1}\frac{\partial}{\partial\tau_-}\right]\times$$

$$\times\partial_t^{-1}\langle\tilde{U}(\sum_{i=1}^{a}l_{+i}+\sum_{j=1}^{b}l_{-j})\rangle \quad , \tag{6.235}$$

which can be summarized as

$$\langle\prod_{i=1}^{a}\tilde{W}_+(l_{+i},\theta_{+i})\prod_{j=1}^{b}\tilde{W}_-(l_{-j},\theta_{+j})\rangle =$$

$$=\prod_{i=1}^{a}(-2\kappa\mathcal{D}_i^+)\prod_{j=1}^{b}(-2\kappa\mathcal{D}_i^-)(-2\kappa\partial_t)^{-1}\langle\tilde{U}(L)\rangle \quad , \tag{6.236}$$

where the differential operators

$$\mathcal{D}_i^{\pm}\equiv\partial_t+\theta_{\pm i}\frac{\partial}{\partial\tau_\pm} \quad , \tag{6.237}$$

satisfy (due to $\theta_i^2=0$)

$$\prod_{i=1}^{n}(-2\kappa\mathcal{D}_i^{\pm})=(-2\kappa)^n\left[\partial_t^n+\Theta_\pm\partial_t^{n-1}\frac{\partial}{\partial\tau_\pm}\right] \quad , \tag{6.238}$$

where

$$\Theta_\pm\equiv\sum_i\theta_{\pm i} \quad , \tag{6.239}$$

is interpreted[84] as a total super length, while

$$L=\sum_i l_{+i}+\sum_j l_{-j} \quad , \tag{6.240}$$

is the usual (even) total length of the loop.

Therefore the superloop correlators depend only on the total length L and total superlengths $\Theta_\pm$, which is also reminiscent of the bosonic case (recall (3.84)). This property was indeed used in ref. [84] to justify, a posteriori, the Ansatz proposed to solve the superloop equations.

6.3.3. Macroscopic Loops Versus Scaling Operators

Also as in Chapter 3, we can use the 1-point functions of the macroscopic and microscopic loop operator to establish a relation[84] between these operators: expanding the r.h.s. of Eq. (6.225) as a power series in the loop length l, and distinguishing the regular part of that expansion, one has

$$\begin{aligned}
\langle\tilde{W}_{\pm}(l,\theta_{\pm})\rangle &= \frac{1}{\kappa}(-\partial_t^{-1}+\tau_+\tau_-\partial_t \pm \theta_{\pm}\partial_t^{-1}\tau_{\mp}\partial_t)\frac{e^{-lu}}{l} \\
&= \frac{1}{\kappa}(-\partial_t^{-1}+\tau_+\tau_-\partial_t \pm \theta_{\pm}\partial_t^{-1}\tau_{\mp}\partial_t)\left(\frac{1}{l}+\sum_{n\geq 0}\frac{(-1)^{n+1}}{(n+1)!}l^n u^{n+1}\right) \\
&= \frac{1}{\kappa}\sum_{n\geq 0}\frac{(-1)^{n+1}}{n!}l^n\left[(-\partial_t^{-1}+\tau_+\tau_-\partial_t \pm \theta_{\pm}\partial_t^{-1}\tau_{\mp}\partial_t)\frac{u^{n+1}}{(n+1)}\right] + (\text{singular terms in } l) \\
&= 2\kappa\sum_{n\geq 0}\frac{(-1)^{n+1}}{n!}l^n\left[\langle\sigma_n\rangle+\theta_{\pm}\langle\nu_n^{\pm}\rangle\right] + (\text{singular terms in } l) \quad , \qquad (6.241)
\end{aligned}$$

where we have recognized the expectation values of scaling operators (6.147, 151) and (6.156). We are thus allowed to use the following operator expansion

$$\tilde{W}_{\pm}(l,\theta_{\pm}) \text{ ``}={}\text{'' } 2\kappa\sum_{n\geq 0}\frac{(-1)^{n+1}}{n!}l^n\left(\sigma_n+\theta_{\pm}\nu_n^{\pm}\right) \quad . \qquad (6.242)$$

The quotation marks have the same meaning as in Eq. (3.101): the above expansion does not reproduce the singularities of the 1-loop function. The relation (6.227) together with our experience with the bosonic models tell us that those singular terms might represent universal analytical terms in the coupling constants, which are important in the study of wave functions (see [156] and end of this Chapter).

So many similarities with the purely bosonic theory foretell us that a naive comparison with the super Liouville approach should be avoided: indeed, if one tries to identify the scaling operators $\{\sigma_n\}$ and $\{\nu_n^{\pm}\}$ with the vertex operators of Neveu-Schwarz and Ramond sectors, as presented in Chapter 5, one comes up against inconsistencies at 1- and 2-point functions already.

Indeed, from momentum conservation in the Coulomb gas framework, one expects that all 1-point functions vanish, except for the area operator expectation value. On the contrary, Eq. (6.204) gives an infinite number of non-vanishing 1-point functions of bosonic scaling operators in the zero-coupling regime.

Also from the results in Chapter 5, one expects an orthogonality among vertex operators: only correlators of the form $\langle\Psi_{NS}(k)\Psi_{NS}(2\alpha_0-k)\rangle$ in the Neveu-Schwarz sector, and $\langle V_{-1/2}(k,1)V_{-1/2}(2\alpha_0-k,-1)\,\Psi_{NS}^{-1}(0)\rangle$ in the Ramond sector, are different from zero. On the other hand the results (6.204) and (6.206) show that neither $\langle\sigma_n\sigma_m\rangle$ nor $\langle\nu_n^+\nu_m^-\rangle$ vanish for any values of n,m.

Therefore we need to proceed as in Chapter 4: studying the dimensions of coupling constants and observing other basis of scaling operators, we shall establish the correct correspondence between the two formulations of the theory.

6.4. Conformal Basis

Here we follow the strategy presented in Chapter 4 to define a suitable set of scaling operators, in both sectors of Ramond and Neveu-Schwarz whose 1- and 2-point functions are compatible with the continuum predictions. These results lead us to a dictionary between the super Liouville and super eigenvalue formulations, as presented at the end of this Chapter.

6.4.1. Orthogonal 2-Point Functions

We shall use the orthogonalization strategy used in the bosonic model, starting from the string equation (which is the same in both bosonic and supersymmetric models) written as

$$u^m = t + \sum_{n\geq 0} t_n u^n \quad . \tag{6.243}$$

In the supersymmetric theory a "physical cosmological constant" should have dimension of inverse of length (see Chapter 5). From (6.213) we learn that $a^{2/m}$ is our length scale, while (6.214) tells us that u has dimension of $[\text{length}]^{-1}$. Therefore

$$[\mu] = [\text{length}]^{-1} = [u] \quad , \tag{6.244}$$

while the coupling constants obey, as follows from (6.243),

$$[t_n] = [u]^{m-n} = [\mu]^{m-n} \quad , \tag{6.245}$$

which is compatible with (6.193). Now we can take (6.194) to read the dimension of the fermionic couplings,

$$[\tau_n^{\pm}] = [\mu]^{m-n\pm 1/2} \quad . \tag{6.246}$$

Therefore t_{m-1} is in general the coupling with dimension of a physical cosmological constant rather than t. As outlined in the bosonic case, it is convenient to shift the bosonic couplings as in (4.7),

$$\begin{aligned} t_0 &\to t_0 - t \quad , \\ t_{m-1} &\to t_{m-1} + \mu \quad , \\ t_n &\to t_n \quad , \quad n \neq 0, m-1 \quad , \end{aligned} \tag{6.247}$$

turning the string equation into

$$u^m = \mu u^{m-1} + \sum t_n u^n \quad , \tag{6.248}$$

which, in the zero coupling limit, given by $t_n \to 0 \quad , \quad \tau_n^{\pm} \to 0$, implies $u \to \mu$.

In order to find an orthogonal basis, we consider the analytical transformations in the space of coupling constants, compatible with the dimensions in (6.245). In analogy to the bosonic model, we first concentrate ourselves on a first-order approximation in the perturbative couplings, given by

$$\begin{aligned} t_n &= \sum_{s\geq 0} A_s^{(n+s)} \mu^s \widehat{t}_{n+s} \quad , \\ \tau_n^{\pm} &= \sum_{s\geq 0} B_s^{(n+s)} \mu^s \widehat{\tau}_{n+s}^{\pm} \quad . \end{aligned} \tag{6.249}$$

Notice that $t_n \to 0 \Longleftrightarrow \widehat{t}_n \to 0$, $\tau_n^{\pm} \to 0 \Longleftrightarrow \widehat{\tau}_n^{\pm} \to 0$, i.e., both zero-coupling regimes imply $u \to \mu$.

Coupled to the new constants we find the following set of operators

$$\begin{aligned} \widehat{\sigma}_n &= \sum_{k\geq 0} \frac{\partial t_k}{\partial \widehat{t}_n} \sigma_k = \sum_{k=0}^{n} A_{n-k}^{(n)} \mu^{n-k} \sigma_k \\ &= \sum_{s=0}^{n} A_s^{(n)} \mu^s \sigma_{n-s} \quad , \end{aligned} \tag{6.250}$$

and analogously, in the Ramond sector,

$$\widehat{\nu}_n^{\pm} = \sum_{s=0}^{n} B_s^{(n)} \mu^s \nu_{n-s}^{\pm} \quad . \tag{6.251}$$

We determine the coefficients $A_s^{(n)}$ and $B_s^{(n)}$ studying the 2-point functions in the zero-coupling (zc) regime: in the bosonic sector, we have

$$\begin{aligned} \langle \widehat{\sigma}_i \widehat{\sigma}_j \rangle_{zc} &= \sum_{s=0}^{i} \sum_{r=0}^{j} A_s^{(i)} A_r^{(j)} \mu^{s+r} \langle \sigma_{i-s} \sigma_{j-r} \rangle_{zc} \\ &= \sum_{s=0}^{i} \sum_{r=0}^{j} A_s^{(i)} A_r^{(j)} \mu^{s+r} \left\{ -\frac{1}{2\kappa^2} \rho_{i+j-s-r+1}[u] \right\}_{zc} \\ &= -\frac{1}{2\kappa^2} \mu^{i+j+1} \sum_{s=0}^{i} \sum_{r=0}^{j} A_s^{(i)} A_r^{(j)} \frac{1}{(i+j-s-r+1)} \\ &= -\frac{1}{2\kappa^2} \mu^{i+j+1} g_{ij}(A) \quad , \end{aligned} \tag{6.252}$$

where the symmetric matrix

$$\begin{aligned} g_{ij}(A) &\equiv \sum_{s=0}^{i} \sum_{r=0}^{j} \frac{A_s^{(i)} A_r^{(j)}}{(i+j-s-r+1)} \\ &= \int_0^1 \mathrm{d}x \left(\sum_{s=0}^{i} A_s^{(i)} x^{i-s} \right) \left(\sum_{r=0}^{j} A_r^{(j)} x^{j-r} \right) \end{aligned} \tag{6.253}$$

is interpreted, as in (4.20), as a metric in the space of couplings. We also find analogous equations for $\langle \widehat{\nu}_i^+ \nu_j^- \rangle$ replacing $A \to B$. Just like in the bosonic model we have reduced our problem to finding orthogonal polynomials with respect to an integral, $\int_0^1 \mathrm{d}x$:

$$\begin{aligned} \pi_i(x) &= \sum_{s=0}^{i} A_s^{(i)} x^{i-s} \quad , \\ \int_0^1 \mathrm{d}x\, \pi_i(x)\pi_j(x) &= h_j \delta_{ij} \quad . \end{aligned} \tag{6.254}$$

We have listed the main properties of the π-polinomials in Appendix E: they can be related to the Legendre polynomials used in the bosonic model. The π-polynomials are therefore the characteristic polynomials of the superconformal basis. We thus take formula (E.13) to define the conformal scaling operators

$$\widehat{\sigma}_n = \sum_{s=0}^{n} \frac{(-1)^s}{s!} \frac{(2n-s)!}{[(n-s)!]^2} \mu^s \sigma_{n-s} \quad , \tag{6.255}$$

which can be inverted to give

$$\sigma_n = [n!]^2 \sum_{s=0}^{n} \frac{(2n+1-2s)}{s!(2n+1-s)!} \mu^s \widehat{\sigma}_{n-s} \quad . \tag{6.256}$$

Using (E.11) we have the following orthogonal 2-point functions

$$\langle \widehat{\sigma}_i \widehat{\sigma}_j \rangle_{zc} = -\frac{1}{2\kappa^2} \frac{\mu^{2i+1}}{(2i+1)} \delta_{ij} \quad . \tag{6.257}$$

Since $\langle \nu_i^+ \nu_j^- \rangle_{zc} = \langle \sigma_i \sigma_j \rangle_{zc}$, we take the coefficients $B_s^{(n)} = A_s^{(n)}$ and find analogous results for the Ramond sector

$$\widehat{\nu}_n^\pm = \sum_{s=0}^{n} \frac{(-1)^s}{s!} \frac{(2n-s)!}{[(n-s)!]^2} \mu^s \nu_{n-s}^\pm \quad , \tag{6.258}$$

$$\langle \widehat{\nu}_i^+ \widehat{\nu}_j^- \rangle_{zc} = -\frac{1}{2\kappa^2} \frac{\mu^{2i+1}}{(2i+1)} \delta_{ij} \quad . \tag{6.259}$$

While the 2-point functions seem to be satisfactory, we must also examine whether the 1-point functions agree with the predictions of the continuum formulation.

To start with we write the string equation in terms of the conformal couplings,

$$\begin{aligned} u^m &= \mu u^{m-1} + \sum_{n\geq 0} \left(\sum_{s\geq 0} A_s^{(n+s)} \mu^s \widehat{t}_{n+s} \right) u^n \\ &= \mu u^{m-1} + \sum_{k\geq 0} \widehat{t}_k \mu^{-k} \left(\sum_{s=0}^{k} A_s^{(k)} \left(\frac{u}{\mu}\right)^{k-s} \right) \\ &= \mu u^{m-1} + \sum_{k\geq 0} \widehat{t}_k \mu^{-k} \pi_k(u/\mu) \quad . \end{aligned} \tag{6.260}$$

From Eq. (6.249) we derive the flows

$$\frac{\partial}{\partial t_n} u = u^{n-m+1} \frac{\partial}{\partial \mu} u \quad , \tag{6.261}$$

and using the transformation (6.249) we arrive at

$$\begin{aligned}
\frac{\partial}{\partial \widehat{t}_n} &= \sum_{k \geq 0} \frac{\partial t_k}{\partial \widehat{t}_n} u^{k-m+1} \frac{\partial}{\partial \mu} u \\
&= \sum_{k=0}^{n} A^{(n)}_{n-k} \mu^{n-k} u^{k-m+1} \frac{\partial}{\partial \mu} u \\
&= \sum_{s=0}^{n} A^{(n)}_{s} \mu^{s} u^{n-s-(m-1)} \frac{\partial}{\partial \mu} u \\
&= \mu^n u^{-(m-1)} \pi_n(u/\mu) \frac{\partial}{\partial \mu} u \quad .
\end{aligned} \tag{6.262}$$

In the zero-coupling(zc)-limit, when $u \to \mu$, we have

$$\frac{\partial}{\partial \widehat{t}_n} u = \mu^{n-(m-1)} \frac{\partial}{\partial \mu} u \quad . \tag{6.263}$$

which is consistent with the association between the $\frac{\partial}{\partial \mu}$ and the insertion of the area-operator (cosmological term) $\widehat{\sigma}_{m-1} \longleftrightarrow \frac{\partial}{\partial \mu}$.

This allows one to calculate 1-point functions integrating the following 2-point functions,

$$\begin{aligned}
\frac{\partial}{\partial \mu} \langle \sigma_i \rangle &= \langle \widehat{\sigma}_{m-1} \sigma_i \rangle \\
&= [i!]^2 \sum_{s=0}^{i} \frac{(2i+1-2s)}{s!(2i+1-s)!} \mu^s \langle \widehat{\sigma}_{m-1} \widehat{\sigma}_{i-s} \rangle \\
&= [i!]^2 \sum_{s=0}^{i} \frac{(2i+1-2s)}{s!(2i+1-s)!} \left(\frac{-1}{2\kappa^2} \frac{\mu^{2m-1}}{2m-1} \right) \delta_{s,i-(m-1)} \\
&= \begin{cases} 0 & i < m-1 \\ \frac{-1}{2\kappa^2} \mu^{i+m} \frac{[i!]^2}{(i-m+1)!(i+m)!} & i \geq m-1 \end{cases} \quad ,
\end{aligned} \tag{6.264}$$

giving as a result

$$\langle \sigma_i \rangle = \begin{cases} 0 & i < m-1 \\ \frac{-1}{2\kappa^2} \mu^{i+m+1} \frac{[i!]^2}{(i-m+1)!(i+m+1)!} & i \geq m-1 \end{cases} \quad . \tag{6.265}$$

Now the definition (6.255) and the above result imply

$$\langle \widehat{\sigma}_i \rangle = \begin{cases} 0 & i < m-1 \\ \frac{-1}{2\kappa^2} \mu^{i+m+1} \sum_{k=m-1}^{i} \frac{(-1)^{i-k}}{(i-k)!} \frac{(i+k)!}{(k-m+1)!(k+m+1)!} & i \geq m-1 \end{cases} \quad . \tag{6.266}$$

Therefore we expect non-vanishing 1-point functions only when $i \geq m-1$, which we parametrize as follows

$$
\begin{aligned}
\langle\hat{\sigma}_{m-1+a}\rangle &= -\frac{1}{2\kappa^2}\mu^{2m+a}\sum_{k=m-1}^{a+m-1}\frac{(-1)^{a+m-1-k}}{(a+m-1-k)!}\frac{(a+m-1+k)!}{(k-m+1)!(k+m+1)!} \\
&= -\frac{1}{2\kappa^2}\mu^{2m+a}\sum_{k=0}^{a}\frac{(-1)^{a-k}}{(a-k)!}\frac{(k+2m+a-2)!}{k!(k+2m)!} \\
&= -\frac{1}{2\kappa^2}\mu^{2m+a}\left\{\left(\frac{\partial}{\partial x}\right)^{a-2}\sum_{k=0}^{a}\frac{(-1)^{a-k}}{k!(a-k)!}x^{k+2m+a-2}\right\}_{x=1} \\
&= -\frac{1}{2\kappa^2}\mu^{2m+a}\left\{\frac{1}{a!}\left(\frac{\partial}{\partial x}\right)^{a-2}x^{2m+a-2}(x-1)^a\right\}_{x=1} \quad . \qquad (6.267)
\end{aligned}
$$

The expression in brackets is obviously zero for $a \geq 2$, while the particular cases $a = 0, 1$ are explicitly calculated below:

$$
\begin{aligned}
\langle\hat{\sigma}_{m-1}\rangle &= -\frac{1}{2\kappa^2}\mu^{2m}\frac{(2m-2)!}{(2m)!} \\
&= -\frac{1}{2\kappa^2}\mu^{2m}\frac{1}{2m(2m-1)} \quad , \qquad (6.268) \\
\langle\hat{\sigma}_m\rangle &= -\frac{1}{2\kappa^2}\mu^{2m}\left\{-\frac{(2m-1)!}{(2m)!}+\frac{(2m)!}{(2m+1)!}\right\} \\
&= +\frac{1}{2\kappa^2}\mu^{2m+1}\frac{1}{2m(2m+1)} \quad . \qquad (6.269)
\end{aligned}
$$

As in the bosonic theory, we find only two non-vanishing 1-point functions: $\langle\hat{\sigma}_{m-1}\rangle$ corresponds to the average "area", as we expected from the continuum approach; the operator $\hat{\sigma}_m$ can be associated[156] to the energy of the system, as in the bosonic model (recall the conclusions of Chapter 4).

The scaling behavior of the free energy can be read now from Eq. (6.268)

$$\frac{\partial}{\partial\mu}\mathcal{F}_S = \langle\hat{\sigma}_{m-1}\rangle = -\frac{1}{2\kappa^2}\mu^{2m}\frac{1}{2m(2m-1)} \quad , \qquad (6.270a)$$

whose integration gives

$$\mathcal{F}_S = -\frac{1}{2\kappa^2}\mu^{2m+1}\frac{1}{(2m+1)2m(2m-1)} = -\frac{\langle\hat{\sigma}_m\rangle}{2m-1} \quad . \qquad (6.270b)$$

From the scaling with respect to the physical cosmological constant μ in (6.270b) we have a family of string susceptibilities $2-\hat{\gamma} = 2m+1, m = 1, 2, 3\cdots$ to be compared with the results of Chapter 4.

One can also calculate higher order correlators in the conformal basis: in principle it is necessary to go beyond the first-order transformation (6.249); nevertheless, the scaling behavior is universal[156], and we find

$$\langle\prod_i \mathcal{O}_i\rangle \sim \mu^{2-\hat{\gamma}+\sum_i \delta_i} \quad , \qquad (6.271)$$

where $\delta_{\hat{\sigma}_{n_i}} = (n_i - m)$ and $\delta_{\hat{\nu}^{\pm}_{n_i}} = (n_i - m \mp \frac{1}{2})$. These exponents are sufficient to study the spectrum of the model.

6.5. Identification of the Model

As in the purely bosonic case we may compare the scaling of the free energy to the partition function of the continuum formulation, in two different regimes, to identify the (dressed) superconformal theory corresponding to the m-th critical super eigenvalue model.

For this purpose we recall some useful formulae: a minimal superconformal model is characterized by a pair of integers (p, q), with $p < q$, so that its central charge $c = \frac{3}{2}\hat{c}$ is parametrized as

$$\hat{c} = 1 - 8\alpha_0^2 \quad , \quad \alpha_0 = \frac{p-q}{2\sqrt{pq}} \quad . \tag{6.272}$$

In Chapter 5 we have also introduced some important parameters, namely $Q = 2\sqrt{1+\alpha_0^2}$ and $\alpha_\pm = -\frac{Q}{2} \pm |\alpha_0|$, which can also be written in terms of (p, q).

We concentrate our attention on the identity operator, with conformal weight $\Delta = 0$, whose gravitational dressing provides the area (or cosmological) operator: its dressing parameter is α_+ and when the scaling is measured with respect to the physical cosmological constant μ, the super Liouville approach predicts the behavior

$$\mathcal{Z} \sim \mu^{-Q/\alpha_+} \quad . \tag{6.273}$$

In terms of the (p, q) indices the above exponent reads

$$-\frac{Q}{\alpha_+} = 1 + \frac{q}{p} \quad . \tag{6.274}$$

When $q = p + 2$ we have the unitary series and $\Delta = 0$ is the minimum weight. However the minimum weight reads in general $\Delta_{min} = \frac{4-(p-q)^2}{8pq}$ and the corresponding operator has $\beta_{min} = \frac{2-(p+q)}{2\sqrt{pq}}$ as its dressing. If t is the coupling constant associated to the minimum weight operator, one expects the following scaling

$$\mathcal{Z} \sim t^{-Q/\beta_{min}} \quad , \tag{6.275}$$

where

$$-\frac{Q}{\beta_{min}} = \frac{2(p+q)}{p+q-2} \quad . \tag{6.276}$$

By comparison with the scalings in (6.182) and (6.270b) obtained in the discrete model, we can determine the indices (p, q) from the following pair of equations

$$\begin{aligned} -\frac{Q}{\beta_{min}} &= \frac{2(p+q)}{p+q-2} = 2 + \frac{1}{m} \quad , \\ -\frac{Q}{\alpha_+} &= 1 + \frac{q}{p} = 2m + 1 \quad , \end{aligned} \tag{6.277}$$

that is, we have the following family of models[84]

$$p = 2 \quad , \quad q = 4m \quad ; \quad m = 1, 2, \cdots$$
$$c = \frac{3}{2}\left[1 - \frac{(2m-1)^2}{m}\right] = 0, -\frac{21}{4}, \cdots \quad . \tag{6.278}$$

The case $m = 1$, $c = 0$ corresponds to the pure supergravity theory and is the only unitary model in (6.278).

Concerning the spectrum of primary operators, we have two sectors (Neveu-Schwarz and Ramond): the total number of primaries is $\frac{1}{2}((p-1)(q-1)+1)$, that makes up $2m$ for the family (6.278), which means m operators in each sector. The conformal weight $\Delta_{rr'}$ of such operators is given by the formula

$$\Delta_{rr'} = \frac{(rp - r'q)^2 - (q-p)^2}{8pq} + \frac{1-(-1)^{r-r'}}{32} \quad , \tag{6.279}$$

with $1 \le r' \le p - 1 = 1$ (i.e. $r' = 1$ in our case) and $1 \le r \le q - 1 = 4m - 1$. When the difference $r - r'$ is even, the operator is in the NS-sector:

$$\Delta^{NS}_{2i+1,1} = \frac{i(i-2m+1)}{4m} \quad , \quad i = 0, 1, \cdots 2m-1 \quad ; \tag{6.280}$$

and for $r - r'$ odd one has the R-sector:

$$\Delta^{R}_{2i+2,1} = \frac{(i+1-m)^2 - (m-1/2)^2}{4m} + \frac{1}{16} \quad , \quad i = 0, 1, \cdots 2m-2 \quad . \tag{6.281}$$

In fact one can restrict the label i to the interval $i = 0, \cdots, m-1$ in (6.280) and (6.281), since the remaining values only duplicate the spectrum of weights.

In the discrete approach (see Eq. (6.271)) each bosonic operator contributes to the scaling of a correlator with a factor

$$\widehat{\sigma}_n \quad \rightarrow \quad \mu^{n-m} \quad , \tag{6.282}$$

while in the continuum one expects

$$\widehat{\sigma}_n \quad \rightarrow \quad \mu^{-\beta/\alpha_+} \quad , \tag{6.283}$$

where the dressing β relates to the conformal weight through the condition $\Delta_n - \frac{\beta(\beta+Q)}{2} = 1/2$. Therefore we find the spectrum corresponding to (6.282):

$$\Delta(\widehat{\sigma}_n) = \Delta^{NS}_n = \frac{(n-m+1)(m+n)}{4m} \quad . \tag{6.284}$$

Redefining $n = m - 1 - i$ we recover the spectrum (6.280): as i runs from 0 to $m-1$ the index n goes from $m-1$ to 0. Therefore we associate

$$\widehat{\sigma}_n \quad \leftrightarrow \quad \text{dressed } \mathcal{O}^{NS}_{2(m-n)-1,1} \quad , \quad n = 0, \cdots, m-1 \quad . \tag{6.285}$$

As for the fermionic operators, we have given in (6.271) the respective scaling contributions,

$$\widehat{\nu}_n^{\pm} \quad \to \mu^{n-m\mp 1/2} \quad , \tag{6.286}$$

while the corresponding vertex operators should scale as in Eq. (6.283) with the dressing β determined in this case by the equation $(\Delta_K - \frac{1}{16}) - \frac{\beta(\beta+Q)}{2} = \frac{1}{2}$ (recall Eq. (5.135)). Taking the set of operators $\widehat{\nu}_n^+$ (which includes the non-degenerate case $\widehat{\nu}_0^+$) we find

$$\Delta(\widehat{\nu}_n^+) = \Delta_n^{R} = \frac{n^2-(m-1/2)^2}{4m} + \frac{1}{16} \quad . \tag{6.287}$$

Renaming $n = m-1-i$ one reproduces the formula (6.281): taking $n = 0, \cdots, m-1$ one spans the values $i = m-1, \cdots, 0$. The correspondence in the Ramond sector is therefore

$$\widehat{\nu}_n \quad \leftrightarrow \quad \text{dressed } \mathcal{O}^{R}_{2(m-n),1} \quad , \quad n = 0, \cdots, m-1 \quad . \tag{6.288a}$$

In order to interpret the odd operators $\widehat{\nu}_n^-$, we recall that in the vertex formulation, the non-vanishing 2-point functions had the form

$$\left\langle V_{-1/2}(k,1)\left[V_{-1/2}(2\alpha_0-k,-1)\mu^2\Psi_{NS}^{(-1)}(0)\right]\right\rangle \sim \mu^{-2\beta/\alpha_+ +1} \quad , \tag{6.289}$$

where the operator $\Psi_{NS}^{(-1)}(0) = \int d^2z\, e^{-\sigma+\alpha_+\phi}$ acts as a screening required by the superselection rule (5.101); the corresponding coupling constant has dimension of $[\text{length}]^{-2}$, therefore the factor μ^2 in (6.289). Altogether its contribution to the scaling behavior reads $\mu^2 \int d^2z\, e^{-\sigma+\alpha_+\phi} \sim \mu^{2-\alpha_+/\alpha_+} = \mu$. The composed operators $V_{1/2}(2\alpha_0-k,1)\mu^2\Psi_{NS}^{(-1)}(0)$ had therefore the scaling $\mu^{-\beta(k)/\alpha_+ +1}$, while $V_{-1/2}(k,1)$ scales as $\mu^{-\beta(k)/\alpha_+}$. This difference of one unit in the exponent of μ is analogous to the one found in the scalings (6.286), and we propose the association

$$\widehat{\nu}_n^- \leftrightarrow \mu^2\Psi_{NS}^{(-1)}(0) \times \text{dressed } \mathcal{O}^{R}_{2(m-n),1} \quad . \tag{6.288b}$$

6.5.1. Wave Functions and Minisuperspace Approximation

In order to establish a super Liouville/discrete model dictionary we shall analyze the wave functions of scaling operators. We begin by considering the bosonic loop 1-point-function, using its expansion in terms of scaling operators: from (6.214) and (6.242) we have

$$\begin{aligned}
\langle U(l)\rangle_{zc} &= \frac{2\kappa}{\sqrt{\pi}}\sum_{n\geq 0}\frac{(-1)^{n+1}}{n!}l^{n+1/2}\langle\sigma_n\rangle_{zc}+\cdots \\
&= \frac{2\kappa}{\sqrt{\pi}}\sum_{n\geq m-1}\frac{(-1)^{n+1}}{n!}l^{n+1/2}\left(-\frac{1}{2\kappa^2}\mu^{n+m+1}\frac{[n!]^2}{(n-m+1)!(n+m+1)!}\right)+\cdots \\
&= \frac{1}{\sqrt{\pi}\kappa}\mu^{m+1/2}(\mu l)^{1/2}\sum_{n\geq m-1}\frac{n!}{(n-m+1)!(n+m+1)!}(-\mu l)^n+\cdots \\
&= \frac{1}{\sqrt{\pi}\kappa}\mu^{m+1/2}(\mu l)^{1/2}\sum_{s\geq 0}\frac{(s+m-1)!}{s!(s+2m)!}(-\mu l)^{s+m-1}+\cdots \\
&= \frac{1}{\sqrt{\pi}\kappa l}\mu^{m-1/2}\sum_{s\geq 0}(-1)^{s+m-1}\frac{(s+m-1)!}{(s+2m)!}\frac{(\mu l)^{s+(m+1/2)}}{s!}+\cdots \\
&= -\frac{(-1)^m}{\sqrt{\pi}\kappa l}\mu^{m-1/2}\frac{\Gamma(m)}{\Gamma(2m+1)}\varphi(m,2m+1;\mu l)+\cdots\quad,
\end{aligned}\tag{6.290}$$

where dots mean singular terms in l and

$$\varphi(\alpha,\gamma;x)\equiv\sum_{s\geq 0}(-1)^s\frac{\Gamma(s+\alpha)}{\Gamma(\alpha)}\frac{\Gamma(\gamma)}{\Gamma(s+\gamma)}\frac{x^{s+\gamma/2}}{s!}$$

are recognized as degenerate hypergeometric functions. The reader can find in Appendix E some useful properties of the functions $\varphi(\alpha,\gamma)$ as well as their relations with other special hypergeometric functions. They also manifest themselves as one expands the superloop operator in terms of the conformal basis, as follows,

$$\begin{aligned}
W_\pm(l,\theta_\pm) &= \frac{2\kappa}{\sqrt{\pi}}\sum_{n\geq 0}\frac{(-1)^{n+1}}{n!}l^{n+1/2}\sum_{s=0}^{n}\frac{(2n+1-2s)}{s!(2n+1-s)!}[n!]^2\mu^s\left(\widehat{\sigma}_{n-s}+\theta_\pm\widehat{\nu}^\pm_{n-s}\right)+\cdots \\
&= \frac{2\kappa}{\sqrt{\pi}}\sum_{i\geq 0}\left\{\sum_{s\geq 0}\frac{(-1)^{i+s+1}}{s!}\frac{(2i+1)(i+s)!}{(2i+s+1)!}l^{i+s+1/2}\mu^s\right\}\left(\widehat{\sigma}_i+\theta_\pm\nu^\pm_i\right)+\cdots \\
&= -\frac{2\kappa}{\sqrt{\pi}}\sum_{i\geq 0}(-1)^i(2i+1)\mu^{-i-1/2}\left(\widehat{\sigma}_i+\theta_\pm\widehat{\nu}^\pm_i\right)\times \\
&\quad\times\frac{i!}{(2i+1)!}(\mu l)^{-1/2}\left\{\sum_{s\geq 0}(-1)^s\frac{(s+i)!}{i!}\frac{(2i+1)!}{(s+2i+1)!}\frac{(\mu l)^{s+i+1}}{s!}\right\}+\cdots \\
&= -\frac{2\kappa}{\sqrt{\pi}}\sum_{i\geq 0}\frac{(2i+1)}{\mu^{i+1/2}}\left(\widehat{\sigma}_i+\theta_\pm\widehat{\nu}^\pm_i\right)(-1)^i\frac{\Gamma(i+1)}{\Gamma(2i+2)}\frac{\varphi(i+1,2i+2;\mu l)}{(\mu l)^{1/2}}+\cdots \\
&= -4\kappa\sum_{i\geq 0}\frac{(2i+1)}{\mu^{i+1/2}}\left(\widehat{\sigma}_i+\theta_\pm\widehat{\nu}^\pm_i\right)\frac{(-1)^i\varphi(i+1,2i+2;\mu l)}{4^{i+1/2}\Gamma(i+3/2)(\mu l)^{1/2}}+\cdots,
\end{aligned}\tag{6.291}$$

Using the 2-point functions (6.257-259) and the property (E.34) that relates $\varphi(\alpha, 2\alpha)$ to Bessel functions, we find

$$W_\pm(l,\theta_\pm) = \frac{2\sqrt{\pi}}{\kappa}\sum_{i\geq 0}\left(\frac{\widehat{\sigma}_i}{\langle\widehat{\sigma}_i^2\rangle} + \theta_\pm\frac{\widehat{\nu}_i^\pm}{\langle\widehat{\nu}_i^+\widehat{\nu}_i^-\rangle}\right)\mu^{i+\frac{1}{2}}e^{-\mu l/2}(-1)^i I_{i+1/2}(\mu l/2) + \cdots \tag{6.292}$$

From this expansion one finds, for instance, the wave functions

$$\langle\widehat{\sigma}_i W_\pm(l,\theta_\pm)\rangle = \frac{4}{\kappa\sqrt{\pi}}\mu^{i+\frac{1}{2}}e^{-\mu l/2}(-1)^i I_{i+1/2}(\mu l/2) + \cdots \quad .$$

In ref. [156] it is argued that, as in the bosonic theory, the neglected singular terms in l are universal and analytical in the coupling constants, and the correct wave functions are obtained by the replacement $(-1)^i I_{i+1/2}(\mu l/2) \longrightarrow \frac{2}{\pi}K_{i+1/2}(\mu l/2)$, where K_n are the Bessel functions of second kind (see also Appendix E). We therefore have the following wave functions

$$\begin{aligned}\psi_i^{NS} &= \langle\widehat{\sigma}_i W_\pm(l,\theta_\pm)\rangle \\ &= \frac{1}{\kappa\sqrt{\pi}}\mu^{i+\frac{1}{2}}\,e^{-\mu l/2}K_{i+1/2}(\mu l/2) \quad ,\end{aligned} \tag{6.293a}$$

$$\begin{aligned}\psi_{i\pm}^{R} &= \langle\widehat{\nu}_i^\pm W_\pm(l,\theta_\pm)\rangle \\ &= \mp\frac{\theta_\mp}{\kappa\sqrt{\pi}}\mu^{i+\frac{1}{2}}\,e^{-\mu l/2}K_{i+1/2}(\mu l/2) \quad ,\end{aligned} \tag{6.293b}$$

which satisfy the following wave equation

$$\left\{-\left(l\frac{\partial}{\partial l}\right)^2 - \mu l^2\frac{\partial}{\partial l} - \frac{\mu l}{2} + \left(i+\frac{1}{2}\right)^2\right\}\psi_i = 0 \quad , \tag{6.294a}$$

or yet

$$\left\{-\left(l\frac{\partial}{\partial l} + \frac{\mu l}{2}\right)^2 + \frac{\mu^2 l^2}{4} + \left(i+\frac{1}{2}\right)^2\right\}\psi_i = 0 \quad . \tag{6.294b}$$

These equations should be compared with the Bessel equations (4.65) that characterized the wave functions in the bosonic theory. They can also be related[156] to a minisuperspace approximation of the Wheeler-DeWitt equation obeyed by the wave functions in the continuum.

6.6. Conclusions

We conclude that the mth critical super eigenvalue model corresponds to the $(2,4m)$ minimal superconformal theories coupled to two-dimensional supergravity; and we find the following association between (conformal) scaling operators and dressed vertices:

$$\hat{\sigma}_n \leftrightarrow \Psi_{NS}(k_{2(m-n)-1,1}) \quad , \tag{6.295a}$$

$$\hat{\nu}_n^+ \leftrightarrow V_{-1/2}(k_{2(m-n),1}) \quad , \tag{6.295b}$$

$$\hat{\nu}_n^- \leftrightarrow \mu^2 \Psi_{NS}^{(-1)}(0) V_{-1/2}(2\alpha_0 - k_{2(m-n),1}) \quad , \tag{6.295c}$$

where $k_{r,r'}$ is the momentum (in the Coulomb gas formulation) corresponding to the weight $\Delta_{r,r'}$ as it stands in (6.279).

As in the bosonic theory (see also discussion in ref. [156]), higher order correlation functions require the study of analytical transformations of couplings at orders higher than the linear approximation (6.249) considered in this Chapter.

7. Correlation Functions in N=2 Super Liouville Theory

7.1. Introduction

Extended supersymmetry plays an important role in string theory. In fact, $N=2$ supersymmetric theories are important objects in the study of integrable models, and string vacua[157,89,158]. Moreover, there is a strong relation between self-duality in four dimensions and integrability[159], a fact that has extrapolated the barrier of dimensionality[160]. Finally, we should mention that there is a deep relation between integrable models and deformations of conformally invariant theories[161], which although very interesting will not concern us in this book, but which might be important for $N = 2$ in order to understand the string vacuum[157].

In fact, the above facts are well connected. First, one should mention that there are close analogies between $N = 2$ strings and two-dimensional gravity, among them we cite the infinite conservation laws, and the their relation to integrable systems[89]. And finally, the analog of the conformal matter in two-dimensions is presumably the self-dual Yang-Mills theory in four dimensions, which on the other hand arises in the heterotic $N = 2$ string. The fact that $N = 2$ strings may be of utter importance in the study of integrable systems, stems from the conjecture, that integrable two-dimensional systems arise from reductions of self-dual objects in four dimensions.

Our present aim is to consider the non-critical $N = 2$ string theory. This might be seen as a generalization of previous efforts to understand string theories away from criticality. We will be concerned with a $N = 2$ matter supermultiplet with central charge $\hat{c} \leq 1\,(c = 3\hat{c})$ in a (super) Coulomb gas representation conformally coupled to a $N = 2$ super Liouville theory. However, as we shall see, the present case contains a number of new technical difficulties, which in part are due to the absence of the so-called "barrier" in the central charge. In fact, both critical points coalesce, and the critical and non-critical theories display a unique amalgamation of their properties, enhancing the difficulties in obtaining closed results.

We have seen in the case of the non-critical bosonic and $N = 1$ supersymmetric string that due to the low dimensionality of space time, the remnants of the excited states of the corresponding critical theories have no room to propagate and they can only appear as poles in the amplitudes for certain discrete values of the momentum. The $N = 2$ critical theory, however, contains only a massless scalar field in its spectrum, in the Neveu Schwarz sector. Scattering amplitudes of this particle have been calculated[89] and the expected simplicity was confirmed by the vanishing of the 4-point function. Those results suggest the study of a possible $N = 2$ *non-critical* string. The correlators must vanish, in the critical case, beyond the 3-point functions, due to consistency reasons. Such vanishing is not trivial, and can only be obtained for the integrated amplitudes, after imposing the on shell conditions, and extensive use must be made out of the (2,2) signature, mandatory in such case. Moreover, such vanishing amplitudes imply a Plebanski equation to be satisfied by fields in the simplest version of $N = 2$ strings[89], equivalent to Ricci-flatness.

7.2. The Critical Theory

We thus start with a brief study of the critical $N = 2$ theory. In the $N = 2$ string, self consistent (anomaly free) propagation requires again a fixed space-time dimensionality. In the $N = 1$ string we have d matter supermultiplets coupled to $N = 1$ supergravity in two dimensions; which consists basically of a zweibein e^a_μ and a gravitino χ_μ. These fields can be completely gauged away in the superconformal gauge, where we have the reparametrization ghosts (b, c) giving a contribution -26 to the central charge and the $N = 1$ supersymmetry ghosts (β, γ) giving $+11$, such that $c_{gh} = -26 + 11 = -15$. Each $N = 1$ matter superfield possess one fermionic $(c = \frac{1}{2})$ and one bosonic $(c = 1)$ degree of freedom. Thus to have a vanishing total central charge $c_T = d(1+\frac{1}{2})-15$ we need a $d = 10$ spacetime. In the $N = 2$ case the supergravity sector contains two important differences with respect to the $N = 1$ case, namely, first the gravitino is now complex, and second we have a $U(1)$ gauge field. Once again these are pure gauge fields and can be eliminated in the $N = 2$ superconformal gauge where we have, besides the (b, c) ghosts, two sets of (β, γ) ghosts (each set for each supersymmetry) and the $U(1)$ ghost whose central charge contribution is -2. Therefore $c_{gh} = -26+2\times 11-2 = -6$. In order to cancel c_{gh} we need 4 bosons $(c_B = 4)$ and 4 fermions $(c_F = 4\times\frac{1}{2} = 2)$. Comparing with the $N = 1$ case we might be led to $d = 4$ real dimensions $(4(1+\frac{1}{2}) = 6)$, but actually this would break the $N = 2$ supersymmetry down to $N = 1$. Indeed, $N = 2$ supersymetry requires a complex Kähler manifold such that each $N = 2$ matter supermultiplet has twice as many degrees of freedom as the $N = 1$ case $(c = 2(\frac{3}{2}) = 3)$, thus we need just $d = 2$ complex dimensions $(c_T = 3 \times d - 6)$ to have a vanishing total central charge. It has also been argued by Ooguri and Vafa[89] that another reason for the (2,2) signature is the fact that, loosely speaking, the critical d-dimensional theory, is equivalent to the subcritical $(d - 1)$-dimensional case due to the fact that the Liouville mode becomes dynamical, and exactly at that dimension it remains free, behaving as a time-like coordinate. For the $N = 2$ case, beyond the Liouville mode, there is also the $U(1)$ gauge field which becomes dynamical, and there are two bosonic degrees of freedom with negative metric, implying the (2,2) signature. Each dimension is associated with a chiral (anti-chiral) superfield $X^\mu(\overline{X}^\mu, i = 1, 2)$. The chirality condition is given by:

$$D_+X^\mu = (\frac{\partial}{\partial\theta^+} + \theta^-\partial)X^\mu = 0 \tag{7.1a}$$

$$\overline{D}_+X^\mu = (\frac{\partial}{\partial\overline{\theta}^+} + \overline{\theta}^-\partial)X^\mu = 0 \tag{7.1b}$$

(analogous for $\overline{X}^\mu$) where $\theta^\pm$ are the two Grassmannian variables satisfying $(\theta^\pm)^\dagger = \overline{\theta}^\mp$. The most general solution of $(7.1a, b)$ is:

$$X^\mu = x^\mu(\mathcal{Z}, \mathcal{Z}^*) + \psi_R{}^\mu(\mathcal{Z}, \mathcal{Z}^*)\theta^- + \psi_L{}^\mu(\mathcal{Z}, \mathcal{Z}^*)\overline{\theta}^- + F(\mathcal{Z}, \mathcal{Z}^*)\theta^-\overline{\theta}^- \tag{7.2}$$

with,

$$\mathcal{Z} = z - \theta^+\theta^- \tag{7.3a}$$

$$\mathcal{Z}^* = \overline{z} - \overline{\theta}^+\overline{\theta}^-. \tag{7.3b}$$

The $N = 2$ string dynamics in the superconformal gauge is defined by the following free-field action

$$S = \frac{1}{2\pi} \int d^2z d^4\theta (X^1\overline{X}^1 - X^2\overline{X}^2). \tag{7.4}$$

with $d^4\theta = d^2\theta^+ d^2\theta^-$. The equations of motion, after eliminating auxiliary fields, are the massless free field equations

$$\overline{\partial}\psi_R^\mu = 0 \; ; \; \partial\psi_L^\mu = 0 \; ; \; \partial\overline{\partial}x^\mu = 0 \quad , \tag{7.5a}$$

$$\overline{\partial}\overline{\psi}_R^\mu = 0 \; ; \; \partial\overline{\psi}_L^\mu = 0 \; ; \; \partial\overline{\partial}\overline{x}^\mu = 0 \quad . \tag{7.5b}$$

If introduced in (7.2), they lead us to the on-shell decomposition[89] (analogously for $\overline{X}^\mu$)

$$X^\mu = x^\mu(z,\overline{z}) + \psi_R^\mu(z)\theta^- + \psi_L^\mu(\overline{z})\overline{\theta}^- - \partial x^\mu\theta^+\theta^- - \overline{\partial}x^\mu\overline{\theta}^+\overline{\theta}^- \quad , \tag{7.6}$$

where $(\theta^\pm)^\dagger = \overline{\theta}^\mp$. The component fields have in our notation the following propagators

$$\langle x^\mu(z)\overline{x}^\nu(w)\rangle = \eta^{\mu\nu}\ln|z-w|^{-2} \quad , \tag{7.7a}$$

$$\langle \psi_R^\mu(z)\overline{\psi}_R^\nu(w)\rangle = \langle \psi^\mu{}_L(\overline{z})\overline{\psi}_L^\nu(\overline{w})\rangle^* = 2\eta^{\mu\nu}(z-w)^{-1} \quad , \tag{7.7b}$$

where $\eta^{\mu\nu} = (+,-)$.

We consider the spectrum of the $N = 2$ critical string. First, it should be noticed that due to the $U(1)$ gauge symmetry we do not have just the R-sector and the NS-sector but rather a one parameter family of sectors. The fermion boundary conditions can be twisted by a continuous phase $e^{2\pi i\theta}$ and fortunately all these sectors are simply related to the NS-sector to which we restrict ourselves henceforth. By calculating the partition function[89] of the theory (for any boundary condition) we see that the transverse oscillator modes cancel out and we are left with a massless scalar particle whose scattering amplitudes we are going to calculate. The vertex operator below represents the massless scalar particle already mentioned,

$$V(k) = \int d^2z d^4\theta \, e^{ik\cdot\overline{X}(z) + i\overline{k}\cdot X(z)} \quad . \tag{7.8}$$

In order that $V(k)$ be a physical operator, analogous to the $N = 1$ case, it must be superconformally invariant and invariant under the residual $U(1)$ gauge invariance generated by the $U(1)$ current J,

$$J = \frac{1}{4} : \overline{\psi}_R^1\psi_R^1 : -\frac{1}{4} : \overline{\psi}_R^2\psi_R^2 : \quad . \tag{7.9}$$

The last requirement is equivalent to imposing that $V(k)$ has vanishing $U(1)$ charge q, where the $U(1)$ charge q of a field ϕ is defined from the short distance expansion

$$J(w)\varphi(z) = \frac{q\varphi}{w-z} + \cdots \quad . \tag{7.10}$$

With the basic assignments $q(\mathrm{d}\theta^+) = \frac{1}{2} = -q(\mathrm{d}\theta^-)$ it is easy to show that $V(k)$, given in (7.8), has vanishing $U(1)$ charge. Supersymmetry of (7.8) is also clear from the superfield notation. With the energy-momentum tensor

$$T = - : \partial x \cdot \partial \overline{x} : + \frac{1}{4} : \overline{\psi}_R \cdot \partial \psi_R : + \frac{1}{4} : \psi_R \cdot \partial \overline{\psi}_R : \quad , \tag{7.11}$$

we can prove that $\Delta(V(k)) = 0$, which guarantees its conformal invariance, if and only if the on-shell condition holds, i.e.,

$$k \cdot \overline{k} = k_1 \overline{k}_1 - k_2 \overline{k}_2 = 0 \quad . \tag{7.12}$$

Now we can calculate n-particle amplitudes

$$\mathcal{A}_n = \langle V_{k_1} \cdots V_{k_n} \rangle = \prod_{i=1}^{n} \int \mathrm{d}^2 z_i \mathrm{d}^4 \theta_i \left\langle \prod_{i=1}^{n} \mathrm{e}^{i(k_i \overline{X}_i + \overline{k}_i X_i)} \right\rangle \quad . \tag{7.13}$$

Integration over the zero-modes x_0^μ, y_0^ν of the first component of the supercoordinate $(X^\nu = x^\nu + iy^\nu)$ leads to momentum and energy conservation,

$$\sum_{j=1}^{n} k_j = 0 = \sum_{j=1}^{n} \overline{k}_j \quad . \tag{7.14}$$

Next we have to fix the residual invariance of the $N = 2$ superconformal gauge, which is now generated by the superalgebra $OSP(2,2)$. Instead of the 5 generators of the superalgebra $OSP(2,1)$ of the $N = 1$ case, the superalgebra $OSP(2,2)$ possesses 8 generators $L_{\pm 1}, L_0, G^{\pm}_{\mp\frac{1}{2}}$, and T_0 which appears in the anticomutators $G^{\pm}_{\mp\frac{1}{2}}, G^{\mp}_{\mp\frac{1}{2}}$. Thus, we have four bosonic $(T_0, L_0, L_{\pm 1})$ and four fermionic $(\{G^{\pm}_{\pm\frac{1}{2}}\})$ generators which permit us to fix the position of 3-particle on the sphere and the two grassmannian variables $\theta^{\pm}$ of two particles; we will fix, e.g., $\theta_1^{(\pm)} = \theta_3^{(\pm)} = 0$ and $z_1 = \infty\,,\,, z_2 = 1\,,\, z_3 = 0$. In this case we have for the 3-particle scattering,

$$\begin{aligned} \mathcal{A}_3 = \Big\langle & \mathrm{e}^{i(k_3 \cdot \overline{x}(0) + \overline{k}_3 \cdot x(0))} \mathrm{e}^{i(k_2 \cdot \overline{x}(1) + \overline{k}_2 \cdot x(1))} \left[ik_2 \cdot \partial \overline{x} - i\overline{k}_2 \cdot \partial x - (k_2 \cdot \overline{\psi}_R)(\overline{k}_2 \cdot \psi_R) \right] \\ & \times \left[ik_2 \cdot \overline{\partial} \overline{x} - i\overline{k}_2 \cdot \overline{\partial} x - (k_2 \cdot \overline{\psi}_L)(\overline{k}_2 \cdot \psi_L) \right] \Big\rangle \quad . \end{aligned} \tag{7.15}$$

Using the propagators (7.7) we reproduce Ooguri and Vafa's result[89],

$$\mathcal{A}_3 = (c_{23})^2 \quad , \tag{7.16}$$

where $c_{ij} = k_i \cdot \overline{k}_j - \overline{k}_i \cdot k_j$.

For the four-particle scattering Ooguri and Vafa have obtained[89] (in the gauge $\theta_1^{(\pm)} = \theta_4^{(\pm)} = 0\,, z_1 = \infty\,, z_2 = 1\,, z_3 = z\,, z_4 = 0$)

$$\mathcal{A}_4 = \int \mathrm{d}^2 z |z|^{2s_{34}} |1-z|^{2s_{32}} \left| \frac{s_{32}(s_{32}-1)}{(1-z)^2} + \frac{c_{12}c_{34}}{z} + \frac{c_{23}c_{41}}{1-z} \right|^2 \quad , \tag{7.17}$$

where $s_{ij} = k_i \cdot \overline{k}_j + \overline{k}_i \cdot k_j$. The above integral can be calculated by generalizing the technique of analytic continuation of Dotsenko[30] implemented originally in the calculation of the simpler integral $\int \mathrm{d}^2 z |z|^{2\alpha} |1-z|^{2\beta}$, that is[89]

$$\begin{aligned} &\int \mathrm{d}^2 z z^{\alpha+n_1} \overline{z}^{\alpha+n_2} (1-z)^{\beta+m_1} (1-\overline{z})^{\beta+m_2} = \\ &= \frac{\sin\pi\alpha \sin\pi\beta}{\sin\pi(\alpha+\beta)} \frac{\Gamma(1+\alpha+n_1)\Gamma(1+\alpha+n_2)\Gamma(1+\beta+m_1)\Gamma(1+\beta+m_2)}{\Gamma(\alpha+\beta+n_1+m_1+2)\Gamma(\alpha+\beta+n_2+m_2+2)} \end{aligned} \quad ,(7.18))$$

which can be understood from the fact that the integral above must reproduce the previous results for $n_1 = n_2, m_1 = m_2$; moreover, for $\alpha = -n_1 + \epsilon - l$, l non-negative integer, one should have simple poles for $n_2 \le n_1 + l$, but not for $n_2 > n_1 + l$. The same reasoning applies to β, respectively m_1, m_2. Therefore, from a detailed analysis of the type used extensively in Appendix C, we arrive at the above result. It is now a question of collecting the several different terms in (7.17), use the fact that the main contribution is of the type of (7.18) times a rational function of momenta, to arrive at

$$\mathcal{A}_4 = -\frac{\pi F^2}{16} \Delta(1+s_{34})\Delta(1+s_{14})\Delta(1+s_{24}) \quad , \tag{7.19}$$

where $\Delta(x) = \Gamma(x)/\Gamma(1-x)$ as before, and

$$F = 1 - \frac{c_{23}c_{41}}{s_{14}s_{24}} - \frac{c_{34}c_{12}}{s_{34}s_{24}} \quad . \tag{7.20a}$$

Above, we have used the identities

$$s_{32} = s_{14} \quad , \quad s_{14} + s_{24} + s_{34} = 0 \quad . \tag{7.20b}$$

It turns out that after use of the on shell condition (7.12) one gets $F = 0$ identically. Indeed, we first write F as

$$F = \frac{1}{s_{24}s_{34}s_{14}} \left[s_{24}s_{34}s_{14} - s_{34}s_{23}s_{41} - s_{14}s_{34}s_{12} \right] \quad , \tag{7.20c}$$

and now eliminate k_4 in the bracket using momentum conservation. We notice that the term is linear in $k_1, \overline{k}_1, k_2, \overline{k}_2, k_3$ and $\overline{k}_3$, and hence proportional to the product of all of them. It is not difficult to convince ourself that it must be proportional to $\Re\mathrm{e}(k_1 \cdot \overline{k}_2 k_2 \cdot \overline{k}_3 k_3 \cdot \overline{k}_1)$, (recall that $k_i \cdot \overline{k}_i = 0$). In fact,

$$F = \frac{2}{s_{24}s_{34}s_{14}} \Re\mathrm{e}(k_1 \cdot \overline{k}_2 k_2 \cdot \overline{k}_3 k_3 \cdot \overline{k}_1) \quad . \tag{7.20d}$$

On the other hand, such real part is zero, as a consequence of the signature, since we can parametrize the momentum as $k_i = K_i(e^{i\theta_i}, e^{i\widetilde{\theta}_i})$, thus, using $\delta_{i,j} = e^{i(\theta_i - \theta_j)}$, $\widetilde{\delta}_{i,j} = e^{i(\widetilde{\theta}_i - \widetilde{\theta}_j)}$, we find

$$\begin{aligned}(k_1 \cdot \overline{k}_2)(k_2 \cdot \overline{k}_3)(k_3 \cdot \overline{k}_1) &= (\delta_{1,2} - \widetilde{\delta}_{1,2})(\delta_{2,3} - \widetilde{\delta}_{2,3})(\delta_{3,1} - \widetilde{\delta}_{3,1})(K_1 K_2 K_3)^2 \\ &= 2i\Im m(\delta_{1,2}\widetilde{\delta}_{1,2})(\delta_{2,3}\widetilde{\delta}_{2,3})(\delta_{3,1}\widetilde{\delta}_{3,1})(K_1 K_2 K_3)^2 \quad .\end{aligned}$$

Therefore,

$$\mathcal{A}_4 = 0 \quad . \tag{7.21}$$

It is expected that higher-point amplitudes also vanish in the same fashion.

Note that formula (7.19) may be checked by looking at the residues of (7.17) even though we did not know how to calculate exactly the complicated integrals (7.17). For instance, let us calculate the residue (R) of $\mathcal{A}_4$ at the first pole of the (34)-channel. Taking $s_{34} = -1 + \epsilon$ we have

$$\frac{R}{\epsilon} = \int d^2 z |z|^{-2+\epsilon} |1-z|^{2s_{32}} \left| \frac{s_{32}(s_{32}-1)}{(1-z)^2} + \frac{c_{12}c_{34}}{z} + \frac{c_{23}c_{41}}{1-z} \right|^2 . \tag{7.22}$$

Next we use the following representation for the distribution $|z|^{-2+\epsilon}$ (see Eq. (C.55), and Appendix C in general)

$$|z|^{-2+\epsilon} = \frac{\pi}{\epsilon}\delta^{(2)}(z) \quad . \tag{7.23}$$

Thus we obtain,

$$\begin{aligned}\frac{R}{\pi} &= (s_{32}(s_{32}-1) + c_{23}c_{41})^2 + (c_{34}c_{12})^2 \int d^2 z \partial_z \partial_{\overline{z}} \delta^{(2)}(z) |1-z|^{2s_{32}} \\ &+ c_{34}c_{12}s_{32}(s_{32}-1) \int d^2 z (-\partial_z \delta^{(2)}(z))(1-z)^{s_{32}} (1-\overline{z})^{s_{32}-2} + h.c. \\ &+ c_{34}c_{12}c_{23}c_{41} \int d^2 z \partial_z \delta^{(2)}(z)(1-z)^{s_{32}}(1-\overline{z})^{s_{32}-1} + h.c. \\ &= [s_{32}(s_{32}-1)]^2 \left(1 + \frac{c_{23}c_{41}}{s_{32}(s_{32}-1)} + \frac{c_{34}c_{12}}{(s_{32}-1)}\right)^2 \quad .\end{aligned} \tag{7.24}$$

Using the identities (7.20) we have:

$$\begin{aligned}R &= \pi[F(s_{34} = -1)]^2 (1 - s_{14})^2 (s_{14})^2 \\ &= \pi[F(s_{34} = -1)]^2 \Delta(1 + s_{14})\Delta(1 + s_{24}) \quad .\end{aligned} \tag{7.25}$$

This is the result expected from (7.19). The analysis of other poles and channels also confirms formula (7.19).

The fact that the coefficient of the amplitude is zero is a solution to the following problem: if the theory does not contain further massive states, how could it be true that the four-point amplitudes contain the poles displayed in (7.19)? Therefore, for consistency, it must have a zero coefficient.

Such problems characterize the critical case, and we are led to verify to which extension such ideas carry over to the non-critical theory.

7.3. The Non-critical Case

Similar to the $N = 0,1$ non-critical strings studied in Chapters 2 and 5 it is natural to consider the coupling of a chiral (anti-chiral) $N = 2$ superfield $X(\overline{X})$ in a Coulomb gas representation with $\hat{c} \leq 1 (c = 3\hat{c})$ to a super Liouville chiral (anti-chiral) superfield $\Phi(\overline{\Phi})$ such that the total action is given by

$$S = S_\Phi + S_X \tag{7.26a}$$

$$S_\Phi = \frac{1}{4\pi} \int d^2w d^4\theta \hat{E} \left(\Phi\overline{\Phi} - Q\hat{Y}(\Phi + \overline{\Phi}) \right) + S_{cosm} \tag{7.26b}$$

$$S_X = \frac{1}{4\pi} \int d^2w d^4\theta \hat{E} \left(X\overline{X} + 2i\alpha_0 \hat{Y}(X + \overline{X}) \right) \quad . \tag{7.26c}$$

The cosmological term for the Liouville sector is given by the sum of the chiral and the anti-chiral contributions,

$$S_{cosm} = \mu \int d^2w d^2\theta^- e^{\overline{\alpha}\Phi} + \mu \int d^2w d^2\theta^+ e^{\alpha\overline{\Phi}} \quad . \tag{7.27}$$

The quantity $\hat{Y}$ stands for the $N = 2$ supercurvature superfield and $\hat{E}$ for the superdeterminant of the superzweibein. In order to introduce the notation we give the on shell decomposition of the chiral field Φ (an analogous for $\overline{\Phi}$ is valid, see also (7.6))

$$\Phi = \varphi(z, \overline{z}) + \xi_R(z)\theta^- + \xi_L(\overline{z})\overline{\theta}^- - \partial\varphi\theta^+\theta^- - \overline{\partial}\varphi\overline{\theta}^+\overline{\theta}^- \quad . \tag{7.28}$$

In fact, the choice of the cosmological term has some arbitrariness, worth commenting on. In the $N = 2$ case we have the possibility of introducing a cosmological term integrating over the whole Grassmann space. However, this could introduce new interactions, and would not correspond to the usual cosmological term one introduces in the 4-dimensional $N = 1$ supersymmetric case[162].

The total energy momentum tensor and $U(1)$ current are calculated by standard means, and their expressions read

$$T = T_\Phi + T_X \quad , \tag{7.29a}$$

$$T_\Phi = -{:}\partial\overline{\varphi}\partial\varphi{:} + \frac{1}{4}{:}\overline{\xi}_R\partial\xi_R{:} + \frac{1}{4}{:}\xi_R\partial\overline{\xi}_R{:} - \frac{Q}{2}\partial^2(\varphi + \overline{\varphi}) \quad , \tag{7.29b}$$

$$T_X = -{:}\partial\overline{x}\partial x{:} + \frac{1}{4}{:}\overline{\psi}_R\partial\psi_R{:} + \frac{1}{4}{:}\psi_R\partial\overline{\psi}_R{:} + i\alpha_0\partial^2(x + \overline{x}) \quad , \tag{7.29c}$$

$$J = J_\Phi + J_X \quad , \tag{7.30a}$$

$$J_\Phi = \frac{1}{4}{:}\overline{\xi}_R\xi_R{:} + \frac{Q}{2}\partial(\varphi - \overline{\varphi}) \quad , \tag{7.30b}$$

$$J_X = \frac{1}{4}{:}\overline{\psi}_R\psi_R{:} - i\alpha_0\partial(x - \overline{x}) \quad . \tag{7.30c}$$

Analogously to Chapter 5 we have the following propagators,

$$\langle x(z)\overline{x}(w)\rangle = \langle\varphi(z)\overline{\varphi}(w)\rangle = \ln|z-w|^{-2} \quad , \tag{7.31a}$$

$$\langle\psi_R(z)\overline{\psi}_R(w)\rangle = \langle\xi_R(z)\overline{\xi}_R(w)\rangle = 2(z-w)^{-1} \quad , \tag{7.31b}$$

$$\langle\psi_L(\overline{z})\overline{\psi}_L(\overline{w})\rangle = \langle\xi_L(\overline{z})\overline{\xi}_L(\overline{w})\rangle = 2(\overline{z}-\overline{w})^{-1} \quad , \tag{7.31c}$$

which can be read from (7.26*b*) and (7.26*c*). Following [50] we fix Q in (7.26*b*) imposing the vanishing of the total central charge

$$c_T = c_{SL} + c_M + c_{ghosts} = 0 \quad , \tag{7.32a}$$

$$c_{SL} = 3(1+2Q^2) \quad , \tag{7.32b}$$

$$c_M = 3\widehat{c},\, \widehat{c} = 1 - 8\alpha_0^2 \quad , \tag{7.32c}$$

$$c_{ghosts} = -6 \quad . \tag{7.32d}$$

Thus we fix the value of the parameter Q to be

$$Q = 2|\alpha_0| \quad , \tag{7.33}$$

where we chose Q to be real; this corresponds to a choice of phases, as one readily verifies.

The constants $\overline{\alpha}$ and α in Eq. (7.27) can be fixed imposing that the operators $\mathrm{e}^{\overline{\alpha}\Phi}$ and $\mathrm{e}^{\alpha\overline{\Phi}}$ have dimension (1/2,1/2) (because of the double integration over the Grassmann variables):

$$\Delta\left(\mathrm{e}^{\overline{\alpha}\Phi}\right) = -\frac{\overline{\alpha}Q}{2} = \frac{1}{2} \ , \tag{7.34a}$$

$$\Delta\left(\mathrm{e}^{\alpha\overline{\Phi}}\right) = -\frac{\alpha Q}{2} = \frac{1}{2} \ , \tag{7.34b}$$

$$\alpha = \overline{\alpha} = -\frac{1}{Q} \quad . \tag{7.34c}$$

Note that the operator $\mathrm{e}^{\alpha\Phi}$ ($\mathrm{e}^{\alpha\overline{\Phi}}$) is chiral (antichiral) since it satisfies the chirality condition $\Delta = +q\,(\Delta = -q)$. It is easy to check that the values of α and $\overline{\alpha}$ in Eq. (7.34*c*) assure vanishing $U(1)$ charge for the action S_{SL} as required. Therefore the solution (7.34*c*) is clearly a consistent one.

Still following ref. [50] the non-critical version of the vertex (7.9) is defined by integration over the tachyonic exponential over the z and θ variables; expanding the superfields we obtain

$$\begin{aligned}V(k,\overline{k}) &= \int \mathrm{d}^2 z \mathrm{d}^4\theta \mathrm{e}^{i(k\overline{X}+\overline{k}X)+\beta\overline{\Phi}+\overline{\beta}\Phi} = \int \mathrm{d}^2 z \mathrm{e}^{i(k\overline{x}+\overline{k}x)+\beta\overline{\varphi}+\overline{\beta}\varphi} \\ &\times \left[ik\partial\overline{x} + \beta\partial\overline{\varphi} - i\overline{k}\partial x - \overline{\beta}\partial\varphi - \overline{k}k\overline{\psi}_R\psi_R + \overline{\beta}\beta\overline{\xi}_R\xi_R + i\beta\overline{k}\overline{\psi}_R\xi_R - i\overline{k}\beta\psi_R\overline{\xi}_R\right] \\ &\times \left[ik\overline{\partial}\overline{x} + \beta\overline{\partial}\overline{\varphi} - i\overline{k}\overline{\partial}x - \overline{\beta}\overline{\partial}\varphi - \overline{k}k\overline{\psi}_L\psi_L + \overline{\beta}\beta\overline{\xi}_L\xi_L + i\beta\overline{k}\overline{\psi}_L\xi_L - i\overline{k}\beta\psi_L\overline{\xi}_L\right] \quad .\end{aligned} \tag{7.35}$$

Notice that we have used on shell expressions (with $\overline{\mu} = 0$) for the superfields $X(\overline{X})$ and $\Phi(\overline{\Phi})$, e.g.,

$$X = x(z,\overline{z}) + \psi_R(z)\theta^- + \psi_L(\overline{z})\overline{\theta}^- - \partial x\theta^+\theta^- - \overline{\partial} x\overline{\theta}^+\overline{\theta}^- \quad , \tag{7.36a}$$

$$\overline{X} = \overline{x}(z,\overline{z}) - \overline{\psi}_R(z)\theta^+ - \overline{\psi}_L(\overline{z})\overline{\theta}^+ + \partial\overline{x}\theta^+\theta^- + \overline{\partial}\overline{x}\overline{\theta}^+\overline{\theta}^- \quad . \tag{7.36b}$$

Imposing vanishing conformal weight and $U(1)$ charge we have two equations for the dressings β and $\overline{\beta}$,

$$\Delta\left(\mathrm{e}^{i(k\overline{X}+\overline{k}X)+\beta\overline{\Phi}+\overline{\beta}\Phi}\right) = 2\left[(\overline{k}-\alpha_0)(k-\alpha_0) - (\overline{\beta}+\frac{Q}{2})(\beta+\frac{Q}{2})\right] = 0 \tag{7.37}$$

$$q\left(\mathrm{e}^{i(k\overline{X}+\overline{k}X)+\beta\overline{\Phi}+\overline{\beta}\Phi}\right) = 2\alpha_0(k-\overline{k}) + Q(\beta-\overline{\beta}) = 0 \quad . \tag{7.38}$$

The equation (7.38) determines the imaginary part of the dressing (for $\alpha_0 \neq 0$) and plugging it in (7.37) we obtain the real part of the dressing (up to a sign),

$$E_\pm = \left(\frac{\beta+\overline{\beta}}{2}\right)_\pm + \frac{Q}{2} = \pm\left|\frac{k+\overline{k}}{2} - \alpha_0\right| \quad , \tag{7.39}$$

where $E_\pm$ is the energy associated with the time direction $(\Phi+\overline{\Phi})$. As discussed before, we take henceforth only positive energy solutions (E_+).

Having defined the vertex operator we can start calculating n-point correlation functions $\mathcal{A}_n = \langle V_{k_1}\cdots V_{k_n}\rangle$. Integrating over the double zero-modes of x and Φ we obtain the momentum and energy conservation (assuming $s = 0 = \overline{s}$) laws, respectively[1)],

$$\sum_{j=1}^{n} k_j = 2\alpha_0 = \sum_{j=1}^{n}\overline{k}_j \quad , \tag{7.40}$$

$$\sum_{1}^{n}\beta_i + Q = 0 = \sum_{1}^{n}\overline{\beta}_i + Q \quad . \tag{7.41}$$

The calculation of the amplitudes is very similar to the critical case and we obtain for the 3-point function,

$$\begin{aligned}\mathcal{A}_3 = \frac{(\ln\mu)^2}{\alpha^2}\Big\langle &\mathrm{e}^{ik_3\cdot x(0)}\mathrm{e}^{ik_2\cdot x(1)}[i k_2\cdot\partial\overline{x} - i\overline{k}_2\cdot\partial x - (k_2\cdot\overline{\psi}_R)(\overline{k}_2\cdot\psi_R)] \\ &\times [ik_2\cdot\overline{\partial}\overline{x} - i\overline{k}_2\cdot\overline{\partial}x - (k_2\cdot\overline{\psi}_L)(\overline{k}_2\cdot\psi_L)]\Big\rangle_{S(\mu=0)} \quad , \\ = \frac{(\ln\mu)^2}{\alpha^2}&(c_{23})^2 \quad ,\end{aligned} \tag{7.42}$$

where

$$\begin{aligned} c_{ij} &= k_i\cdot\overline{k}_j - \overline{k}_i\cdot k_j \\ &= k_i\overline{k}_j - \overline{k}_i k_j - \beta_i\overline{\beta}_j + \overline{\beta}_i\beta_j \quad .\end{aligned} \tag{7.43}$$

---

1) We have used that on the sphere $\int \mathrm{d}^2z\sqrt{\hat{g}}\hat{R} = 8\pi$.

The factor $(\ln \mu)^2$ in (7.42) is the finite part of $\left(\lim_{\bar{s},s\to 0} \mu^{s+\bar{s}}\Gamma(-s)\Gamma(-\bar{s}) \right)$. In order to rewrite $\mathcal{A}_3$ in a more suggestive form we need some facts about the kinematics. First of all it is easy to show (using (7.37)-(7.41) and (7.43)) that

$$\sum_{j=1}^{n} c_{ij} = 2\alpha_0(k_i - \bar{k}_i) + (\beta_i - \bar{\beta}_i)Q = 0 \quad . \tag{7.44}$$

The vanishing of $\sum_{j=1}^{n} c_{ij}$ holds in the critical case as a consequence of the momentum conservation and the on shell condition $k \cdot \bar{k} = k_1\bar{k}_1 - k_2\bar{k}_2 = 0$. It is remarkable that (7.44) holds also in the non-critical case as a consequence of the zero $U(1)$ current condition (7.38).

We assume[2)] from now on that $\alpha_0 < 0$ in this case we have from (7.37), (7.38) and (7.39),

$$\beta(k) = \begin{cases} k & , \text{if } \Re e k = \frac{k+\bar{k}}{2} > \alpha_0 \quad , \\ 2\alpha_0 - \bar{k} & , \text{if } \Re e k < \alpha_0 \quad , \end{cases} \tag{7.45}$$

therefore

$$k \cdot \bar{k} = k\bar{k} - \beta\bar{\beta} = \begin{cases} 0 & , \text{if } \Re e k > \alpha_0 \quad , \\ 4\alpha_0(\Re e k - \alpha_0) & , \text{if } \Re e k < \alpha_0 \quad . \end{cases} \tag{7.46}$$

Using all these kinematic relations we may write $\mathcal{A}_3$ from Eq. (7.42) in the region $\Re e k_2$, $\Re e k_3 < \alpha_0$, $\Re e k_1 > \alpha_0$ in a factorized form,

$$\mathcal{A}_3 = \frac{(\ln \mu)^2}{\alpha^2} \frac{(\Im m k_1)^2}{\alpha_0^2} (k_2 \cdot \bar{k}_2)(k_3 \cdot \bar{k}_3) \quad , \tag{7.47}$$

where $\Im m k = (-i)\frac{(k-\bar{k})}{2}$. The amplitude vanishes for any other kinematic region, where at least two momenta satisfy $\Re e k > \alpha_0$ and are therefore "on shell" $(k \cdot \bar{k} = 0)$ in the critical sense. It should be stressed that the amplitude $\mathcal{A}_3$ in the critical case[89] has the same form (7.42) but it cannot be written in a factorized form as in (7.47).

Now the important difference with respect to the critical case comes from the non-analytical structure of the dispersion relation (7.39) which allows us to eliminate completely, in a given kinematic region, the real part of the momentum of one of the scattered particles and to rewrite $\mathcal{A}_3$ in a factorized form. For instance, in the region above we have,

$$\Re e k_1 = 0 \quad ; \quad k_1 \cdot \bar{k}_1 = 0 \tag{7.48}$$

and (7.47) can be written as[87]

$$\mathcal{A}_3 = (\ln \mu)^2 \left(\frac{|k_1|}{\alpha_0} \right)^2 (k_2 \cdot \bar{k}_2)(k_3 \cdot \bar{k}_3) \quad . \tag{7.49}$$

Now we come to the computation of $\mathcal{A}_4$ which corresponds to

$$\mathcal{A}_4 = \frac{(\ln \mu)^2}{16\alpha^2} \int d^2z |z|^{-s} |1-z|^{-t} \left(\frac{t(t+2)}{(1-z)^2} + \frac{4c_{12}c_{34}}{z} + \frac{4c_{23}c_{41}}{1-z} \right) \times (\text{h.c.}) \tag{7.50}$$

2) Calculations for $\alpha_0 > 0$, are completely analogous.

where[3] $s = -2s_{34}$, $t = -2s_{23}$ and $s_{ij} = k_i \cdot \overline{k}_j + \overline{k}_i \cdot k_j$. The "hermitian conjugated" term of (7.50) corresponds to the previous term inside the brackets with $\overline{z}$ instead of z. Note that it is not really the hermitian conjugated expression since $\overline{k}_i$ is not the complex conjugated of k_i.

After performing the integrals in (7.50) using formula (7.18) and making algebraic manipulations which are consequence of kinematic relations common to the critical and non-critical cases we have

$$\mathcal{A}_4 = -\frac{\pi(\ln \mu)^2}{\alpha^2} F^2 \Delta(s_{34})\Delta(s_{14})\Delta(s_{24}) \quad , \tag{7.51}$$

where

$$F = [(k_1 \cdot \overline{k}_2)(k_2 \cdot \overline{k}_3)(k_3 \cdot \overline{k}_1) + \text{“h. c.”}] \tag{7.52}$$

and $\Delta(x) = \Gamma(x)/\Gamma(1-x)$. The expression (7.51) is essentially the same one derived in critical case, the difference now comes from the fact that after fixing the kinematic region $\Re e k_1$, $\Re e k_2$, $\Re e k_3 < \alpha_0$, $\Re e k_4 > \alpha_0$ we have the following kinematic identities,

$$c_{ij} = -\frac{i}{\alpha_0}[\Im m \beta_i (k_j \cdot \overline{k}_j) - \Im m \beta_j (\overline{k}_i \cdot \overline{k}_i)] \quad , \tag{7.53}$$

$$s_{i4} = -k_i \cdot \overline{k}_i \quad , \tag{7.54}$$

$$s_{ij} = k_i \cdot \overline{k}_i + k_j \cdot \overline{k}_j \quad , \quad i,j = 1,2,3 \quad , \tag{7.55}$$

which permit us to derive a very simple expression for F,

$$F = -\frac{((\Im m k_4)^2 + \alpha_0^2)}{4\alpha_0^2} \prod_{i=1}^{3} (k_i \cdot \overline{k}_i) = \frac{1}{4}\left(\frac{|k_4|}{\alpha_0}\right)^2 \prod_{i=1}^{3} s_{i4} \quad . \tag{7.56}$$

Using equations (7.53)-(7.56) we can finally write $\mathcal{A}_4$ in a factorized form

$$\mathcal{A}_4 = \frac{\pi(\ln \mu)^2}{16}\left(\frac{|k_4|}{\alpha_0}\right)^4 \prod_{1}^{3} \Delta(1 - k_i \cdot \overline{k}_i) \quad . \tag{7.57}$$

It's important to remark that in any kinematic region where at least two particles satisfy $\Re e k_i \geq \alpha_0$ $(\rightarrow k_i \cdot \overline{k}_i = 0)$ both amplitudes $\mathcal{A}_3$ and $\mathcal{A}_4$ vanish.

We can start now the analysis of the results (7.49) and (7.57) observing that when we take $\alpha_0 = 0$ $(\hat{c} = 1)$ the $U(1)$ charge of the vertex operator vanishes identically (see (7.38)) and we have no restrictions on the imaginary part of the dressing $(\beta - \overline{\beta})$. In particular, this means that the dispersion relation (7.39) does not apply to the $\alpha_0 = 0$ case. Therefore the factorized results that we have obtained so far are only true, strictly speaking, for $\hat{c} < 1$ $(\alpha_0 \neq 0)$. For $\alpha_0 = 0$ the requirement of vanishing conformal weight (see (7.37)) just reproduces the on shell condition $k \cdot \overline{k} = k\overline{k} - \beta\overline{\beta} = 0$ and we recover the critical case whose amplitudes have been already calculated by Ooguri and Vafa[89]. The discontinuous nature of the $\hat{c} \rightarrow 1^-(\alpha_0 \rightarrow 0)$ limit can be also seen from other points of view. Note, for

[3] Our definition of s and t correspond to twice of ref. [89] because their propagators correspond to half of ours (see (7.7))

instance, that contrary to the $N = 0,1$ non-critical strings, in the $N = 2$ case it is impossible to obtain the $\hat{c} = 1$ non-critical theory by an appropriate rotation of the $\hat{c} < 1$ model (see (7.29a)-(7.30c) and (7.33)). One can also take the factorized expression (7.57) for $\mathcal{A}_4$ in the limit $\alpha_0 \to 0$ to see that the result diverges with $\frac{1}{\alpha_0}$ (we absorb the factor $1/\alpha^2$ in the definition of the measure, which is necessary to have a finite result for $\mathcal{A}_3$ in the $\alpha_0 \to 0$ limit), which shows the non-existence of discrete states in the $\hat{c} \to 1^-$ limit, as expected. For $\hat{c} < 1$ the interesting models are the minimal ones for which the functions $\Delta(1 - k \cdot \overline{k})$ have no poles or zeroes. Thus, as in the $N = 0,1$ non-critical strings, these functions have a mild effect and can be absorbed through renormalizations of the vertices.

Concluding, we must say that there are still many aspects of $N = 2$ non-critical strings to be understood in the continuum which might be useful in developing super matrix models. In particular it is not known how to continue for other kinematic regions the results that we have derived in a given region. To realize this aim it may be useful to calculate higher point functions, as well as, to properly include the cosmological terms[163] ($s, \overline{s} \neq 0$) to understand the space time picture behind the amplitudes that we have obtained.

8. Final Remarks and Outlook

8.1. Comments on Continuum Results

Before entering into the conclusions, it is useful to comment on the validity of the results. First, the zero mode technique requires that the parameter s be continued to non-integer values, a practice that has been achieved following different procedures. Although none of them fully justifies the calculation, a comparison with the matrix model results and KdV, or the existing supersymmetric discrete results, show excellent agreement, as far as they can be compared. The technique used in the present work to continue the amplitudes from positive integer s to arbitrary values is related to the continuation to an arbitrary kinematic region (arbitrary momentum) of the amplitudes obtained inside the specific convergence region of the integrals involved. The latter continuation was based on the space-time interpretation of the tachyon S-matrix for $\mu = 1$. The amplitudes thus continued contain cuts, which prevent a naive analytic continuation, at specific values of the momenta and they appear due to the non-conservation of the Liouville momentum for arbitrary s. Such cuts have also been seen[90] in the discrete approach for $c = 1$ and the agreement between the continuum and the discrete approaches is a very strong indication of the correctness of the continuation in s that we have used and the space-time picture behind it. Although our examples of continued amplitudes (see e.g. $\mathcal{A}_4$ and $\mathcal{A}_5$) were restricted to the bosonic case the continuation procedure can be generalized to the NS-sector of the $N = 1$ superstring (see [79] for an example) for which we also have a clear spacetime picture. Such continuation also shows the presence of cuts in the momentum space. It is important to remark that for the $N = 2$ and the Ramond-sector of the $N = 1$ non-critical superstrings the continuation for an arbitrary kinematic region from the specific region used so far is not yet possible since we still do not have in those cases access to their full S-matrix and consequently their spacetime picture is still incomplete.

The main result in the fermionic sector, which concerns the computation of the N-point correlator for the Neveu-Schwarz sector, shows a striking similarity with the bosonic case. This shows that this sector has little further information with respect to the bosonic sector, and the complications contained in the fermionic determinant of the super Liouville theory are probably encoded in the Ramond sector, which is also much more difficult to be obtained in general.

8.2. Higher Genus in the Continuum

The set of results obtained up to now for (super) Liouville theory on genus-zero surfaces and (super) matrix models are rather encouraging, but the generalization of such results to higher genus, from the continuum point of view, poses some technical difficulties which can be presumably handled. In fact, for the $c = 1$ purely bosonic case, the theory has been shown by Mukhi and Vafa[164] to be equivalent to a twisted $N = 2$ supersymmetric topological model (see also [165]). The core of that observation relies on two basic facts apparently disconnected from the present problems. First, Distler and Vafa addressed the question of using Hermitian matrix models in order to compute the Euler characteristic of moduli space of Riemann surfaces, a problem already studied by the mathematicians [166]. They found that the problem is solved by the $c = 1$ string at the self dual radius, by arguing that the string susceptibility vanishes, upon comparison of the genus-g partition function of moduli space with the Euler character of moduli space of genus-g Riemann surfaces with n punctures. The second piece of information comes from the fact that the Euler characteristic of moduli space is also given by a twisted $N = 2$ superconformal $SU(2)/U(1)$ coset model at $k = -3$ (or equivalently, $SL(2,R)/U(1)$ at $k = 3$). Indeed, oversimplifying the argument, the twisted model above is a topological theory and the correlators are position independent, and Witten has shown that it is given by the above Euler invariant.

On the top of the above fact, one also knows that bosonic string theories have an $N = 2$ superconformal symmetry related to the stress tensor, ghost number current, b ghost and BRST current, with a further ingredient, namely, we consider a scalar field $\eta(z)$ with a background charge Q_η; the $N = 2$ twisted superconformal algebra is generated by

$$
\begin{aligned}
&T(z) = T^M(z) + T^{gh}(z) \quad , \quad J(z) = c(z)b(z) - y\partial\eta(z) \quad , \\
&G^+(z) = c(z)T^M(z) + \frac{1}{2}c(z)T^{gh}(z) + x\partial^2 c(z) + y\partial\left(c(z)\partial\eta(z)\right) \quad , \\
&G^-(z) = b(z) \quad , \\
&\qquad x = \frac{1}{2}\left(3 + Q_\eta y\right) \quad , \quad y = \frac{1}{2}\left(-Q_\eta + \sqrt{Q_\eta^2 - 8}\right) \quad .
\end{aligned}
$$

Mukhi and Vafa used the matter field X to play the role of the scalar field; it has $c = 1$, and no background charge, therefore $x = 3/2$, $y = i\sqrt{2}$ and the topological central charge of the theory is $\hat{c} = 3$, corresponding to the superconformal $SU(2)/U(1)$ model with $k = -3$.

Relevant physical quantities might be computed using topological methods; so far some specific correlators have been obtained, namely the N-point correlator of the cosmological term (for arbitrary genus) and the 3- and 4-tachyons scattering on the sphere ($h = 0$) which we have also computed in Chapter 2. The results completely agree with matrix models.

8.3. Matrix Models Beyond the Planar Approximation

Matrix models turn out to be a fruitful approach to the problem of the sum over geometries in two-dimensional quantum gravity. Indeed they can be solved to all orders in the genus expansion, in this way surpassing the results obtained from the Liouville formulation. In this section we summarize some of these achievements from the matrix model description and their connection with integrable systems.

8.3.1. The Method of Orthogonal Polynomials

In Chapter 6 we have seen how the one matrix problem can be transformed into an eigenvalue model, by eliminating the angular matrices from the integrals. In this case the evaluation of any correlation function of $O(N)$-invariant operators is reduced to an N-dimension integral over the eigenvalues $\{\lambda_i\}$.

We have also mentioned, along Chapters 3 and 6, that the double scaling limit, which constraints the large N limit to the critical point $\Lambda \to \Lambda_c$, is a suitable definition for the continuum limit of the model including the sum over the complete topological expansion.

Concerning large N problems, one of the most powerful methods is based on the use of orthogonal polynomials. As it enables one to solve the models for finite N, the double scaling limit can be carefully controlled. We shall apply this technique to solve the eigenvalue model given by Eq. (6.16).

Let us begin by introducing the orthogonal polynomials $P_n(\lambda)$ w.r.t. the measure $dw(\lambda) \equiv \exp\{-\frac{N}{\Lambda}V(\lambda)\}\, d\lambda$, which satisfy

$$\int d\lambda\, e^{-\frac{N}{\Lambda}V(\lambda)} P_i(\lambda) P_j(\lambda) = h_i \delta_{ij} \quad , \tag{8.1}$$

where $V(\lambda)$ is the potential function and h_i are normalization constants. We have normalized the coefficient of λ^i in $P_i(\lambda)$ to one, that is

$$P_i(\lambda) = \lambda^i + \sum_{j=1}^{i} a_i^{(j)} \lambda^{i-j} \quad . \tag{8.2}$$

Using their orthogonality property, one can show that these polynomials satisfy the following two-term recursion relation:

$$\lambda P_i(\lambda) = P_{i+1}(\lambda) + S_i P_i(\lambda) + R_i P_{i-1}(\lambda) \quad . \tag{8.3}$$

The coefficients S_i and R_i characterize completely the set of polynomials. While S_i vanishes for symmetric potentials, the R_i coefficients can be written in terms of the normalization constants h_i, as follows

$$\begin{aligned}
h_{i+1} &= \int \mathrm{d}\lambda\, \mathrm{e}^{-\frac{N}{\Lambda}V(\lambda)} P_{i+1}(\lambda) P_{i+1}(\lambda) \\
&= \int \mathrm{d}\lambda\, \mathrm{e}^{-\frac{N}{\Lambda}V(\lambda)} P_{i+1}(\lambda)[\lambda P_i(\lambda) + \cdots] \\
&= \int \mathrm{d}\lambda\, \mathrm{e}^{-\frac{N}{\Lambda}V(\lambda)} [\lambda P_{i+1}(\lambda)] P_i(\lambda) \\
&= \int \mathrm{d}\lambda\, \mathrm{e}^{-\frac{N}{\Lambda}V(\lambda)} [P_{i+2} + S_{i+1}P_{i+1} + R_{i+1}P_i] P_i \\
&= R_{i+1} h_i \quad ,
\end{aligned}$$

implying

$$R_{i+1} = \frac{h_{i+1}}{h_i} \tag{8.4}$$

and therefore

$$h_{i+1} = h_0^{i+1} \prod_{j=1}^{i+1} R_j \quad . \tag{8.5}$$

Back to the eigenvalue model (6.16), we notice that the Vandermonde determinant can be written as a sum of products of orthogonal polynomials,

$$\begin{aligned}
\Delta(\lambda) = \prod_{i<j}(\lambda_i - \lambda_j) &= \begin{vmatrix} 1 & 1 & \cdots & 1 \\ \lambda_1 & \lambda_2 & \cdots & \lambda_N \\ \vdots & \vdots & \cdots & \vdots \\ \lambda_1^{N-1} & \lambda_2^{N-1} & \cdots & \lambda_N^{N-1} \end{vmatrix} \\
&= \begin{vmatrix} P_0(\lambda_1) & P_0(\lambda_2) & \cdots & P_0(\lambda_N) \\ P_1(\lambda_1) & P_1(\lambda_2) & \cdots & P_1(\lambda_N) \\ \vdots & \vdots & \cdots & \vdots \\ P_{N-1}(\lambda_1) & P_{N-1}(\lambda_2) & \cdots & P_{N-1}(\lambda_N) \end{vmatrix} \quad .
\end{aligned} \tag{8.6}$$

Expanding $\Delta^2(\lambda)$ in the partition function (6.16), then integrating over λ_i using the orthogonality of the P_i polynomials, we obtain

$$\begin{aligned}
\mathcal{Z}_B &= N! \prod_{i=1}^{N-1} h_i = N!\, h_0^N \prod_{i=1}^{N-1} R_i^{N-i} \\
&= N!\, h_0^N \exp\left\{ \sum_{i=1}^{N-1} (N-i) \ln R_i \right\} \quad .
\end{aligned} \tag{8.7}$$

Hence all the "physics" of the model has been concentrated in the R_i and S_i parameters. They can be determined, in terms of the coupling constants, from the following relations: using simple integrations by parts and the orthogonality property, one finds

$$\begin{aligned}\frac{N}{\Lambda}\int \mathrm{d}\lambda\, \mathrm{e}^{-\frac{N}{\Lambda}V(\lambda)} P_n(\lambda)V'(\lambda)P_n(\lambda) &= \int \mathrm{d}\lambda\, \mathrm{e}^{-\frac{N}{\Lambda}V(\lambda)} \frac{\mathrm{d}}{\mathrm{d}\lambda}(P_n(\lambda)P_n(\lambda)) \\ &= \int \mathrm{d}\lambda\, \mathrm{e}^{-\frac{N}{\Lambda}V(\lambda)} 2P_n(\lambda)[nP_{n-1}(\lambda)+\cdots] \\ &= 0\end{aligned}$$

and

$$\begin{aligned}\frac{N}{\Lambda}\int \mathrm{d}\lambda\, \mathrm{e}^{-\frac{N}{\Lambda}V(\lambda)} P_n(\lambda)V'(\lambda)P_{n-1}(\lambda) &= \int \mathrm{d}\lambda\, \mathrm{e}^{-\frac{N}{\Lambda}V(\lambda)} \frac{\mathrm{d}}{\mathrm{d}\lambda}(P_n(\lambda)P_{n-1}(\lambda)) \\ &= \int \mathrm{d}\lambda\, \mathrm{e}^{-\frac{N}{\Lambda}V(\lambda)} \left[P_n'(\lambda)P_{n-1}(\lambda) + P_n(\lambda)P_{n-1}'(\lambda)\right] \\ &= \int \mathrm{d}\lambda\, \mathrm{e}^{-\frac{N}{\Lambda}V(\lambda)} \left[(nP_{n-1}(\lambda)+\cdots)P_{n-1}(\lambda) + P_n(\lambda)((n-1)P_{n-2}(\lambda)+\cdots)\right] \\ &= nh_{n-1} \quad ,\end{aligned}$$

or briefly

$$\int \mathrm{d}\lambda\, \mathrm{e}^{-\frac{N}{\Lambda}V(\lambda)} P_n(\lambda)V'(\lambda)P_n(\lambda) = 0 \quad , \tag{8.8}$$

$$\frac{n}{N}\Lambda = \frac{1}{h_{n-1}}\int \mathrm{d}\lambda\, \mathrm{e}^{-\frac{N}{\Lambda}V(\lambda)} P_{n-1}(\lambda)V'(\lambda)P_n(\lambda) \quad . \tag{8.9}$$

Let us restrict ourselves to even potentials. In that case, Eq. (8.8) is immediately satisfied, and we can concentrate our attention on Eq. (8.9). If $V'(\lambda)$ is a polynomial of degree $(2a-1)$ in λ, we have to apply the recursion relation (8.3) (with $S_i = 0$) $(2a-1)$ times onto $P_n(\lambda)$, in order to find the coefficient of $P_{n-1}(\lambda)$ in the product $V'(\lambda)P_n(\lambda)$. If we define a matrix Q as

$$Q_{ij} = \delta_{j,i+1} + R_i\delta_{j,i-1} \quad , \tag{8.10}$$

the λ-multiplication relation (8.3) becomes

$$\lambda P_n(\lambda) = \sum_j Q_{nj}P_j(\lambda) \tag{8.11}$$

and therefore

$$\begin{aligned}V'(\lambda)P_n(\lambda) &= \sum_i 2ig_{2i}\lambda^{2i-1}P_n(\lambda) \\ &= \sum_i 2ig_{2i}\sum_j (Q^{2i-1})_{nj}P_j(\lambda) \\ &= \sum_j [V'(Q)]_{nj}P_j(\lambda) \quad .\end{aligned} \tag{8.12}$$

We thus obtain

$$\int d\lambda\, e^{-\frac{N}{\Lambda}V(\lambda)} P_{n-1}(\lambda)V'(\lambda)P_n(\lambda) = h_{n-1}[V'(Q)]_{n,n-1} \quad . \tag{8.13}$$

This means that the fundamental Eq. (8.9) can be expressed in terms of the matrix Q as

$$\frac{n}{N}\Lambda = [V'(Q)]_{n,n-1} \quad . \tag{8.14}$$

In particular, for $n = N$ we find the discrete version of the string equation,

$$\Lambda = [V'(Q)]_{N,N-1} \quad , \tag{8.15}$$

relating the R_i coefficients (encoded in the matrix Q) to the coupling constants and Λ. For polynomial potentials the r.h.s. of (8.15) only depends on a finite number of R_n's, with n of the order of N.

Let us study the first non-trivial example, given by the potential

$$V(\lambda) = \left(\frac{4\Lambda_c}{R_c}\right)\lambda^2 - \left(\frac{4\Lambda_c}{3R_c^2}\right)\lambda^4 \quad , \tag{8.16}$$

corresponding to Kazakov's $m = 2$ critical potential (recall (3.65)). Using the definition (8.10) we verify that

$$\begin{aligned} (Q)_{n,n-1} &= R_n \quad , \\ (Q^3)_{n,n-1} &= R_n(R_{n-1} + R_n + R_{n+1}) \quad , \end{aligned} \tag{8.17}$$

implying that

$$\begin{aligned} [V'(Q)]_{n,n-1} &= \left(\frac{4\Lambda_c}{R_c}\right) 2(Q)_{n,n-1} - \left(\frac{4\Lambda_c}{3R_c^2}\right) 4(Q^3)_{n,n-1} \\ &= 2\Lambda_c\left(\frac{4R_n}{R_c}\right) - \frac{1}{3}\Lambda_c\frac{16R_n}{R_c}\left(\frac{R_{n-1}}{R_c} + \frac{R_n}{R_c} + \frac{R_{n+1}}{R_c}\right) \quad . \end{aligned} \tag{8.18}$$

In this case, Eq. (8.14) becomes

$$\frac{n}{N}\frac{\Lambda}{\Lambda_c} = 2\left(\frac{4R_n}{R_c}\right) - \frac{1}{3}\frac{4R_n}{R_c}\left(\frac{4R_{n-1}}{R_c} + \frac{4R_n}{R_c} + \frac{4R_{n+1}}{R_c}\right) \quad ,$$

or

$$\left(1 - \frac{\Lambda}{\Lambda_c}\frac{n}{N}\right) = \left(1 - \frac{4R_n}{R_c}\right)^2 + \frac{1}{3}\frac{4R_n}{R_c}\left(\frac{4R_{n-1}}{R_c} - 2\frac{4R_n}{R_c} + \frac{4R_{n+1}}{R_c}\right) \quad . \tag{8.19}$$

We are interested in n close to N, in the limits $N \to \infty$ and $\Lambda \to \Lambda_c$. Therefore it is convenient to define some scaling variables: as in the planar approximation, we use a parameter a to control the critical point limit,

$$1-\frac{\Lambda}{\Lambda_c}=a^2t \quad , \tag{8.20}$$

i.e. $a \to 0$ implies $\Lambda \to \Lambda_c$; then we define a scaling variable x related to n, as suggested by Eq. (8.19),

$$1-\frac{\Lambda}{\Lambda_c}\frac{n}{N}=a^2x \quad , \tag{8.21}$$

so that $n \to N$ corresponds to $x \to t$; finally we define $u(x)$ as the scaling function related to R_n,

$$1-\frac{4R_n}{R_c}=a\,u(x) \quad . \tag{8.22}$$

Notice that the a-dependence was adjusted according to the leading terms in Eq. (8.19). Assuming that $u(x)$ is a smooth function of x, we can take the following Taylor expansions

$$\begin{aligned}\frac{4R_{n\pm1}}{R_c}&=1-a\,u(x\mp\frac{\Lambda}{\Lambda_c}\frac{1}{Na^2})\\&=1-a\left\{u(x)\mp\frac{\Lambda}{\Lambda_c}\frac{1}{Na^2}u'(x)+\frac{1}{2}\frac{\Lambda^2}{\Lambda_c^2}\frac{1}{N^2a^4}u''(x)+\cdots\right\}\end{aligned} \tag{8.23}$$

and substitute them back in Eq. (8.19), which now reads

$$\begin{aligned}a^2x&=a^2u^2(x)-\frac{a}{3}(1-au(x))\left\{\frac{\Lambda^2}{\Lambda_c^2}\frac{1}{N^2a^4}u''(x)+\mathcal{O}\left(\frac{1}{N^3a^6}\right)\right\}\\&=a^2\left\{u^2(x)-\frac{1}{3}\frac{\Lambda^2}{\Lambda_c^2}\frac{1}{N^2a^5}\left(u''-auu''\right)+\mathcal{O}\left(\frac{1}{N^3a^7}\right)\right\} \quad ,\end{aligned} \tag{8.24}$$

If the limit $N \to \infty$ is taken before (and independently from) the scaling limit $a \to 0$, one finds $x=u^2(x)$. In particular, for $x \to t$, we recover the planar string equation for $m=2$, $t=u^2$. But if we constrain those limits imposing that

$$Na^{5/2}=\kappa^{-1}=\text{constant} \quad , \tag{8.25}$$

which implies

$$\begin{aligned}&\frac{\Lambda^2}{\Lambda_c^2}\frac{1}{N^2a^5}\overset{\text{DSlim}}{\longrightarrow}\kappa^2 \quad ,\\&\frac{1}{N^ja^{2j+1}}=\left(\frac{1}{Na^{5/2}}\right)^j a^{j/2-1}=a^{(j-2)/2}\kappa^j\overset{\text{DSlim}}{\longrightarrow}0 \quad , \quad j\geq 3 \quad ,\end{aligned} \tag{8.26}$$

a second derivative term also survives the limit in (8.24), and for $x=t$ we obtain

$$t=u^2-\frac{1}{3}\kappa^2u'' \quad . \tag{8.27}$$

This is the Painlevé equation of the first kind. It admits a (formal) perturbative solution for small κ (or equivalently, for large t):

$$u = t^{1/2} - \frac{1}{24}\frac{\kappa^2}{t^2} - \cdots \quad . \tag{8.28}$$

Remember that small κ means large N: κ plays the role of the renormalized constant parametrizing the topological expansion. Hence the solution (8.28) corresponds to the genus expansion of the $m = 2$ model: $t^{1/2}$ is the sphere contribution, the next term comes from the torus, and so on.

Now we return to Eq. (8.7) to analyze the double scaling limit of the free energy. We are interested in the scaling behavior with respect to the constant t. Since h_0 is t-independent and $N!h_0^N$ only gives a non-interesting additive constant to the free energy, we may write

$$\begin{aligned} F &= \text{const.} + \sum_{i=1}^{N-1}(N-i)\ln R_i \\ &= \text{const.} + N\sum_{i=1}^{N-1}\left(1-\frac{i}{N}\right)\left\{\ln\left(\frac{4R_i}{R_c}\right) - \ln\frac{4}{R_c}\right\} \\ &= \text{const}'. + N\sum_{i=1}^{N-1}\left(1-\frac{i}{N}\right)\ln\left(\frac{4R_i}{R_c}\right) \quad . \end{aligned} \tag{8.29}$$

Using the scaling variables defined in (8.20, 21, 22) we can approximate

$$\begin{aligned} N\sum_{i=1}^{N-1} &\to N\left(-N\frac{\Lambda_c}{\Lambda}a^2\right)\int_{1/a^2}^{t} \mathrm{d}x \\ \left(1-\frac{i}{N}\right) &\to \frac{\Lambda_c}{\Lambda}a^2(x-t) \\ \ln\left(\frac{4R_i}{R_c}\right) &\to \ln(1-a\,u(x)) \to -a\,u(x) \end{aligned} \tag{8.30}$$

and therefore

$$\begin{aligned} F &\to \text{const.} - \left(\frac{\Lambda_c^2}{\Lambda^2}N^2a^5\right)\int_t^{1/a^2} \mathrm{d}x\,(x-t)u(x) \\ &\xrightarrow{\text{DSlim}} \text{const.} + \frac{1}{\kappa^2}\int_t^{\infty} \mathrm{d}x\,(t-x)u(x) \quad . \end{aligned} \tag{8.31}$$

We can eliminate non-interesting constants and analytical terms in t by taking the second derivative of F w.r.t. t, thus defining the (renormalized) free energy $\mathcal{F}$ as

$$\frac{\partial^2}{\partial t^2}\mathcal{F} = -\frac{1}{\kappa^2}u \quad . \tag{8.32}$$

Therefore the universal function u is interpreted as the specific heat, like in the planar approximation. Eq. (8.32) together with the string equation (8.27) obeyed by the specific heat provide a non-perturbative description of the $m = 2$ model, which corresponds to the pure gravity theory according the the conclusions of Chapter 4.

In principle we could follow the same algorithm used for $m = 2$ to define the continuum limit of the general m-th critical model: the double scaling condition (8.25) is generalized as $Na^{2+1/m} = \kappa^{-1}$; the r.h.s. of the definition (8.22) is replaced by $a^{2/m}u$, as in the planar approximation; we verify that the relation between free energy and the function u stays the same as in Eq. (8.32), but u obeys different string equations. For instance, when $m = 3$, Eq. (8.27) is replaced by

$$t = u^3 - \kappa^2 uu'' - \frac{1}{2}\kappa^2(u')^2 + \frac{1}{10}\kappa^4 u^{(4)} \quad , \tag{8.33}$$

while the 4-th critical string equation reads

$$\begin{aligned} t = u^4 - 2\kappa^2 u(u')^2 - 2\kappa^2 u^2 u'' + \frac{3}{5}\kappa^4 (u'')^2 \\ + \frac{4}{5}\kappa^4 u' u^{(3)} + \frac{2}{5}\kappa^4 uu^{(4)} - \frac{1}{35}\kappa^6 u^{(6)} \quad . \end{aligned} \tag{8.34}$$

Using different methods the authors in ref. [70] showed that the general m-th critical string equation is given by

$$t = \frac{m!}{(2m-1)!!} R_m[u] \quad , \tag{8.35}$$

where $R_m[u]$ are the Gelfand-Dikii polynomials obtained from the recursion relation

$$R_{n+1}[u] = \left(u + D^{-1}uD - \frac{1}{2}\kappa^2 D^2\right) R_n[u] \quad , \tag{8.36a}$$

$$R_0 = 1 \quad , \quad D = \frac{\partial}{\partial t} \quad . \tag{8.36b}$$

Later on we shall briefly review Douglas' method, which can be generalized to multi-matrix models, the result (8.36) being obtained as a particular case.

Notice that in the limit $\kappa \to 0$ the relation (8.36) generates the monomials

$$R_n[u] = \frac{(2n-1)!!}{n!} u^n \tag{8.37}$$

and Eq. (8.35) tends to the planar m-th critical string equation $t = u^m$. This is in agreement with the interpretation of κ as the genus counting parameter, so that as $\kappa \to 0$ all contributions cancel except the one from genus zero.

8.3.2. On the Integrability of Matrix Models

We have described in Chapters 3 and 4 the appearance of Virasoro constraints obeyed by the partition function of the 1-matrix model. One can show[131,132] that, in the continuum (double scaling) limit, the *square root* of the partition function also satisfies an infinite set of linear constraints, as a consequence of the KdV equations.

The theory is defined by the partition function $\mathcal{Z} = e^{-\mathcal{F}}$, where the free energy $\mathcal{F}$ relates to the specific heat u through Eq. (8.32)

$$-\kappa^2 D^2 \mathcal{F} = u \quad , \quad D \equiv \frac{\partial}{\partial t} \quad . \tag{8.38}$$

The specific heat, on the other hand, satisfies the string equation (8.35) for the m-th model. In a general massive model the string equation reads

$$t = -\sum_{n\geq 1}(2n+1)t_n R_n[u] \quad . \tag{8.39}$$

Compare this equation with Eq. (3.70): the monomials u^n have been replaced by the Gelfand-Dikii polynomials, and we have normalized the couplings t_n so that $t_m = -1/(2m+1)\,,\, t_n = 0\,,\, n \neq m$ corresponds to the m-th multi-critical point, $t = R_m[u]$.

As in the planar approximation, the scaling operators σ_n couple to the sources t_n, so that the insertion of σ_n corresponds to differentiation with respect to t_n, and is identified with the n-th KdV flow of u,

$$\frac{\partial}{\partial t_n} u = D R_{n+1}[u] \quad , \tag{8.40}$$

(recall Eq. (3.71)). Using the recursion relation (8.36), the above equation implies

$$\frac{\partial}{\partial t_{n+1}} u = \left[2u + (Du)D^{-1} + \frac{1}{2}\kappa^2 D^2\right] \frac{\partial}{\partial t_n} u \quad , \tag{8.41}$$

which becomes a recursion relation for the free energy, as follows from Eq. (8.38),

$$D^2 \frac{\partial}{\partial t_{n+1}} \mathcal{F} = \left[2uD^2 + (Du)D + \frac{1}{2}\kappa^2 D^4\right] \frac{\partial}{\partial t_n} \mathcal{F} \quad . \tag{8.42}$$

On the other hand, the string equation (8.39) is equivalent to a constraint on $\mathcal{F}$: indeed, from Eqs. (8.38-40) we have

$$\begin{aligned} 0 &= \sum_{n\geq 1}(2n+1)t_n D R_n[u] + 1 \\ &= \sum_{n\geq 1}(2n+1)t_n \frac{\partial u}{\partial t_{n-1}} + 1 \\ &= -\kappa^2 D^2 \left\{ \sum_{n\geq 1}(2n+1)t_n \frac{\partial \mathcal{F}}{\partial t_{n-1}} \right\} + 1 \\ &= 4\kappa^2 D^2 \left\{ \sum_{n\geq 1}\left(n+\frac{1}{2}\right) t_n \frac{\partial}{\partial t_{n-1}}\left(-\frac{\mathcal{F}}{2}\right) + \frac{t^2}{8\kappa^2} \right\} \quad , \end{aligned} \tag{8.43}$$

which can be integrated (as shown in [132], the integration constants are zero), giving as a result

$$\sum_{n\geq 1}\left(n+\frac{1}{2}\right)t_n\frac{\partial}{\partial t_{n-1}}\left(-\frac{\mathcal{F}}{2}\right)+\frac{t^2}{8\kappa^2}=0 \quad , \tag{8.44a}$$

which can otherwise be written as

$$e^{\mathcal{F}/2}\left\{\sum_{n\geq 1}\left(n+\frac{1}{2}\right)t_n\frac{\partial}{\partial t_{n-1}}+\frac{t^2}{8\kappa^2}\right\}e^{-\mathcal{F}/2}=0 \quad . \tag{8.44b}$$

Now we use the recursion relation (8.42) to generate other constraints from Eq. (8.44): applying the operator $[2uD^2+(Du)D+\frac{1}{2}\kappa^2D^4]$ on both sides of (8.44a) we have

$$\begin{aligned}
0 &= \left[2uD^2+(Du)D+\frac{1}{2}\kappa^2D^4\right]\left[\sum_{n\geq 1}\left(n+\frac{1}{2}\right)t_n\frac{\partial}{\partial t_{n-1}}\left(-\frac{\mathcal{F}}{2}\right)+\frac{t^2}{8\kappa^2}\right] \\
&= \sum_{n\geq 1}\left(n+\frac{1}{2}\right)t_nD^2\frac{\partial}{\partial t_n}\left(-\frac{\mathcal{F}}{2}\right)+\frac{1}{2\kappa^2}u+\frac{1}{4\kappa^2}tDu \\
&= D^2\sum_{n\geq 1}\left(n+\frac{1}{2}\right)t_n\frac{\partial}{\partial t_n}\left(-\frac{\mathcal{F}}{2}\right)-\frac{1}{2}D^2\mathcal{F}-\frac{1}{4}tD^3\mathcal{F} \\
&= D^2\sum_{n\geq 1}\left(n+\frac{1}{2}\right)t_n\frac{\partial}{\partial t_n}\left(-\frac{\mathcal{F}}{2}\right)+D^2\left(\frac{t}{2}D\left(-\frac{\mathcal{F}}{2}\right)\right) \\
&= D^2\left\{\left[\sum_{n\geq 1}\left(n+\frac{1}{2}\right)t_n\frac{\partial}{\partial t_n}+\frac{1}{2}t\frac{\partial}{\partial t}\right]\left(-\frac{\mathcal{F}}{2}\right)\right\} \quad .
\end{aligned} \tag{8.45}$$

Calling $t=t_0$, we have

$$\sum_{n\geq 1}\left(n+\frac{1}{2}\right)t_n\frac{\partial}{\partial t_n}\left(-\frac{\mathcal{F}}{2}\right)+C_0+C_1t=0 \quad . \tag{8.46}$$

As shown in ref. [132], we have $C_1=0$ and $C_0=1/16$, and the constraint (8.46) may be rewritten as follows

$$e^{\mathcal{F}/2}\left\{\sum_{n\geq 0}\left(n+\frac{1}{2}\right)t_n\frac{\partial}{\partial t_n}+\frac{1}{16}\right\}e^{-\mathcal{F}/2}=0 \quad . \tag{8.47}$$

One can proceed generating another equation for the free energy, combining Eqs. (8.42) and (8.47); by repeating this procedure, one finds the following recursion relation[132]

$$D^2\left(\mathrm{e}^{\mathcal{F}/2}L_{n+1}\mathrm{e}^{-\mathcal{F}/2}\right)=\left(2uD^2+(Du)D+\frac{1}{2}\kappa^2D^4\right)\left(\mathrm{e}^{\mathcal{F}/2}L_n\mathrm{e}^{-\mathcal{F}/2}\right) \tag{8.48}$$

where L_{-1} is the operator in (8.44*b*), corresponding to the string equation. After taking care of the integration constants, one has

$$L_{-1}=\sum_{n\geq1}\left(k+\frac{1}{2}\right)t_k\frac{\partial}{\partial t_{k-1}}+\frac{t_0^2}{8\kappa^2}\quad, \tag{8.49a}$$

$$L_0=\sum_{n\geq0}\left(k+\frac{1}{2}\right)t_k\frac{\partial}{\partial t_k}+\frac{1}{16}\quad, \tag{8.49b}$$

$$L_n=\sum_{n\geq0}\left(k+\frac{1}{2}\right)t_k\frac{\partial}{\partial t_{k+n}}+\frac{\kappa^2}{2}\sum_{k=1}^{n}\frac{\partial^2}{\partial t_{k-1}\partial t_{n-k}} \tag{8.49c}$$

and since $L_{-1}\mathrm{e}^{-\mathcal{F}/2}=0$, we obtain

$$L_n\tau=0\quad,\quad\tau=\mathrm{e}^{-\mathcal{F}/2}=\sqrt{\mathcal{Z}}\quad. \tag{8.49d}$$

Since $u=-\kappa^2D^2\mathcal{F}=2\kappa^2D^2\ln\tau$ is a solution of the KdV equations (8.40), we say that τ is a τ-function of the KdV hierarchy. It is nevertheless a special τ-function: it satisfies the constraints (8.49), which are consequences of the string equation obeyed by u.

The generators (8.49) satisfy a closed algebra, namely

$$[L_n,L_m]=(n-m)L_{n+m}\quad, \tag{8.50}$$

the same Virasoro algebra truncated to $n\geq-1$ found in the discrete model (recall (3.29)). However, notice that the generators are written in terms of different sets of coupling constants (compare (3.28) and (8.49)); moreover, the constraints are imposed on the discrete partition function Z in the finite N model, as stated in Eq. (3.27), while Eqs. (8.49) are linear relations on the square root of the continuum $\mathcal{Z}$.

For a rigorous analysis of the continuum levels we suggest ref. [133]. Also in ref. [134] the integrability properties of matrix models are clarified using discrete integrable systems. In particular, for even potentials, one can show the correspondence between the one matrix model and the Volterra hierarchy[134].

8.4. Multi-Matrix Models

As discussed in Chapters 3 and 4, the planar limit of the m-th critical Hermitian 1-matrix model generates the spectrum of anomalous dimensions of the $(2, 2m-1)$ series of minimal models coupled to gravity. In section 8.3 we have extended the results to all orders in genus expansion. In order to generate other models from the (p, q) series - in particular, unitary minimal theories - we resort to other matrix models.

8.4.1. Ising Model on a Random Lattice

The partition function of the Ising model with a magnetic field on random ϕ^4 lattices turns out to be equal[55] (up to an unimportant additive term) to the free energy of an exactly soluble model of *two* Hermitian matrices, whose partition function reads

$$Z = \int \mathcal{D}X\mathcal{D}Y \mathrm{e}^{-\beta W(X,Y)} \quad , \tag{8.51a}$$

$$W(X,Y) = \mathrm{tr}\ \left[(X^2 + Y^2 - 2cXY) - g(\mathrm{e}^H X^4 + \mathrm{e}^{-H} Y^4)\right] \quad , \tag{8.51b}$$

where H is the magnetic field, and c is related to the Ising temperature[55]. As in the 1-matrix case, one can integrate[126] over the angular degrees of freedom of the Hermitian matrices X and Y, reducing (8.51) to the following integral over eigenvalues

$$Z = \mathrm{const.} \int \prod_{i=1}^{N} \mathrm{d}x_i \mathrm{d}y_i \Delta(x)\Delta(y) \mathrm{e}^{-\beta \sum_i W(x_i, y_i)} \quad , \tag{8.52a}$$

$$W(x_i, y_i) = (x_i^2 + y_i^2 - 2cx_iy_i) - g(\mathrm{e}^H x_i^4 + \mathrm{e}^{-H} y_i^4) \quad , \tag{8.52b}$$

where $\{x_i\}$ (resp. $\{y_i\}$) is the set of eigenvalues of the matrix X (resp. Y); $\Delta(x) = \prod_{i<j}(x_i - y_i)$ is the Vandermonde determinant. The partition function (8.52) thus factorizes as a product of double integrals: defining orthogonal polynomials $P_i^+(x)$ and $P_i^-(y)$ with respect to the two-dimensional measure in (8.52)

$$\int \mathrm{d}x\mathrm{d}y \mathrm{e}^{-\beta W(x,y)} P_i^+(x) P_j^-(y) = h_i \delta_{ij} \tag{8.53}$$

and expanding the Vandermonde determinants as products of these polynomials, one obtains

$$Z \propto \prod_{i=1}^{N} h_i \quad , \tag{8.54}$$

to be compared with the result (8.7). Therefore we need to calculate h_i to solve the model.

For the quartic potential (8.52b), the orthogonality (8.53) implies that the polynomials $P_i^{\pm}(x)$ obey a three-term recursion relation

$$xP_i^{\pm}(x) = P_{i+1}^{\pm}(x) + r_i^{\pm} P_{i-1}^{\pm}(x) + s_i^{\pm} P_{i-3}^{\pm}(x) \quad . \tag{8.55}$$

From (8.53) one easily derives the following equations for the coefficients $r^{\pm}, s^{\pm}$:

$$cr_i^{\mp} = R_i \left(1 + 2ge^{\mp H} \left(r_{i+1}^{\pm} + r_i^{\pm} + r_{i-1}^{\pm}\right)\right) \tag{8.56a}$$

$$cR_i + \frac{i}{2\beta} = r_i^{\pm} + 2ge^{\pm H} \left(\left(s_{i+2}^{\pm} + s_{i+1}^{\pm} + s_i^{\pm}\right) + r_i^{\pm} \left(r_{i+1}^{\pm} + r_i^{\pm} + r_{i-1}^{\pm}\right)\right) \tag{8.56b}$$

$$cs_{i+1}^{\pm} = 2ge^{\mp H} R_{i+1} R_i R_{i-1} \quad , \tag{8.56c}$$

where we have introduced the parameters

$$R_i \equiv \frac{h_i}{h_{i-1}} \quad , \tag{8.57}$$

in analogy to the 1-matrix model (recall Eq. (8.4)). The free energy is therefore given as

$$F = \sum_{i=1}^{N} (N - i) \ln R_i \quad , \tag{8.58}$$

up to non-interesting constants.

Now we can apply the treatment of the 1-matrix model in the double-scaling limit: as $N \to \infty$, one can replace i/β by a continuous variable x; substituting $r_i^{\pm} \to r^{\pm}(x)$, $s_i^{\pm} \to s^{\pm}(x)$, $R_i \to R(x)$ one transforms (8.56) into differential equations for the functions $r^{\pm}, s^{\pm}$ and R. Moreover the free energy tends to $\mathcal{F} \sim \int_0^x dy (x - y) \ln R(y)$.

In order to preserve the contribution from higher-genus surfaces, we have to approach the critical values of the parameters g and c, and take $N \to \infty$, $\beta/N \to 1$ as prescribed by the double scaling limit.

Using this procedure for the the magnetic field $H = i\pi/2$, which corresponds to the Yang-Lee edge singularity[61], one finds[56,64] the equation

$$t = u^3 - \kappa^2 u u'' - \frac{1}{2}\kappa^2 (u')^2 + \frac{1}{10}\kappa^4 u^{(4)} \quad , \tag{8.59}$$

where t is the scaling variable corresponding to x, κ is the renormalized coupling constant and $u(t)$ is the scaling function corresponding to $R(x)$. This is precisely the string equation (8.32) of the $m = 3$ multi-critical one-matrix model. On the other hand, when the magnetic field vanishes, the system (8.56) implies the equation[56]

$$t = u^3 - \kappa^2 u u'' - \frac{1}{2}\kappa^2 (u')^2 + \frac{2}{27}\kappa^4 u^{(4)} \quad . \tag{8.60}$$

The difference between the coefficients in (8.59) and (8.60) is more consequent that it seems[56,64]: expanding the solution of Eq. (8.60) in inverse power of $t^{7/3}$, one finds that all terms are positive[56]; in turn the series obtained from (8.59) becomes negative at high genus (recalling the non-unitarity of the $m = 3$ multi-critical model).

8.4.2. Chains of Matrices

The partition function (8.51) exemplifies the fact that by considering multi-matrix models one may span a larger spectrum of minimal models[63]. Indeed, let us take the partition function

$$Z = \int \prod_{t=1}^{n} \mathcal{D}M(t) \mathrm{e}^{-\beta W} \quad , \tag{8.61a}$$

$$W = \mathrm{tr} \left\{ \sum_{t=1}^{n} V_t(M(t)) + \sum_{t=1}^{n-1} c_t M(t) M(t+1) \right\} \quad , \tag{8.61b}$$

thus generalizing (8.51) to the case of a chain of Hermitian matrices $\{M(t), t = 1, \cdots n\}$ with arbitrary potentials $V_t(x)$. The matrices $M(t)$ can be simultaneously diagonalized in (8.61), using the result[126]

$$\int \mathcal{D}A \mathrm{e}^{-\beta \mathrm{tr}\,[V(A)+2cAB]} = \mathrm{const.} \int \prod_i \mathrm{d}a_i \mathrm{e}^{-\beta \sum_i [V(a_i)+2ca_ib_i]} \frac{\Delta(a)}{\Delta(b)} \quad , \tag{8.62}$$

where $\{a_i\}$ (resp. $\{b_i\}$) is the set of eigenvalues of A (resp. B), and $\Delta(a) = \prod_{i<j}(a_i - a_j)$ is the Vandermonde determinant.

After discarding unimportant multiplicative factors, we reduce (8.61) to the following eigenvalue model

$$Z = \int \prod_{i,t} \mathrm{d}\lambda_i(t)\, \mathrm{e}^{-\beta W(\lambda)} \Delta(\lambda(1)) \Delta(\lambda(n)) \quad , \tag{8.63a}$$

$$W(\lambda) = \sum_{i,t} [V_t(\lambda_i(t)) + c_t \lambda_i(t) \lambda_i(t+1)] \quad , \tag{8.63b}$$

where $\{\lambda_i(t)\}$ represents the set of eigenvalues of $M(t)$. To evaluate (8.63) we define orthogonal polynomials $p_i^{(1)}(x)$ and $p_i^{(n)}(x)$ with respect to the multiple integral

$$\int \prod_t \mathrm{d}\lambda(t) \mathrm{e}^{-\sum_t [V_t(\lambda(t)) + c_t \lambda(t)\lambda(t+1)]} p_i^{(1)}(\lambda(1)) p_j^{(n)}(\lambda(n)) = h_i \delta_{ij} \quad . \tag{8.64}$$

Expanding the determinants in (8.63a) and integrating, the partition function becomes a product of the coefficients h_i, as in Eq. (8.54). The problem is therefore reduced to finding such coefficients.

Recall the 1-matrix results (8.8) and (8.9): those equations were used to determine h_i in terms of coefficients appearing in the recursion relations (8.3) among orthogonal polynomials. To derive such equations, one basically needs to know how to expand $\lambda p_i(\lambda)$ and $\frac{\mathrm{d}}{\mathrm{d}\lambda} p_i(\lambda)$ in terms of the orthogonal polynomials (as in Eq. (8.11)). Following ref. [63] we define the operators $Q(t)$ and $P(t)$, which represent the insertion of $\lambda(t)$ and $\frac{\mathrm{d}}{\mathrm{d}\lambda(t)}$ in the integral (8.64). In particular, the (discrete) equations of motion of the model imply that $P(t)$ and $Q(t)$ are determined by $P(1)$ and $Q(1)$. These equations of motion, together with the condition

$$[P(t), Q(t)] = 1 \quad , \tag{8.65}$$

are sufficient to determine the operators P and Q. However, we are interested in the double scaling limit of the theory: therefore, instead of solving the equation (8.65) immediately, we can take the large N limit first. For polynomial potentials $V_t(x)$, the operators P and Q have non-zero matrix elements $\mathcal{O}_{ij}$ only for $|i-j| \leq k$, for some finite integer k. This implies that, in the large N limit, P and Q become differential operators of finite order[63].

After tuning the coupling constants and suitably normalizing the polynomials, one approaches the critical point and take $N \to \infty$: the operator $Q(1)$ is expected[63] to assume the general form

$$Q = D^p + \sum_{i=0}^{p-2} u_i D^i \quad , \tag{8.66}$$

where u_i are functions to be determined (in particular, one can show that $D^2\mathcal{F} \propto u_{p-2}$, indicating that u_{p-2} is the specific heat), $D = \frac{\mathrm{d}}{\mathrm{d}x}$, and x is a continuous variable defined from the discrete one in a way similar to Eq. (8.21). Therefore Q is a differential operator of order p.

In the same way we assume that the couplings can be tuned to produce a differential operator P of order q. For p and q relatively prime (and supposing $p < q$ without loss of generality), the solution of Eq. (8.65) is[63]

$$P = Q_+^{q/p} \quad , \tag{8.67}$$

where $Q_+^{q/p}$ is the true differential piece of the pseudo-differential operator $Q^{q/p}$. The equation

$$[Q_+^{q/p}, Q] = 1 \tag{8.68}$$

implies a system of $(p-1)$ non-linear differential equations for the $(p-1)$ functions u_i in (8.66). The free energy can be written in terms of u_{p-2}, and its scaling behavior with respect to x (which is supposed to couple to the most negative dimension operator), gives[63] the string susceptibility γ

$$2 - \gamma = \frac{2(p+q)}{p+q-1} \quad , \tag{8.69}$$

which agrees with the continuum value for the (p,q) minimal model (see Eq. (4.50)).

This elegant formulation in terms of differential operators also provides the scaling operators[63]: they are given by $Q_+^{(mq-np)/q}$, for integer values of m, n. In ref. [83] this technique is applied to the unitary models ($p = n$, $q = n+1$) and to the D series. Some correlation functions are computed: one finds the precise fusion rules of the underlying conformal theory.

If the results obtained from the multi-matrix model (8.61) seem to be more than satisfactory, one can look for a more economic construction. Indeed by taking *two* matrix models and appropriately adjusting the potentials of both matrices, one can get[167,168] the critical regimes corresponding to the various (p,q) - rational models. By studying the invariance of the two-matrix partition function under transformations similar to (3.21), one can show [128] that there is a universal $W^{(N)}$ - symmetry in the two-matrix model, which is also an indication that it describes all minimal conformal theories coupled to two-dimensional gravity.

8.5. The Kontsevich Model and the Virasoro Constraints

Consider the partition function defined by the following integral over $N \times N$ Hermitian matrices

$$Z = \int D\phi\, e^{-\mathrm{tr}\left(\frac{1}{3}g\phi^3 + A\phi^2\right)} \quad , \tag{8.70}$$

where A is a positive Hermitian matrix. It corresponds to a particular case of (8.62), when $V(\phi)$ is taken as a cubic potential. As we can read from (8.62), the partition function (8.70) is a function of the N eigenvalues a_i of the matrix A.

In his remarkable paper [169], Kontsevich showed that the expansion of Z in terms of Feynman graphs can be interpreted as a decomposition of the moduli space of Riemann surfaces. This means that the correlation functions of two-dimensional topological gravity[67] can be generated by (8.70). On the other hand, one can show[170–173] that Z satisfies the Virasoro constraints of the 1-matrix model, thus completing the proof of the conjecture[67] that matrix models are equivalent to topological gravity.

Here we outline the demonstration given by ref. [171]. Using the transformation of variables

$$\phi \to \widehat{\phi} = \phi + \frac{A}{2g} \quad , \tag{8.71}$$

the partition function (8.70) can be written as

$$Z = e^{\mathrm{tr}\left(\frac{A^3}{12g^2}\right)}\widehat{Z} \quad , \tag{8.72}$$

where

$$\widehat{Z} = \int D\widehat{\phi}\, e^{-\mathrm{tr}\left(\frac{g}{3}\hat{\phi}^3 - X\hat{\phi}\right)} \quad , \tag{8.73a}$$

$$X = \frac{A^2}{4g} \quad . \tag{8.73b}$$

$\widehat{Z}$ represents a Hermitian matrix model in an external field X, whose Schwinger-Dyson equations of motion can be written[171] in terms of the eigenvalues $x_i = a_i^2/4g$ of X as

$$\left\{\frac{\partial^2}{\partial x_i^2} + \sum_{j\neq i}\frac{1}{x_i - x_j}\left(\frac{\partial}{\partial x_i} - \frac{\partial}{\partial x_j}\right) + \frac{x_i}{g}\right\}\widehat{Z} = 0 \quad . \tag{8.74}$$

Factorizing Z in the form

$$Z = \prod_{i,j} (a_i + a_j)^{-1/2} Y \quad , \tag{8.75}$$

the equations (8.72) and (8.74) imply that Y obeys the constraints

$$K_i Y = 0 \quad , \tag{8.76a}$$

$$\begin{aligned} K_i = & \frac{1}{4} \sum_{k,l=0} \frac{1}{a_i^{2(k+l+3)}} \frac{\partial^2}{\partial t_k \partial t_l} + \frac{1}{8g^2} \sum_{k=0}^{\infty} \frac{1}{a_i^{2(k+1)}} \frac{\partial}{\partial t_k} \\ & + \frac{1}{2} \sum_{k=0}^{\infty} \left(\frac{(2k+3)}{a_i^2} t_{k+1} + \frac{(2k+1)}{a_i^4} t_k + \cdots + \frac{t_0}{a_i^{2(k+2)}} \right) \frac{\partial}{\partial t_k} \\ & + \frac{1}{16 a_i^4} + \frac{t_0^2}{4 a_i^2} \quad , \end{aligned} \tag{8.76b}$$

where we have defined the couplings

$$t_k = -\frac{1}{2k+1} \sum_{i=1}^{N} \frac{1}{a_i^{2k+1}} \quad . \tag{8.77}$$

Notice that the t_k's become truly independent couplings only in the limit $N \to \infty$. Choosing $g = i/2$ and rearranging (8.76) in powers of $1/a_i^2$ one obtains the equations

$$L_n Y = \frac{1}{2} \frac{\partial}{\partial t_{n+1}} Y \quad , \tag{8.78}$$

where

$$L_{-1} = \sum_{k=1}^{\infty} \left(k + \frac{1}{2} \right) t_k \frac{\partial}{\partial t_{k-1}} + \frac{1}{4} t_0^2 \quad , \tag{8.79a}$$

$$L_0 = \sum_{k=0}^{\infty} \left(k + \frac{1}{2} \right) t_k \frac{\partial}{\partial t_k} + \frac{1}{16} \quad , \tag{8.79b}$$

$$L_n = \sum_{k=0}^{\infty} \left(k + \frac{1}{2} \right) t_k \frac{\partial}{\partial t_{k+n}} + \frac{1}{4} \sum_{k=1}^{n} \frac{\partial^2}{\partial t_{k-1} \partial t_{n-k}} \quad , \quad n \geq 1 \quad , \tag{8.79c}$$

i.e., the Virasoro generators (8.49) found in the continuum (double scaling) limit of the 1-matrix model. The remarkable point is that, starting from the simple cubic potential in (8.70) and taking the large N limit (without using the sophistications of the double scaling limit), one can have access to the whole set of multicritical potentials.

More details about the demonstration and the analysis of other matrix ensembles can be found in [171].

8.6. Double Scaling Limit and Non-Perturbative Solution of the Super Virasoro Constraints

The solution of the discrete superloop equations or equivalently the discrete super-Virasoro constraints, Eqs. (6.52-53), to all orders in the genus expansion was found by K. Becker and M. Becker [154]. They first considered the partition function (6.44) in the absence of fermionic couplings ($\xi_{k+1/2} = 0$), i.e.

$$Z_S^{2N}(g_k, \xi_{k+1/2} = 0) = \int \prod_{n=1}^{2N} \mathrm{d}\lambda_n \mathrm{d}\theta_n \mathrm{e}^{-\frac{2N}{\Lambda}\sum_k \sum_{i=1}^{2N} g_k \lambda_i^k} \prod_{i<j}^{2N} (\lambda_i - \lambda_j - \theta_i \theta_j) \quad . \tag{8.80}$$

In this case, the fermionic variables can be easily integrated, giving rise to the following effective (bosonic) eigenvalue model

$$Z_S^{2N}(g_k, 0) = \int \prod_{n=1}^{2N} \mathrm{d}\lambda_n \mathrm{e}^{-\frac{2N}{\Lambda}\sum_k \sum_{i=1}^{2N} g_k \lambda_i^k} Pf^{2N}(\lambda_{ij}^{-1}) \prod_{i<j}^{2N} (\lambda_i - \lambda_j) \quad , \tag{8.81}$$

where $Pf^{2N}(\lambda_{ij}^{-1})$ is the Pfaffian of the antisymmetric matrix

$$\lambda_{ij}^{-1} \equiv \frac{1}{\lambda_i - \lambda_j} \quad , \quad i \neq j \quad , \tag{8.82}$$

$$Pf(M) = \sqrt{\det M} = \frac{1}{2^N N!} \epsilon^{a_1 a_2 \cdots a_{2N}} M_{a_1 a_2} \cdots M_{a_{2N-1} a_{2N}} \quad . \tag{8.83}$$

Then, making use of the following identity

$$Pf^{2N}(\lambda_{ij}^{-1}) \prod_{i<j}^{2N} (\lambda_i - \lambda_j) = \frac{1}{2^N} \sum_{I<J} S^{I,J} \Delta^2(I) \Delta^2(J) \quad , \tag{8.84}$$

where $I = (\lambda_{i_1}, \cdots, \lambda_{i_n})$, $i_1 < \cdots i_N$ is a subset containing half of the $2N$ eigenvalues variables (same for J) and $S^{I,J}$ means symmetric permutations between the elements of I and J, one concludes that

$$Z_S^{2N}(g_k, 0) = \frac{1}{2^N} \binom{2N}{N} \left[Z_B^N(2g_k, 0) \right]^2 \quad , \tag{8.85}$$

where $Z_B^N(2g_k)$ is the partition function (6.14) of the purely bosonic theory. In terms of the free energies

$$F_S^{(0)} = \frac{1}{(2N)^2} \ln Z_S^{2N}(g_k, 0) \quad , \tag{8.86a}$$

$$F_B = \frac{1}{N^2} \ln Z_B^N(2g_k) \quad . \tag{8.86b}$$

Eq. (8.85) implies that, up to non-important additive constants,

$$F_S^{(0)} = \frac{1}{2} F_B \quad . \tag{8.87}$$

In order to include the fermionic couplings, one uses the super Virasoro constraints (6.32). In terms of the free energy, the solution is[154]

$$F_S = \left\{ 1 - \sum_{k,l} \xi_{k+1/2}\xi_{l+1/2} \frac{\partial^2}{\partial g_{k+1}\partial g_l} \right\} \frac{F_B}{2} \quad . \tag{8.88}$$

Remarkably the free energy depends on two fermionic couplings at most, as it was first observed in the planar approximation[84].

The continuum (double scaling) limit of the super Virasoro constraints was obtained in [152,153] and read

$$G_{n+1/2}\mathcal{Z}_s(t_k, \tau_k) = 0 \quad , \quad n \geq -1 \quad , \tag{8.89a}$$

$$G_{n+1/2} = \frac{t_0\tau_0}{4\kappa^2}\delta_{n,-1} + \sum_{k\geq 0}\left(k+\frac{1}{2}\right) t_k \frac{\partial}{\partial \tau_{n+k+1}} + \frac{\tau_0}{2}\frac{\partial}{\partial t_n}$$
$$+ \sum_{k\geq 0} \tau_{k+1} \frac{\partial}{\partial t_{n+k+1}} + \kappa^2 \sum_{k=0}^{n} \frac{\partial^2}{\partial t_k \partial \tau_{n-k}} \quad . \tag{8.89b}$$

The bosonic sector of constraints is obtained from the anti-commutation of the generators $G_{n+1/2}$. The coupling constants $\{t_k, \tau_k\}$ are the renormalized couplings related to the scaling operators. The equations that describe the theory in the double scaling limit are determined by[154] the free energy

$$\kappa^2 D^2 \mathcal{F}_S = u - 2\tau D^2 \tau \quad , \tag{8.90}$$

where $D = \frac{\partial}{\partial t}$, t is the renormalized cosmological constant, u is the bosonic piece of the specific heat and satisfies the string equation

$$t = -\sum_{k\geq 1}(2k+1)t_k R_k[u] \tag{8.91}$$

and the fermionic function $\tau(t_k, \tau_k)$ is given by

$$\tau = -\sum_{k\geq 0} \tau_k R_k[u] \quad . \tag{8.92}$$

$R_k[u]$ are the Gel'fand-Dikii polynomials. In particular (8.91) coincides with the string equation (8.39) of the bosonic theory. The relation (8.92) should be compared to its planar approximation (6.154). To verify that (8.90) gives a solution to the constraints (8.89) it suffices to show that $G_{-1/2}$ and $G_{3/2}$ annihilate $e^{-\mathcal{F}_s}$, since the algebra of $G_{n+1/2}$'s guarantees the remaining constraints.

Using the fact that the function u satisfies the KdV-flows

$$\frac{\partial}{\partial t_n} u = D R_{n+1}[u] \quad , \tag{8.93}$$

one shows[154] that the fermionic free energy can be written in terms of the purely bosonic one

$$\mathcal{F}_S = \frac{1}{2}\left\{1 - \sum_{k,l} \tau_k \tau_l \frac{\partial^2}{\partial t_k \partial t_{l-1}}\right\} \mathcal{F}_B \quad . \tag{8.94}$$

Notice the similarities between the discrete (8.88) and continuum (8.94) relations. Compare also Eq. (8.94) with the planar result (6.208): the expansion is at most bilinear in the fermionic couplings.

One concludes that the bilinear dependence on the fermionic couplings is a property that holds to all orders in the genus expansion. If this result seems incompatible with the non-vanishing 4-point functions of Ramond vertex operators, as obtained in the continuum formulation of the theory, the solution probably comes from choosing the appropriate (conformal) basis of scaling operators, and it involves analytic transformations among the coupling constants[156].

To finish this section, we comment on the relation between the super eigenvalue model and integrable systems. As a consequence of the first non-trivial flow in Eq. (8.93) and the definition (8.92), we have the following set of equations

$$\frac{\partial u}{\partial t_1} = D(3u^2 + \kappa^2 D^2 u) \quad , \tag{8.95a}$$

$$\frac{\partial \tau}{\partial t_1} = (6uD + \kappa^2 D^3)\tau \quad . \tag{8.95b}$$

The first one is the ordinary KdV equation. On the other hand, one verifies[154] that these equations are invariant under the following global supersymmetric transformations

$$\delta u = \epsilon D\tau \quad , \quad \delta\tau = \epsilon u \quad , \tag{8.96}$$

where ϵ is an anti-commuting constant parameter.

Equations (8.95) turn out to be a new supersymmetric extension of the KdV equation[154,155] in which the bosonic flows satisfy the recursion relations

$$\begin{pmatrix} \frac{\partial u}{\partial t_{n+1}} \\ \frac{\partial \tau}{\partial t_{n+1}} \end{pmatrix} = \begin{pmatrix} \kappa^2 D^2 + 2u + 2DuD^{-1} & 0 \\ 2D\tau D^{-1} + 2D^{-1}D\tau & \kappa^2 D^2 + 2u + 2D^{-1}uD \end{pmatrix} \begin{pmatrix} \frac{\partial u}{\partial t_n} \\ \frac{\partial \tau}{\partial t_n} \end{pmatrix} \tag{8.97}$$

and the fermionic flows read

$$\frac{\partial u}{\partial \tau_n} = 0 \quad , \tag{8.98a}$$

$$\frac{\partial \tau}{\partial \tau_{n+1}} = (\kappa^2 D^2 + 2u + 2D^{-1}uD)\frac{\partial \tau}{\partial \tau_n} \quad . \tag{8.98b}$$

This super-KdV hierarchy thus plays the same role that the ordinary KdV plays in the purely bosonic theory as regards its integrability.

8.7. String Theory of Two-Dimensional QCD (SCD_2)

The most important application of non-critical strings that we can foresee concerns the study of quantum chromodynamics. In fact, string theory was born out of the problem of strong interactions. By the end of the fifties, quantum electrodynamics was a well established theory. There had been also several important developments of general quantum field theories in the framework of the LSZ formalism[174], as well as by means of the axiomatic approach, where some non-perturbative results could be derived from general principles.

However, dynamical calculations were restricted to perturbation theory, rendering the results concerning strong interactions unreliable. In particular, information about the spectrum of the theory was only accessible within approximative non-perturbative, and often non-unitary schemes, as the Bethe Salpeter equation[175] in the ladder approximation. As a result quantum field theory had fallen into stagnation and even discredit in the late fifties.

These difficulties provided, in particular, the motivation for a new approach to the strong interactions, which became known as S-matrix theory[176], which was to play a dominant role in the sixties. The predictive power of the theory turned out to be very low, since it was entirely based on kinematical principles and analyticity supplemented by the bootstrap idea. An underlying dynamical principle was lacking. In fact, the idea of having a Lagrangian was put aside[177], since it was thought that the bootstrap idea could substitute the dynamical principles and provide a more fundamental formulation, implying a very radical position towards conventional developments at that time. This led in particular to the concept of duality, expressing the possibility of representing a given scattering amplitude as a sum over poles in crossed channels.

The explicit realization of these concepts by the remarkable Veneziano[178] formula led to the full development of dual models[2].

Although several important developments followed, there have been quite a number of incorrect results that led physicists to discard them. In particular, there was the high energy behavior, which turned out to be extremely well described by perturbative QCD, which grew in importance, being the central issue in the early seventies. On the other hand, dual models, after several reinterpretations, first as a string model, later with fermions - when super gauge transformations were introduced leading to the important concept of supersymmetry, and finally the superstring theory, which turned into a possible candidate to fulfill the unification aim.

On the other hand the problem of strong interactions could not advance for the understanding of low energy phenomena. In principle this problem should be addressed using a non-perturbative method. In fact, several properties concerning hadrons are understandable by means of the concept of string like flux tubes, being consistent with linear confinement, and linear Regge trajectories, as well as the approximate duality of hadronic scattering amplitudes, which are usual concepts of the string idea.

It is well known that the large N limit of QCD is smooth, and provides a picture of the string in the Feynman diagram space[53,54]. The large N limit is expected to provide most of the qualitative pictures of the low energy limit of the theory. When studied in low dimensional systems, the $1/N$ expansion turns out to be *the* correct expansion for models with problematic infrared behavior, such as CP^{N-1} and Gross-Neveu models, where properties such as confinement and spontaneous mass generation are straightforwardly derived within the scheme, and the exact S-matrix of the theory can be explicitly checked.

The above issues indicate that the understanding of the theory of strong interactions presumably requires the study of the large N limit of QCD. This is a formidable problem that evaded solution in the last two decades. Aiming at that problem, we have first to understand the relation which exists between the large N limit of chromodynamics and string theory. In any case, even the expected string theory cannot be in the same phase as that studied before due to the phase transition at $c = 1$, beyond which the string is in a crumpled phase[51].

There are models rather similar to QCD, displaying the dynamical structure one expects for the low energy theory. The first model is two-dimensional QED - or the Schwinger model[179,180]. It has been completely solved by Lowenstein and Swieca[181], and a rather detailed account of the model is given in [11].

Confinement, θ vacuum structure and the spontaneous breakdown of $U(1) \otimes \widetilde{U}(1)$ symmetry makes it an extraordinary laboratory for four-dimensional QCD. However, one important issue is missing for the above mentioned string interpretation: the non-abelian character of the fundamental field. Indeed, lowest order string scattering must correspond to an infinite summation of Feynman diagrams, which should correspond to a summation of the type expected in the $1/N$ expansion.

Therefore we consider two-dimensional QCD instead of the Schwinger model. In contrast to the latter, the interaction with massless fermions does not imply an exactly solvable theory. Nevertheless, one can derive some exact properties of the theory, and later relate its partition function to the one corresponding to a string theory.

Two-dimensional chromodynamics is a super renormalizable theory with finite mass and coupling constant renormalization. The first attempt to obtain its particle spectrum dates back to 1974, and was based on the $1/N$ expansion[182,183]. The axial gauge eliminates the gauge field self coupling term from the Lagrangian. In the large N limit only planar diagrams without fermion loops contribute.

The model was studied by t'Hooft[53], who found that the fermion self-energy obeyed the integral equation

$$\sum(p) = -\frac{ie^2}{2\pi^2}\int \frac{dk_-}{k_-^2} dk_+ \frac{(k_- + p_-)}{M^2 + (k+p)^2 + (k_- + p_-)\sum(k+p) - i\epsilon} \quad . \tag{8.99}$$

The light cone gauge ($A_- = 0$) was adopted in the computation, as well as the large N approximation. The $i\epsilon$ prescription is ambiguous. One can use it in order to compute first the k_+ integration, and later integrate over k_- with an infrared cut off $\lambda \leq |k_-|$. For gauge invariant quantities this is equivalent to interpret the integrals as principal values, and the solution turns out to be

$$\sum(p) = \frac{e^2}{\pi^2}\frac{1}{p_-} \tag{8.100}$$

and the fermion propagator is given by

$$S_F(p) = \frac{\not p + M + \frac{e^2}{2\pi p_-}\gamma_-}{p^2 - M^2 + e^2/\pi} \quad . \tag{8.101}$$

However, the use of the principal value prescription is very ambiguous[184]. Observe that for $M^2 < e^2/\pi$ we have tachyonic poles in the quark propagator[185]. However the simplicity of the result (8.100) is due to the fact that in the large N limit only the second order self energy diagram contributes. By reducing the sum over ladder graphs for the fermion four-point function to an integral equation for the corresponding Bethe Salpeter wave function[175], t'Hooft found an expression for the quark-antiquark bound state spectrum: the masses lie on a nearly linear Regge trajectory! This is different from the $U(1)$ case - quantum electrodynamics, where only one mesonic state exists.

Nevertheless, t'Hooft's solution can be criticized due to the aforementioned ambiguity, as well as to the severe infrared problems one finds. The ambiguity can in any case be tamed by performing the computation in Euclidian space, and returning upon look at the final result, and one finds for the self energy the expression[184]

$$\sum(p) = \frac{1}{2p_-}\left\{M^2 - p^2 - \sqrt{(M^2 - p^2) + \frac{4e^2}{\pi}p^2}\right\}^{1/2} \quad . \tag{8.102}$$

The presence of the anomalous branch cut reflects the fact that in Euclidian space every planar rainbow diagram contribute to the self-energy. This result is however very doubtful for a series of reasons[11].

There have also been claims[186,187] in the sense that the theory may have two phases. The first, given by t'Hooft's solution, describes the large N limit, where gluons remain massless, since fermion loops do not contribute for the leading term in $1/N$. In this case the mesonic spectrum is described by a Regge trajectory. There should be another phase, where the gluon requires a mass via the intrinsic Higgs mechanism as in the $U(1)$ case. In this strong coupling regime - or Higgs phase - the original $SU(N)$ symmetry would be broken down to the maximal abelian subgroup of $SU(N)$. Support in favour of this idea was given in [188,189]. In all cases, it is an open question whether this mechanism might explain that QCD is at the same time a string theory, but with high energy scattering processes showing powerlike behavior in the momenta (up to calculable logarithmic corrections).

Although very simple, the theory already displays several important features. It has very little more than a topological field theory. Indeed, after gauge fixing almost no interaction remains, as we see choosing the axial gauge. We will verify also that observables depend only on the topology and on its area.

We will consider the Wilson loop for pure QCD_2

$$W(C) = \mathrm{tr}_R P e^{\oint A^\mu \mathrm{d}x_\mu} \quad , \tag{8.103}$$

where C is a non-self intersecting contour. In the axial gauge $A_1 = 0$, the Lagrangian is quadratic in the remaining field, i.e.

$$\mathcal{L} = \frac{1}{2e}\mathrm{tr}\, E^2 \quad , \quad E = \partial_1 A_0 \quad . \tag{8.104}$$

The Wilson loop consists of a pair of charged particles propagating in time. From our knowledge of the $U(1)$ case, these sources produce a constant electric field E^a in two dimensions. Therefore, the Wilson loop can be explicitly computed, and is given by the exponential of the energy of the pair, integrated over the space time, and in terms of the quadratic Casimir for the given representation R, namely $C_2(R)$, the area of the loop - A and the coupling constant e, we find

$$W(C) = \mathrm{e}^{-e^2 C_2(R) A} \quad . \tag{8.105}$$

From the clue we learned from t'Hooft's work about the relation between large N QCD_2 and string theory concerning Regge behavior for the meson masses, our next step is to look for a string theory for QCD_2. We thus consider the simple case of pure gauge theory, which should correspond to a theory of closed strings. The partition function of an $SU(N)$ gauge theory defined on an Euclidian manifold $\mathcal{M}$ is

$$\mathcal{Z}_{\mathcal{M}} = \int \mathcal{D}A_\mu \mathrm{e}^{-\frac{1}{4e^2}\int_{\mathcal{M}} \mathrm{d}^2 x \sqrt{g} \mathrm{tr}\, F_{\mu\nu} F_{\mu\nu}} \quad . \tag{8.106a}$$

In general, the free energy will depend on the manifold $\mathcal{M}$, and we restrain to that dependence in our study of the above partition function, in order to look for the relation between the partition function above and that of a string model. We suppose that $\mathcal{Z}_{\mathcal{M}}$ depends on the topology and area of $\mathcal{M}$. Therefore we can write the partition function as

$$\mathcal{Z}_{\mathcal{M}} = \mathcal{Z}[G, e^2, A, N] = \mathcal{Z}[G, e^2 A, N] \quad , \tag{8.106b}$$

where G is the genus of the manifold $\mathcal{M}$, and scaling properties require that $\mathcal{Z}_{\mathcal{M}}$ depends on the adimensional parameter $e^2 A$, where e is the coupling constant. It is also clear that although the canonical formulation implies a theory without degrees of freedom, we can not gauge away the gluons if the Wilson loops (8.103) is non-trivial (not unit).

The important point to consider about (8.106) is that it is invariant under area-preserving diffeomorphysms, the group W_∞, since in two dimensions

$$F_{\mu\nu} = \epsilon_{\mu\nu} f \tag{8.107}$$

and the action is

$$S = \int \mathrm{d}^2 x\, \mathrm{tr}\, f^2 \sqrt{g} \quad . \tag{8.108}$$

This is also a symmetry of the Nambu action! Let us see what the relation is, if any, between the QCD_2 free energy and the partition function of some theory with coupling $g_{str} = 1/N$, and string tension $\alpha = e^2 N$, that is

$$\ln \mathcal{Z}_{\mathcal{M}}[G, e^2 A, N] \doteq \mathcal{Z}^{str}\left[g_{str} = \frac{1}{N}, \alpha' = e^2 N, A \right] \quad , \tag{8.109}$$

with $\mathcal{Z}^{str}$ obtained from the Nambu string as

$$\mathcal{Z}^{str} = \sum_h g_{str}^{2(h-1)} \int \mathcal{D}x^\mu(\xi)\, \mathrm{e}^{-\alpha \int \mathrm{d}^2\xi \sqrt{g}} \quad , \tag{8.110}$$

where h is the genus of the string.

The Nambu-Goto action is almost topological. Indeed, if the Jacobian $\left|\frac{\partial x^\mu}{\partial \xi^\alpha}\right|$ never vanishes, the Nambu-Goto action is the oriented area, and measures how many times one covers the target space. The true area differs from this topological number by the contribution of folds. These are the important entities to be studied in this case.

In order not to wander through complicated mathematical concepts, we restrain to a study of the QCD_2 partition function for large N and its consequences in the interpretation as a string theory.

We use a lattice regularization. For an arbitrary triangularization of the manifold we find

$$\mathcal{Z}_{\mathcal{M}} = \int \prod_L \mathrm{d}U_L \prod_{\text{plaquettes}} \mathrm{e}^{-\frac{1}{e^2}\mathrm{tr}\,(U_P+U_P^\dagger)} \quad , \tag{8.111}$$

where L are the links, and for a plaquette P

$$U_P = \prod_{L\in P} U_L \quad . \tag{8.112}$$

Migdal[190] has shown that the functional

$$K_P = \mathrm{e}^{\beta N \mathrm{tr}\,(U_P+U_P^\dagger)} \quad , \tag{8.113}$$

called here the heat-kernel functional, can be expanded in terms of the irreducible representations R of the gauge group as [191]

$$K_P = \sum_R d_R \lambda_R(\beta N)\chi_R(U) \quad , \tag{8.114}$$

where $\chi_R(U)$ is the character of R, $d_R = \chi_R(1)$ is the dimension of χ_R and the coefficients λ_R are given by

$$\lambda_R(\beta N) = d_R^{-1} \int \mathrm{d}U\, \chi_R(U)\, \mathrm{e}^{\beta N \mathrm{tr}\,(U_P+U_P^\dagger)} \quad . \tag{8.115}$$

In a region with several small plaquettes, we can show that[191]

$$\lambda_R(\beta N) = \mathrm{e}^{-e^2 C_2(R) A} \quad , \tag{8.116}$$

where A is the area, that is, we write K in terms of the Wilson loop

$$K_P = \sum_R d_R\, \chi_R(U_P)\, \mathrm{e}^{-e^2 C_2(R) A_P} \quad . \tag{8.117}$$

The important property of the heat-kernel functional above is that it is additive, that is, upon integration over each link the functional remains the same; in formulae, we have

$$\int \mathrm{d}U K_{P_1}(U_a U_b U) K_{P_2}(U^+ U_c U_d) = K_{P_1+P_2}(U_a U_b U_c U_d) \quad , \tag{8.118}$$

which is another way of expressing a property of

$$K_P = \langle U|\mathrm{e}^{-e^2 A\Delta}|1\rangle \quad , \tag{8.119}$$

where Δ is the Laplacian on the group.

Thus one can use the simplest triangulation, namely describing $\mathcal{M}$ in terms of a $4G$-gon $aba^{-1}b^{-1}\cdots a_G\, b_G\, a_G^{-1}\, b_G^{-1}$, we can introduce in (8.119) unit in the form of a sum over representations $\sum_R |R\rangle\langle R|$, using

$$\langle U|R\rangle = \chi_R(U) \quad \text{and} \quad \langle R|1\rangle = d_R \quad , \tag{8.120}$$

as well as the orthogonality of the characters, and the relation

$$\int \mathcal{D}U \chi_a(AUBU^+) = \frac{1}{d_a}\chi_a(A)\chi_a(B) \quad , \tag{8.121}$$

to obtain[191]

$$\mathcal{Z}_{\mathcal{M}_g} = \sum_R d_R^{2-2G}\, \mathrm{e}^{-\frac{\lambda A}{N}C_2(R)} \quad , \tag{8.122}$$

where $\lambda = e^2 N$ is kept fixed. This formula may be used to express the large N limit of the QCD_2 partition function in a very simple way. Moreover it makes the bridge between this theory and the string formulation, which as we know, is such that the partition function depends on analogous geometry. For fixed genus G, we only have to expand d_R and $C_2(R)$ in powers of $1/N$.

Indeed, let us suppose that the basis manifold is the torus, i.e. $G = 1$. We have to compute

$$\mathcal{Z}_{\mathcal{M}_1} = \sum \mathrm{e}^{-\frac{\lambda A}{N}C_2(R)} \quad . \tag{8.123}$$

Now, for $SU(N)$, the representations are labelled by Young tableaux indexed by $m \le N$, the number of rows of boxes of length $n_1 \ge n_2 \ge \cdots \ge n_m \ge 0$, and the Casimir is given by

$$C_2^{SU(N)}(R) = N\sum_{i=1}^m n_i + \sum_{i=1}^m n_i(n_i + 1 - 2i) - \frac{(\sum_i n_i)^2}{N} \quad . \tag{8.124}$$

For $N \to \infty$, we have

$$\begin{aligned}\mathcal{Z}_{\mathcal{M}_1}(N\to\infty) &= \sum \mathrm{e}^{-\lambda A\sum_i n_i} = \sum_{k_i n_{i+1}-n_i\ge 0} \mathrm{e}^{-\lambda A\sum_n nK_n}\\ &= \prod_{n=1}^N \frac{1}{1-\mathrm{e}^{-n\lambda A}} \equiv \eta(\mathrm{e}^{-\lambda A}) \quad ,\end{aligned} \tag{8.125}$$

where $\eta(x)$ is the Dedekind function, well known in string theory. Further corrections may be computed without difficulty.

If one expands $\ln\mathcal{Z}$ in terms of $\mathrm{e}^{-\lambda A}$, one sees that the coefficient of $\mathrm{e}^{-n\lambda A}$ counts the number of different maps of a torus onto a torus n times.

Finally, we briefly review[192] the expectation we have concerning the $1/N$ expansion comparison with the corresponding string theory. The expansion of the free energy is

$$\ln \mathcal{Z}_{\mathcal{M}_G} = \sum_{h=0}^{\infty} N^{2-2h} f_h^G(\lambda A) \quad , \tag{8.126}$$

where $f_h^G(\lambda A) \sim \mathrm{e}^{-\lambda n A}$, and n is the winding number of a map from a surface $\mathcal{M}_h$ to the surface $\mathcal{M}_G$, given by $(\xi \to x)$

$$n = \int \mathrm{d}^2\xi \, \det \frac{\partial x^\mu}{\partial \xi^\alpha} \quad , \tag{8.127}$$

which differs from the Nambu action

$$S_{Nambu} = \int \mathrm{d}^2 x \left| \det \frac{\partial x^\mu}{\partial \xi^\alpha} \right| \quad , \tag{8.128}$$

by the folds of the surface onto itself. In fact, the expectation is that

$$f_h^G(\lambda A) = \sum_n \begin{cases} 0 & \text{if } h-1 < n(G-1) \quad , \\ w_{h,G}^n(A)\, \mathrm{e}^{-\lambda n A} & \text{otherwise} \quad , \end{cases} \tag{8.129}$$

where $w_{h,G}^n(A)$ are coefficients related to the above maps (see [192] for further details). In order to have a more convincing and last relation between QCD_2 and string theory, we study a collective Hamiltonian for the string QCD interaction, which will be related with the Das-Jevicki Hamiltonian[193] for the $c = 1$ matrix model.

Consider QCD_2 compactified on a circle, in such a way that states consist of interacting strings that wind around the circle[194]. The winding states are mapped into momentum states. The theory is defined by the partition function

$$\mathcal{Z} = \sum_R \mathrm{e}^{-\frac{\lambda A}{N} C_2(R)} \tag{8.130}$$

and the Casimir is

$$C_2(R) = N\left(\sum n_i + \frac{1}{N} \sum n_i(n_i - 2i + 1) - \left(\frac{\sum n_i}{N} \right)^2 \right) \quad , \tag{8.131}$$

where α depends on the $U(1)$ coupling (for $U(N)$ $\alpha = 0$). The area $A = \beta L$, and every representation of the gauge group corresponds to a physical state of the theory with energy

$$E = \frac{\lambda L}{N} C_2(R) \quad . \tag{8.132}$$

The string picture for the cylinder with Wilson loops is realized by a chiral representation R with n boxes, being equivalent to a linear combination of string states that wrap around the compact dimension n times, or several strings wrapping a total amount of n times around the compactified dimension. Nevertheless, we suppose that two strings of opposite winding will not join to form a single string.

Considering a state with n winding in one direction means that a string state can be described by an element of the permutation group with n elements, and inequivalent states are given by the conjugacy classes of S_n. The states can be obtained by creation and annihilation operators $a_i^\dagger$ operating on the vacuum

$$\prod_{i=1}^{n} \left(a_i^\dagger\right)^{n_i} |0\rangle \quad , \tag{8.133}$$

with the commutation relations[194]

$$\left[a_i, a_j^\dagger\right] = i\delta_{ij} \quad . \tag{8.134}$$

The leading term in the energy (8.132) (see also (8.131)) leads to the zeroth order Hamiltonian

$$\mathcal{H}_0 = \frac{1}{2}\lambda^2 L \sum_{n\neq 0} a_n^\dagger a_n \quad . \tag{8.135}$$

In order to compute the next term Minahan and Polychronakos argued[194] with the operator in S_n which, once one computes its matrix elements between two states s and s' and sum over all such operators, are non-zero, i.e.,

$$\sum_{p\in S_n} \langle s'|p|s\rangle = \sum_{t\in S_n\,, p\in S_n'} \delta_{s'p,tst^{-1}} \tag{8.136}$$

and arrive at the second term of the Hamiltonian, which is cubic, and "conserves" the total number n, being given by

$$\mathcal{H} = \frac{\lambda^2 L}{2N} \sum_{\substack{n,n'>0\\ n,n'<0}} \left(a_{n+n'}^\dagger a_n a_{n'} + \text{c.c.}\right) \quad , \tag{8.137}$$

corresponding to the second term in (8.131), while the last term should be simply

$$\mathcal{H} = \alpha \frac{\lambda^2 L}{2N^2} \left[\sum_{n>0} \left(a_n^\dagger a_n - a_{-n}^\dagger a_{-n}\right)\right]^2 \quad , \tag{8.138}$$

but can be made to vanish fine tuning $\alpha = 0$. Therefore we obtain the Das-Jevicki Hamiltonian for collective states

$$\mathcal{H} = \frac{1}{2}\lambda^2 L \sum_{n\neq 0} a_n^\dagger a_n + \frac{\lambda^2 L}{2N} \sum_{\substack{n,n'>0\\ n,n'<0}} \left(a_{n+n'}^\dagger a_n a_{n'} + \text{c.c.}\right) \quad . \tag{8.139}$$

The fact that a collective field theory can be constructed using the rules derived from QCD_2 suggests also that there exists a free fermion picture of QCD_2.

8.8. Conclusion

We have seen, by means of a few representative examples, that the scope of applications of non-critical strings is vast. Problems of unsurmountable importance, such as quantum chromodynamics, three-dimensional Ising model, and the higher genera results in string theory. These problems are in the core of questions in theoretical physics, and the answer almost unavoidable passes through the understanding of the theory of non-critical strings.

Appendix A: Notation and Conventions

In this Appendix we summarize our notation and conventions. In most of the book we work in Euclidian two-dimensional space, exception is made in Chapter 1, from Eq. (1.77) to Eq. (1.152), where we fix the light-cone gauge, in which case Minkowski space is unavoidable.

The two-dimensional metric is defined by the tensor $g_{\mu\nu}$, with determinant

$$g = \det g_{\mu\nu} \quad . \tag{A.1}$$

In Chapter 1 we use the modulus $|g|$ inside the square root to avoid confusion between Minkowski and Euclidian notation, but it is dropped in subsequent Chapters, being superfluous.

In two dimensions the scalar curvature R can be written as

$$\sqrt{|g|}R = \partial_\mu \left(\frac{1}{\sqrt{|g|}} \partial_\nu (g g^{\mu\nu}) \right) + \frac{1}{4\sqrt{|g|}} \varepsilon^{\mu\nu} \varepsilon_{\alpha\beta} g^{\beta\gamma} \partial_\mu (\varepsilon_{\gamma\rho} g g^{\rho\sigma}) \partial_\nu (\varepsilon_{\sigma\delta} g g^{\delta\alpha}) \tag{A.2}$$

$\varepsilon^{01} = 1$, and the Einstein action is a topological invariant

$$S_E = \frac{1}{4\pi} \int \mathrm{d}^2x \sqrt{g} R = 2(1-h) \quad , \tag{A.3}$$

where h is the Euler characteristic (genus) of the surface.

The light cone variables are defined by

$$x^{\pm} = x^0 \pm x^1 \quad ; \quad \partial_{\pm} = \frac{\partial_0 \pm \partial_1}{2} \quad , \tag{A.4}$$

and the light-cone gauge is

$$g^{\mu\nu} = \begin{pmatrix} g^{++} & g^{+-} \\ g^{-+} & g^{--} \end{pmatrix} = \begin{pmatrix} 0 & 2 \\ 2 & -4g_{++} \end{pmatrix} \quad ; \quad \sqrt{-g} = \frac{1}{2} \quad . \tag{A.5}$$

The conformal gauge is given by the choice

$$g_{\mu\nu} = \hat{g}_{\mu\nu} \mathrm{e}^{2\sigma} \quad , \quad g^{\mu\nu} = \hat{g}^{\mu\nu} \mathrm{e}^{-2\sigma} \quad , \tag{A.6}$$

where $\hat{g}_{\mu\nu}$ is called the fiducial metric.

The graviton/gravitino pair, is described in the light cone gauge as

$$g^{\mu\nu} = \begin{pmatrix} 0 & 2 \\ 2 & -4g_{++} \end{pmatrix} \quad , \quad \Psi_- = 0 \quad \Psi_+ = \begin{pmatrix} \psi_+ \\ 0 \end{pmatrix} \quad , \tag{A.7}$$

with the following representation of γ-matrices

$$\gamma_+ = 2i\begin{pmatrix} 0 & 0 \\ 1 & 0 \end{pmatrix} \quad , \quad \gamma_- = -2i\begin{pmatrix} 0 & 1 \\ 0 & 0 \end{pmatrix} \quad . \tag{A.8}$$

The algebra $SL(2,R)$ is used extensively: its Killing form is taken as

$$\eta^{ab} = \begin{pmatrix} 0 & 0 & 2 \\ 0 & -1 & 0 \\ 2 & 0 & 0 \end{pmatrix} \quad ; \quad \eta_{ab} = \begin{pmatrix} 0 & 0 & 1/2 \\ 0 & -1 & 0 \\ 1/2 & 0 & 0 \end{pmatrix} \quad . \tag{A.9}$$

and $f^{abc} = 2\varepsilon^{abc}$, $\varepsilon^{-0+} = 1$ defines its structure constants.

In the n-point correlator (recall Eq. (1.89)) of the field g_{++} we use the notation

$$\partial_{\pm_i} \equiv \frac{\partial}{\partial x_i^{\pm}} \quad . \tag{A.10}$$

In order to obtain short distance expansions, we must get rid of the delta functions. We need the representation

$$\partial_- \frac{1}{x^+} = 2i\pi\delta^{(2)}(x) \quad , \tag{A.11}$$

or, as a consequence of this,

$$\frac{1}{4\pi i}\partial_-^3 \frac{(x^-)^2}{x^+} = \delta^{(2)}(x) \quad . \tag{A.12}$$

In fact, one has to use the prescription $\frac{1}{x^+} \to \frac{1}{x^+ - i\epsilon \mathrm{sign}(x^-)}$.

The Möbius transformation on τ is defined as

$$\mathcal{S}_\tau = \frac{a\tau + b}{c\tau + d} \quad , \quad ac - bd = 1 \quad . \tag{A.13}$$

Throughout this review $\mathrm{d}^2x = \mathrm{d}x^1\mathrm{d}x^2$. In principle, after changing the variables (x^1, x^2) to $(z, \overline{z})$, one has to take into account a factor 1/2. Whenever this factor leads to an unimportant overall constant we systematically overlook it.

The relations between the central charge c and the parameters α_0 , $\alpha_\pm$, Q, in the bosonic case studied in Chapter 2, are

$$Q = 2\sqrt{2 + \alpha_0^2} \quad , \tag{A.14}$$

$$\alpha_\pm = -\frac{Q}{2} \pm |\alpha_0| \quad , \tag{A.15}$$

$$d_\pm = \alpha_0 \pm \frac{Q}{2} \quad , \tag{A.16}$$

$$c = 1 - 12\alpha_0^2 \quad , \tag{A.17}$$

$$\alpha_+\alpha_- = 2 \quad . \tag{A.18}$$

The Liouville central charge is

$$c_L = 26 - c = 1 + 3Q^2 \quad , \quad Q = \sqrt{\frac{25-c}{3}} \quad . \tag{A.19}$$

In the supersymmetric case, we have

$$Q = 2\sqrt{1+\alpha_0^2} \quad , \tag{A.20}$$

$$\alpha_\pm = -\frac{Q}{2} \pm |\alpha_0| \quad , \tag{A.21}$$

$$d_\pm = \alpha_0 \pm \frac{Q}{2} \quad , \tag{A.22}$$

$$\hat{c} = \frac{2}{3}c = 1 - 8\alpha_0^2 \quad , \tag{A.23}$$

$$\alpha_+\alpha_- = 1 \quad , \tag{A.24}$$

and the super Liouville central charge reads

$$\hat{c}_{SL} = \frac{2}{3}c_{SL} = 1 + 2Q^2 = \quad , \quad Q = \sqrt{\frac{9-\hat{c}}{2}} \quad . \tag{A.25}$$

The ghosts have conformal dimension

$$\Delta\left(e^{a\sigma}\right) = -\frac{1}{2}a(a+2) \quad , \tag{A.26}$$

in such a way that

$$\Delta\left(e^{-\frac{1}{2}\sigma}\right) = \frac{3}{8} \quad , \tag{A.27}$$

$$\Delta\left(e^{\frac{1}{2}\sigma}\right) = -\frac{5}{8} \quad . \tag{A.28}$$

In the (p,q) minimal conformal theories, whose central charge reads

$$c = 1 - 6\frac{(p-q)^2}{pq} \quad , \quad p,q \text{ coprimes} \quad , \tag{A.29}$$

the conformal weights of the primary operators $\mathcal{O}_{rr'}$ are given by

$$\Delta_{rr'} = \frac{(rp-r'q)^2-(q-p)^2}{4pq} \quad , \tag{A.30a}$$

with

$$1 \le r' \le p-1 \quad , \quad 1 \le r \le q-1 \quad . \tag{A.30b}$$

On the other hand the (p,q) minimal superconformal series, characterized by the central charges

$$c = \frac{3}{2}\left\{1 - 2\frac{(p-q)^2}{pq}\right\} \quad |p-q| \ge 2 \quad , \tag{A.31}$$

exhibit the following spectrum of conformal weights

$$\Delta_{rr'} = \frac{(rp - r'q)^2 - (q-p)^2}{8pq} + \frac{1-(-1)^{r-r'}}{32} \quad , \tag{A.32a}$$

where

$$1 \leq r' \leq p-1 \quad , \quad 1 \leq r \leq q-1 \quad , \tag{A.32b}$$

so that when $r - r'$ is even (respectively odd) the operator $\mathcal{O}_{rr'}$ is in the Neveu-Schwarz (respectively Ramond) sector.

The string susceptibility γ_h is defined in terms of the scaling behavior of the partition function (or free energy) of fixed genus h according to

$$\mathcal{Z}_h \propto \mu^{2-\gamma_h} \quad , \tag{A.33}$$

where μ is the cosmological constant.

Appendix B: Conformally Invariant Field Theory in Two Dimensions[1)]

Conformally invariant quantum field theories were first considered in the sixties[199], and used to understand the small distance behavior of field operator products[200]; their consideration was also motivated by the fact that in the high-energy limit, renormalizable theories are approximately conformally invariant[201], thus allowing for the perturbative computation of several high-energy processes[202] involving QCD. Considerable effort has been dedicated to gaining a deeper understanding of conformally invariant theories in any space-time dimension[185]. However they have not found extensive applications. In spite of the importance of conformal invariance in the study of the high-energy behavior of quantum field theory, the main application of conformal quantum theory has still been in the area of statistical mechanics. Indeed, classical scale invariance is a quite general feature shared by many field theories, having no dimensional parameter, a property generally met by statistical systems at criticality[203]. Polyakov and Migdal were the first to use extensively these ideas[204]. The theoretical framework of the renormalization group provided a satisfactory description of the theory at criticality[203]. On the other hand, the computational power of such methods is rather limited, and there is no general framework based on these methods, to study general results, except for some properties of specific models[205].

The fact that a number of two-dimensional models exhibit conformal invariance is of great importance, since the conformal group in two dimensions is infinite dimensional. The recognition that in general, scale invariant theories are also conformally invariant[206] is a major development in recent years. Polyakov thus proposed to use conformal invariance as an essential ingredient in the study of criticality[204].

The conformal algebra

Consider the coordinate transformations $x \to x'(x)$, with the property that the metric tensor $g_{\mu\nu}(x)$ is left invariant except for a local multiplicative factor (Weyl transformation[197]):

$$g_{\mu\nu}\partial_\lambda x'^\mu \partial_\rho x'^\nu = \lambda(x) g_{\lambda\rho} \quad , \tag{B.1}$$

with $\lambda(x)$ a continuous, non-vanishing function of x.

In order to determine the corresponding infinitesimal transformation we look for solutions of the Killing Cartan equation, obtained from $(B.1)$ by the substitutions $x'_\mu = x_\mu + \xi_\mu$, and $\lambda = 1 + \delta\lambda$, such that

$$\partial_\mu \xi_\nu + \partial_\nu \xi_\mu = (\delta\lambda) g_{\mu\nu} \quad . \tag{B.2}$$

[1)] This Appendix is a simplified and corrected version of a section of ref. [11]. We thank K.D. Rothe for discussions

Expressing $\delta\lambda$ in terms of ξ by contracting $(B.2)$ with $g^{\mu\nu}$, we obtain

$$\partial_\mu \xi_\nu + \partial_\nu \xi_\mu = \frac{2}{D} \partial^\rho \xi_\rho g_{\mu\nu} \quad . \tag{B.3}$$

If $D \neq 2$ we find the following Killing vectors[197], that is, solutions of Eq. $(B.3)$[2)] for translations,

$$t^\alpha_\mu = -\delta^\alpha_\mu \quad ; \tag{B.4a}$$

for Lorentz transformations,

$$\xi^\alpha_{\mu\nu} = x_\nu \delta^\alpha_\mu - x_\mu \delta^\alpha_\nu \quad , \tag{B.4b}$$

$(\xi_\mu = \xi^\alpha_{\mu\nu} \epsilon^\nu_\alpha)$; for dilatations,

$$d^\alpha = -x^\alpha \quad , \tag{B.4c}$$

$(\xi_\mu = -\epsilon x_\mu)$; and, for special conformal transformations,

$$\xi^\alpha_\mu = 2x_\mu x^\alpha - x^2 \delta_{\mu\alpha} \quad , \tag{B.4d}$$

$(\xi_\mu = \xi^\alpha_\mu \epsilon_\alpha)$. This completes the set of Killing vectors, in D space-time dimensions.

Holomorphic realization

When restricting the dimension of space time to $D = 2$, Eq. $(B.3)$ may be rewritten, in the Euclidean case[3)], as:

$$(\partial_1 + i\partial_2)(\xi_1 + i\xi_2) = 0 \tag{B.5a}$$

$$(\partial_1 - i\partial_2)(\xi_1 - i\xi_2) = 0 \quad . \tag{B.5b}$$

Therefore, any analytic (anti-analytic) transformation $z = x_1 + ix_2 \rightarrow f(z)$, $(\overline{z} = x_1 - ix_2 \rightarrow \overline{f}(\overline{z}))$ is allowed. Expanding $f(z)$ in a Laurent series in z as

$$f(z) = z + \sum_{-\infty}^{\infty} \epsilon_n z^{n+1} \tag{B.6a}$$

$$\overline{f}(\overline{z}) = \overline{z} + \sum_{-\infty}^{\infty} \overline{\epsilon}_n \overline{z}^{n+1} \quad , \tag{B.6b}$$

with ϵ_n small parameters, we can write

$$f(z) = z + \sum \epsilon_n (L^c_n z) \quad . \tag{B.6c}$$

The classical generators of the transformations $(B.6)$ are thus,

$$L^c_n = z^{n+1} \frac{\mathrm{d}}{\mathrm{d}z} \quad , \tag{B.7a}$$

$$\overline{L}^c_n = \overline{z}^{n+1} \frac{\mathrm{d}}{\mathrm{d}\overline{z}} \quad . \tag{B.7b}$$

---

2) The infinitesimal parameters ξ_μ in Eq. $(B.3)$ are obtained multiplying the Killing vectors in Eqs. $(B.4)$ by appropriate infinitesimal parameters.

3) In the Euclidean formulation $x^E_2 = ix_0$, $x^E_1 = x^1$, $\partial^E_1 = \partial_1$; we drop the index E here.

They obey the classical loop algebra

$$[L_n^c, L_m^c] = -(n-m)L_{n+m}^c \quad , \tag{B.8a}$$

$$[\overline{L}_n^c, \overline{L}_m^c] = -(n-m)\overline{L}_{n+m}^c \quad , \tag{B.8b}$$

as one easily obtains from Eq. $(B.7)$.

Notice that $L_0^c, L_{+1}^c, L_{-1}^c$ (together with $\overline{L}_0^c, \overline{L}_{+1}^c, \overline{L}_{-1}^c$) correspond to the generators of the special conformal group. In two dimensions, this group is described by $SO(2,2) \approx SL(2,R)_+ \otimes SL(2,R)_-$, each connected piece being generated by 3 operators, conveniently labeled as L_0, L_{+1}, L_{-1} (and of course $\overline{L}_0, \overline{L}_{+1}, \overline{L}_{-1}$ for the antiholomorphic component).

The definition of the generators of translations, dilatations and special conformal transformations are, respectively ($x^\pm = x^0 \pm x^1$)

$$P_- = \int \theta(x^-)\mathrm{d}x^- \tag{B.10b}$$

$$\mathcal{D}_- = \int x^-\theta(x^-)\mathrm{d}x^- \tag{B.10c}$$

$$\mathcal{K}_- = \int (x^-)^2\theta(x^-)\mathrm{d}x^- \quad , \tag{B.10d}$$

with corresponding expressions for x^+; $\theta(x^-)$ is the left moving component of the energy-momentum tensor. The generators above satisfy commutation relations summarized by the operators

$$L_0 = P + \frac{1}{4}\mathcal{K} \tag{B.11a}$$

$$L_{\pm 1} = P \pm i\mathcal{D} - \frac{1}{4}\mathcal{K} \quad . \tag{B.11b}$$

These operators ($L_{0,\pm 1}$), satisfy the $SL(2,R)$ Virasoro algebra; P_μ generates the infinitesimal translations, D the infinitesimal dilatations, and K_μ the infinitesimal holomorphic special conformal transformations.

A representation of this algebra is given by the restriction of $(B.7)$ to $n = 0, \pm 1$, with analogous expressions for the antiholomorphic fields.

Conformal dimensions

In a conformally invariant field theory, the transformation of the elementary field is defined by its conformal dimension Δ, which naively corresponds to the scale dimension.

As we already mentioned, in two-dimensional space-time we have a simplification due to the fact that the conformal group factorizes in terms of the $SL(2,R)$ group of Möbius transformations, $G = SL(2,R) \bigotimes SL(2,R)$. Correspondingly there exists a special class of factorizable fields, that transforms in the following simple way[44]

$$\phi(z,\bar{z}) \longrightarrow \left(\frac{\mathrm{d}\zeta}{\mathrm{d}z}\right)^{\Delta} \left(\frac{\mathrm{d}\bar{\zeta}}{\mathrm{d}\bar{z}}\right)^{\bar{\Delta}} \phi(\zeta(z),\bar{\zeta}(\bar{z})) \quad . \tag{B.12a}$$

For Möbius transformations, namely $\zeta(z) = \frac{az+b}{cz+d}$, $ad - bc = 1$, this equation reads

$$\phi(z,\bar{z}) \longrightarrow (cz+d)^{-2\Delta}(\overline{cz}+\bar{d})^{-2\bar{\Delta}}\phi(\zeta(z),\bar{\zeta}(\bar{z})) \quad . \tag{B.12b}$$

These are called primary fields[44], and are the building blocks of the construction of two-dimensional conformally invariant quantum field theory.[4)]

From this point on it is convenient to choose new complex coordinates for the problem, mapping the usual (Euclidean) space-time into a space such that constant "Euclidean" time corresponds to circles. Thus, if we take the map[207] $\frac{1+\frac{i}{2}x^-}{1-\frac{i}{2}x^-}$, with inverse $x^- = 2i\frac{1-z}{1+z}$, we are mapping the real axis into the circle $|z| = 1$, $-\pi \leq \arg z \leq \pi$. Such transformations, in which lines of constant Euclidean time are mapped into circles around the origin are very useful in the study of conserved generators. Another example is

$$z = \mathrm{e}^{(x_2 - ix_1)} \quad , \tag{B.13}$$

where $x_2 = -\infty$ corresponds to the origin, $x_2 = +\infty$ to the circle at infinity, and a line of constant time is still a circle. Since the theory is expected to respect the symmetry generated by holomorphic and anti-holomorphic transformations, as given by ($B.7$), we are always entitled to perform these transformations.

Conserved currents are frequently analyzed in quantum field theory, and appear a number of times in the problems considered here. In the framework of a conformally invariant quantum field theory, if we consider a conserved current, we know from general arguments of conformal quantum field theory, that its scale dimension is given by the naive value[208]. Indeed, as we shall explicitly see, the current obeys a current algebra with a Schwinger term[209]

$$[J_0^a(t,x^1), J_1^b(t,y^1)] = if^{abc}J_1^c\delta(x^1-y^1) + k\frac{i}{\pi}\delta^{ab}\delta'(x^1-y^1) \quad . \tag{B.14}$$

We shall see an explicit example in terms of free fields although this is not the most general case.

If we make a rescaling, we readily see that the Schwinger term matches the remaining terms in the equation only if $\dim J = 1$. In general, if the current is a primary field as in Eq. ($B.12$), we have the transformation property

$$J(z,\bar{z}) \longrightarrow \left(\frac{\mathrm{d}\zeta}{\mathrm{d}z}\right)^{\Delta_J} \left(\frac{\mathrm{d}\bar{\zeta}}{\mathrm{d}\bar{z}}\right)^{\bar{\Delta}_J} J(\zeta,\bar{\zeta}) \quad , \tag{B.12c}$$

---

4) For the time being, we are only considering Möbius transformations; a field obeying ($B.12b$) only, but not ($B12.a$) is actually called quasi-primary.

For dimension one, we have for the holomorphic and antiholomorphic fields,

$$J(z) \longrightarrow \frac{d\zeta}{dz} J(\zeta) \tag{B.15a}$$

$$J(\overline{z}) \longrightarrow \frac{d\overline{\zeta}}{d\overline{z}} J(\overline{\zeta}) \quad . \tag{B.15b}$$

Thus we obtain for the charge

$$Q = \int J(x^0, x^1) dx^1 \rightarrow \frac{1}{2\pi i} \oint_{|z|=const} J(z) dz + \frac{1}{2\pi i} \oint_{|z|=const} J(\overline{z}) d\overline{z} \quad , \tag{B.16}$$

where the integration over constant radius (z) corresponds to an integration at constant time, as dictated by the new time interpretation following Eq. (B.12). Using Eq. (B.15) we see that the charge is invariant under holomorphic (antiholomorphic) transformations.

Two-, three- and four-point correlators

Conformal invariance can be used to compute exactly the two- and three-point function of a scalar theory, up to some dimensionless parameters. These functions are given by simple expressions; for the two-point function, Poincaré and scale invariance imply

$$\langle 0|\phi_i(z)\phi_j(w)|0\rangle = \lambda[(z-w)]^{-2\Delta} \tag{B.17a}$$

$$\langle 0|\phi_i(\overline{z})\phi_j(\overline{w})|0\rangle = \lambda[(\overline{z}-\overline{w})]^{-2\overline{\Delta}} \quad , \tag{B.17b}$$

where we separated holomorphic and antiholomorphic fields (λ is an arbitrary constant)[5)].

Indeed, Poincaré invariance implies that it is a function of the difference of the coordinates; scale invariance implies that it is a homogeneous function; the scale transformation fixes the power behavior (notice that the generator L_0 of scale transformations allows only for powers of the argument). Finally, special conformal transformations generated by L_1 force the two scaling parameters to be equal ($\Delta_i = \Delta_j = \Delta$). For the three-point function we have[195,197]:

$$\begin{aligned}&\langle 0|\phi_1(z_1)\phi_2(z_2)\phi_3(z_3)|0\rangle = \\ &=\lambda[(z_{12})^2]^{\frac{1}{2}(-\Delta_3+\Delta_1+\Delta_2)}[(z_{13})^2]^{\frac{1}{2}(-\Delta_2+\Delta_1+\Delta_3)}[(z_{23})^2]^{\frac{1}{2}(-\Delta_1+\Delta_2+\Delta_3)}\end{aligned} \tag{B.18}$$

where $z_{ij} = z_i - z_j$, and λ is an arbitrary constant.

When considering a four-point function, we have also the Möbius invariant

$$z = \frac{z_{12} z_{34}}{z_{13} z_{24}} \quad . \tag{B.19}$$

---

5) In what follows we use the notation $z, \overline{z}$ also for holomorphic and antiholomorphic, unless otherwise explicitly stated. However, in general, the light-cone notation implies that we are working in Minkowski space.

Indeed, under the transformation

$$\delta_f \phi = \left\{ \Delta_i \frac{\partial f(z_i)}{\partial z_i} + f(z_i) \frac{\partial}{\partial z_i} \right\} \phi \quad , \tag{B.20}$$

with $f(z) = 1$ for Poincaré transformations, $f(z) = z$ for scale transformations and $f(z) = z^2$ for special conformal transformations, we have for the 4-point function the following identity

$$\sum_{i=1}^{4} \left(\Delta_i \frac{\partial f_i}{\partial z_i} + f_i \frac{\partial}{\partial z_i} \right) \langle 0 \mid \phi_1(z_1)\phi_2(z_2)\phi_3(z_3)\phi_4(z_4) \mid 0 \rangle = 0 \quad . \tag{B.21}$$

One easily verifies that $\sum_i f_i \frac{\partial}{\partial z_i} z = 0$, for $f_i = 1, z_i, z_i^2$, i.e., functions of z are Möbius invariant. Therefore the most general conformally invariant 4-point function is given by

$$\langle 0|\phi_1(z_1)\phi_2(z_2)\phi_3(z_3)\phi_4(z_4)|0\rangle = \prod_{i<j} z_{ij}^{-(\Delta_i + \Delta_j - \frac{1}{3}\sum_k \Delta_k)} \mathcal{F}\left(\frac{z_{12} z_{34}}{z_{13} z_{24}} \right) \quad , \tag{B.22}$$

where $\mathcal{F}$ is an arbitrary function depending on the (Möbius invariant) harmonic ratio z.

The BPZ construction

We next verify that the field algebra is tremendously enlarged in a two-dimensional field theory. First notice that Poincaré invariance implies the conservation of the energy momentum tensor,

$$\partial^\mu T_{\mu\nu} = 0 \quad , \tag{B.23}$$

while scale invariance implies its tracelessness

$$T^\mu_\mu = 0 \quad . \tag{B.24}$$

Moreover, the following current is conserved

$$\mathcal{J}_\mu = T_{\mu\nu} \epsilon^\nu \quad , \tag{B.25}$$

whenever $\xi_\mu = \epsilon_\mu$ satisfies Eq. (B.3), and $T_{\mu\nu}$ is traceless.

Conservation and tracelessness of the energy momentum tensor in (Euclidean) 2-dimensional space-time imply that the combination

$$\overline{T} = T_{\bar{z}\bar{z}} = T_{11} + iT_{12} \quad , \tag{B.26}$$

satisfies

$$(\partial_1 - i\partial_2)\overline{T} = 0 \quad . \tag{B.27}$$

Therefore $\overline{T}$ is a function of $\overline{z} = x_1 - ix_2$ only. Analogously:

$$T = T_{zz} = T_{11} - iT_{12} \tag{B.28}$$

is a function of $z = x_1 + ix_2$. However, $T_{\mu\nu}$ generates a space-time symmetry, since each of the above components can be expanded in a Laurent series as

$$T(z) = \sum_{n=-\infty}^{\infty} z^{-n-2} L_n \tag{B.29a}$$

and

$$\overline{T}(\overline{z}) = \sum_{n=-\infty}^{\infty} \overline{z}^{-n-2} \overline{L}_n \quad . \tag{B.29b}$$

It will be very useful to work with complex variables. In this case, quantization has to be effected in a way already indicated previously. This is useful, since the definition of a *time* direction would break the independence of z and $\overline{z}$ variables. We map space-time into circles, through the change of variables $(B.12)$. This means that constant *time* is associated with a constant radius: a circle. *Time* flows from the origin outwards, the origin being the far past, as previously explained. In this case, the conserved charge, analogous to $\int \mathcal{J}(x)dx^1$, is now given by the angular integration of the radial component of the current. This prescription is known as radial quantization[210]. We obtain, for the infinitesimal generator of diffeomorphic transformations $(B.16)$,

$$Q_\epsilon = \frac{1}{2\pi i} \int \mathrm{d}x^1 T_{0\nu} \epsilon^\nu \quad , \tag{B.30a}$$

or, when expressed in terms of the holomorphic variables z and $\overline{z}$,

$$Q_\epsilon = \frac{1}{2i\pi} \oint T(z)\epsilon(z)\mathrm{d}z + \frac{1}{2i\pi} \oint \overline{T}(\overline{z})\overline{\epsilon}(\overline{z})\mathrm{d}\overline{z} \quad , \tag{B.30b}$$

where the integration contour encircles the origin. The commutator of Q_ϵ with some field $\phi(z)$, is then obtained by using the following rule: the integration contour around the origin of the complex ξ-plane is chosen such that z lies inside (outside) of the contour if $\phi(z)$ appears on the right (left) of Q_ϵ. This corresponds to the usual time-ordering prescription in quantum field theory, as we mentioned before (see also discussion after Eq. (B.12)). Indeed, when we continue a given Greens function from Minkowski to Euclidean space[211], and use $\langle \phi(x_1) e^{-H(\tau_1-\tau_2)} \phi(x_2) e^{-H(\tau_2-\tau_3)} \ldots \phi(x_n) \rangle$ with the Hamiltonian H bounded from below, the above expression converges only for $\tau_i > \tau_{i+1}$. An explicit computation of the function $\langle :e^{i\alpha\varphi(z)}::e^{-i\alpha\varphi(\omega)}: \rangle$, with $\varphi(z) = \sum \frac{a_n}{n} z^{-n}$, $a_{-n} = a_n^\dagger$, and the Wick product defined as

$$:e^{i\alpha\varphi}:= e^{i\alpha\varphi^{(+)}} e^{i\alpha\varphi^{(-)}} \quad , \tag{B.30c}$$

where $\varphi^{(+)}(\varphi^{(-)})$ is the creation (annihilation) part of the field $\varphi(x)$, shows that the sum is well defined only for $| z |>| \omega |$. Indeed, using

$$e^A e^B = e^B e^A e^{[A,B]} \tag{B.30d}$$

which holds when $[A,B]$ is a c-number, we find for the two-point function

$$e^{-\alpha^2\sum\frac{1}{n}\left(\frac{\omega}{z}\right)^n}=e^{\alpha^2\ln\left(1-\frac{\omega}{z}\right)} \quad . \tag{B.30e}$$

This is clearly valid only for $|z|>|\omega|$.

Using the above prescription one obtains for an infinitesimal transformation of $\phi(z)$, generated by Q_ϵ,

$$\delta\phi(z)=[Q_\epsilon,\phi(z)]=\frac{1}{2i\pi}\oint T(\xi)\phi(z)\epsilon(\xi)\mathrm{d}\xi \quad , \tag{B.31}$$

where ξ is integrated around z.

The energy-momentum tensor has the Laurent expansion $(B.29)$ which may be inverted to give

$$L_n=\frac{1}{2i\pi}\oint T(z)z^{n+1}\mathrm{d}z \quad . \tag{B.32}$$

We compute now the commutator algebra satisfied by the operators L_n. This algebra is obtained by considering the operator product expansion of the energy momentum tensor. In order to obtain this expansion, we first discuss the transformation properties of $T(z)$ under the conformal group. For an object of dimension γ, one usually has (see Eq. (B.20))

$$\delta_f J=f\frac{\mathrm{d}J}{\mathrm{d}z}+\gamma f'J \quad . \tag{B.33}$$

For conserved quantities, the dimension is equal to the naive value[208]. Hence T has dimension $\gamma=2$. We write

$$\delta_f T(z)=\frac{c}{12}f'''(z)+f(z)\frac{\mathrm{d}T(z)}{\mathrm{d}z}+2\frac{\mathrm{d}f(z)}{\mathrm{d}z}T(z) \quad , \tag{B.34}$$

where we allowed for a term of dimension 3 with arbitrary coefficient[209]. This implies that the transformation law for the energy momentum tensor for a finite transformation is given by the expression

$$T(z)=\left(\frac{\mathrm{d}\zeta}{\mathrm{d}z}\right)^2T(\zeta(z))+\frac{c}{12}\{\zeta,z\} \quad , \tag{B.35}$$

where[6)]

$$\{\zeta,z\}=\frac{\mathrm{d}^3\zeta/\mathrm{d}z^3}{\mathrm{d}\zeta/\mathrm{d}z}-\frac{3}{2}\left(\frac{\mathrm{d}^2\zeta/\mathrm{d}z^2}{\mathrm{d}\zeta/\mathrm{d}z}\right)^2 \quad . \tag{B.36}$$

---

6) This is obtained from the fact that invariance under finite transformations is a consequence of the closure of the algebra; thus one considers a transformation $T\to T^{(1)}\to T^{(2)}$, where $z\to\zeta\to\omega$, which satisfies

$$T^{(2)}(z)=\left(\frac{\mathrm{d}\zeta}{\mathrm{d}z}\right)^2T^{(1)}(\zeta)+\frac{c}{12}\{\zeta,z\}=\left(\frac{\mathrm{d}\zeta}{\mathrm{d}z}\right)^2\left\{\left(\frac{\mathrm{d}\omega}{\mathrm{d}\zeta}\right)^2T(\omega)+\frac{c}{12}\{\omega,\zeta\}\right\}+\frac{c}{12}\{\zeta,z\}$$

$$=\left(\frac{\mathrm{d}\omega}{\mathrm{d}z}\right)^2T(\omega)+\frac{c}{12}\{\omega,z\} \quad ,$$

from which one obtains[212] $\{\omega,z\}=\left(\frac{\mathrm{d}\zeta}{\mathrm{d}z}\right)^2\{\omega,\zeta\}+\{\zeta,z\}$, thus defining $(B.36)$.

is the Schwarzian derivative of ζ with respect to z.

In particular, for an inversion, $\zeta = \frac{1}{z}$, we obtain the behaviour of the energy momentum tensor for large argument, $T(z) \sim \frac{1}{z^4}$. Comparing with the expression (B.31) with $\phi(z) = T(z)$, we obtain

$$T(z)T(w) = \frac{\frac{c}{2}}{(z-w)^4} + \frac{T(z)+T(w)}{(z-w)^2} \quad . \tag{B.37}$$

In particular, the transformation law (B.35) implies that T is not a primary field in the sense of (B.12). However, it is *quasiprimary* in the sense that (B.12) is obeyed by $T(z)$ if we restrict ourselves to Möbius transformations, for which the inhomogeneous term in (B.35) can be shown to vanish.

The constant c is called the central charge. It plays a fundamental role in the Virasoro algebra implied by (B.37),

$$[L_n, L_m] = (n-m)L_{n+m} + \frac{c}{12}n(n^2-1)\delta_{n,-m} \quad . \tag{B.38}$$

Realization of the conformal group

We display the general properties of a theory which is invariant under the conformal transformations. It will be based upon scalar fields $\phi_{\Delta,\overline{\Delta}}(z,\overline{z})$, defined by their anomalous dimensions Δ and $\overline{\Delta}$. At this point some remarks are in order. Local fields transforming under the conformal group have an anomalous dimension γ, defined via dilatations by

$$A(x) \rightarrow A'(\lambda x) = \lambda^{-\gamma} A(x) \quad , \tag{B.39}$$

with $\gamma \geq 0$. In terms of Δ and $\overline{\Delta}$ the dimension γ and the spin s are given by $\gamma = \Delta + \overline{\Delta}$, and $s = \Delta - \overline{\Delta}$. [7)]

We suppose there is a complete set of fields A_i spanning the representation of the conformal algebra. By "complete" we mean that any state may be obtained by application of a suitable linear combination of those fields on the vacuum. This implies also that a product of two of them may be written as a linear combination with finite, single valued c-number coefficients $C^k_{ij}(\xi)$ as[44]

$$A_i(z)A_j(0) = \sum C^k_{ij}(z)A_k(0) \tag{B.40}$$

(all fields actually depend on z and $\overline{z}$, but because of conformal invariance, it is sufficient to consider holomorphic fields; full results may be obtained at the end of the calculation by forming products of holomorphic and antiholomorphic contributions). Conformal invariance imposes severe constraints on the coefficients $C^k_{ij}(z)$, as we see later. We shall now restrict ourselves to an important subset of fields to

---

[7)] γ is the mass dimension of the theory. The above definition of the spin arises from the Lorentz transformations properties of a spin-s field, that is $\phi(e^{\lambda}z, e^{-\lambda}\overline{z}) = e^{\lambda s}\phi(z,\overline{z})$.

which we have already alluded, having particularly simple transformation laws, (see Eq. (B.12)).

We will consider the transformation of conformal fields under the substitution

$$z \to \zeta(z) = z + \epsilon(z) \quad .$$

For purely holomorphic fields we have

$$\phi(z) \longrightarrow \left(\frac{\mathrm{d}\zeta}{\mathrm{d}z}\right)^{\Delta} \phi(\zeta) \quad . \tag{B.41}$$

Such a field will be called primary, and will play a central role in the sequel.

Let us discuss the Ward identities following from conformal invariance. For a general quantum field theory we have the transformation law (B.31) for $A(w)$. For a local field A, its variation must be a linear combination of a finite number of derivatives of ϵ, so that we write

$$\delta_\epsilon A_j(z) = \sum_{k=0}^{\nu_j} B_j^{k-1}(z) \left(\frac{\mathrm{d}}{\mathrm{d}z}\right)^k \epsilon(z) \quad , \tag{B.42}$$

where $B_j^{(k-1)}(z)$ are fields which are local function of the A_j, and ν_j are integers. The first two coefficients are fixed by the translation and dilatation properties of the fields, such that

$$B_j^{-1}(z) = \frac{\partial}{\partial z} A_j(z) \quad , \tag{B.43a}$$

and

$$B_j^0(z) = \Delta_j A_j(z) \quad . \tag{B.43b}$$

The dimension of the field B_j^{k-1} can be read off expression (B.42), to be

$$\Delta_j^{(k-1)} = \Delta_j + 1 - k, \quad k = 0, 1, ..., \nu_j \quad . \tag{B.44}$$

Thus, the set of fields B_j^k have dimensions which are equally spaced by integers, $\Delta_n^k = \Delta_n - k$. It is also clear that for a primary field [8]

$$\delta_\epsilon \phi(z) = \epsilon(z) \frac{\partial \phi(z)}{\partial z} + \Delta \epsilon'(z) \phi(z) \quad . \tag{B.45}$$

As a result, the operator product $T(z)\phi(w)$ is given by the following series involving only non-negative values of k

$$T(z)\phi(w) = \sum_{k=0}^{\infty} (z-w)^{-2+k} \phi^{(-k)}(w) \quad . \tag{B.46}$$

[8] The above equation implies that there is no operator $B^{(k)}$ with dimension smaller than Δ, (in analogy with the lower bound of the magnetic quantum number). Thus Δ plays the role of the highest root in representation theory.

Comparison of $(B.46)$ with $(B.45)$ shows, using $(B.31)$, that

$$\phi^{(0)} = \Delta\phi \quad , \quad \phi^{(-1)} = \frac{\partial\phi}{\partial z} \quad . \tag{B.47}$$

The remaining fields $\phi^{(-k)}, k = 2, 3, \dots$ are local fields generated by the application of L_{-n} on ϕ, and are called the "descendants" of the "ancestor" ϕ, or simply secondary fields

$$L_{-n}\phi_a(z) = \phi_a^{(-n)}(z) \quad . \tag{B.48}$$

We also define

$$\phi_a^{(-n_1,\dots,-n_A)}(z) = L_{-n_1}(z)\dots L_{-n_A}(z)\phi_a(z) \quad , \tag{B.49}$$

where

$$L_{-n}(z) = \oint \mathrm{d}\xi \frac{T(\xi)}{(\xi - z)^{n-1}} \quad , \tag{B.50}$$

with the inverse transformation $T(z) = \sum \frac{L_n(w)}{(z-w)^{n+2}}$.

It is clear that the fields $(B.49)$ are in general not primary since the application of L_n, with $n > 0$, does not annihilate them; they are eigenstates of L_0 with eigenvalue

$$\Delta_a^{n_1 \cdots n_A} = \Delta_a + n_1 + \cdots + n_A \quad . \tag{B.51}$$

We call $\Delta_a^{n_1 \cdots n_A} - \Delta_a$ the *level* of a given state.

We are able to learn more about the operator product expansion of a local field with the energy-momentum tensor. In view of $(B.31)$ and $(B.42)$ we have

$$\begin{aligned}\langle 0|T(z)A_{j_1}(z_1)\dots A_{j_n}(z_n)|0\rangle &= \sum_{l-1}^{n}\sum_{k=0}^{\nu_l} k!(z - z_l)^{k-1} \times \\ &\langle 0|A_{j_1}(z_1)\dots A_{j_{l-1}}(z_{l-1})B_{j_l}^{(k-1)}(z_l)A_{j_{l+1}}(z_{l+1})\dots A_{j_n}(z_n)|0\rangle \quad . \end{aligned} \tag{B.52}$$

Going back to the primary fields, we shall use them in order to build a representation of the conformal group. This representation is constructed in terms of the ancestor together with its descendants, a construction similar to the one well known in group theory, where the representation is constructed by starting with a state with the highest weight[45]. Consider the state vector[9)]

$$|a\rangle = \phi_a(0)|0\rangle \quad , \tag{B.53}$$

where the vacuum $|0\rangle$ is defined by

$$L_n|0\rangle = 0 \qquad n \geq -1 \quad , \tag{B.54}$$

[9)] Recall that $z = 0$ is related to the fact that $|a>$ is an *in* state, at $t = -\infty$ (see comment after Eq. $(B.29)$).

which means that the vacuum is conformally invariant. From $(B.31)$, $(B.45)$ and $(B.46)$, we obtain

$$[L_n, \phi_a(z)] = z^{n+1}\frac{\partial}{\partial z}\phi_a(z) + \Delta_a(n+1)z^n\phi_a(z) \quad . \tag{B.55}$$

Thus we have, for the primary states $|a\rangle$,

$$L_n|a\rangle = 0 \quad \text{for} \quad n > 0 \tag{B.56a}$$

$$L_0|a\rangle = \Delta_a|a\rangle \quad . \tag{B.56b}$$

The above set of states define a huge Hilbert space. We now restrict the fields to a "physical" space, by imposing constraints.

Physical Hilbert space and the Kac determinant

The classification of the conformal representations is obtained by imposing constraints in the form of relations which some of the operators in $(B.48)$ must obey.

The constraints involve only states of a given level. At level one, there is only one operator, namely L_{-1}, and imposing the condition $L_{-1}\phi = 0$ gives a trivial result a constant field. At level 2 there are two operators, L_{-2} and L^2_{-1}, and we require

$$(L_{-2} + aL^2_{-1})\phi = 0 \quad . \tag{B.57}$$

This condition must be invariant under applications of $L_n, n > 0$; this determines a and the dimension of ϕ. Let us work out this case in detail; we have

$$L_1(L_{-2} + aL^2_{-1})\phi = (3 + 2a + 4a\Delta)L_{-1}\phi = 0 \tag{B.58}$$

and

$$L_2(L_{-2} + aL^2_{-1})\phi = (4\Delta + \frac{c}{2} + 6a\Delta)\phi = 0 \quad , \tag{B.59}$$

where equation $(B.38)$ has been used; for $n \geq 3$, L_n annihilates the state on the left hand side of $(B.57)$ identically. Equation $(B.58)$ implies

$$a = \frac{-3}{2(2\Delta + 1)} \tag{B.60a}$$

which can be used to calculate c from $(B.59)$:

$$c = \frac{2\Delta(5 - 8\Delta)}{(2\Delta + 1)}, \quad \text{or} \quad \Delta = -\frac{c-5}{16} \pm \frac{1}{16}\sqrt{(c-1)(c-25)} \quad . \tag{B.60b}$$

This is a special example out of a series of very profound results obtained by Kac[45]. States such as ϕ above constitute the so-called *Verma module*[213]. The condition for obtaining a well defined finite representation fixes the conformal dimension Δ to take the following values, labeled by two integers, p and q,

$$\Delta_{p,q} = \frac{c-1}{24} + \left(\frac{1}{2}p\alpha_+ + \frac{1}{2}q\alpha_-\right)^2 \quad , \tag{B.61}$$

where

$$\alpha_\pm = \frac{\sqrt{-c+1} \pm \sqrt{(-c+25)}}{\sqrt{24}} \quad . \tag{B.62}$$

Compare the above results with Eq. (1.128).

The equations above show the first steps towards defining the so-called degenerate representation of the conformal family. In general there is degeneracy at a given level, and one looks for the *singular vectors*, which are linear combinations of states of a given level, perpendicular to any given state.

The argument leading to the above formulae is sketched in [11]. One sees from here the appearance of the constants $\alpha_\pm$ as in (B.62). Generalizations from the above procedure to theories with non-abelian symmetries as well as with supersymmetries can be similarly obtained.

The central charge in unitary representations

The Kac determinant has no zeros for $c > 1$, and $\Delta > 0$. If the determinant vanishes, there exist linear combinations of states at a given level which are zero, or a null state. We obtain linear differential equations for the correlators as a consequence of this property.

A detailed study of the Kac determinant done by[214] Friedan, Qiu and Shenker reveals that unitarity imposes stringent restrictions. By looking for the zeros of the Kac determinant in the plane Δ *versus* c, they studied the positions of the zero norm states, separating regions of positive and negative norm states, and proved that in order to have unitarity, for $c < 1$, we should have

$$c = 1 - \frac{6}{n(n+1)} \quad , \tag{B.63}$$

with $n = 3, 4, 5 \ldots$

In this case, we have for the conformal dimension the expression

$$\Delta_{p,q} = \frac{(p(n+1) - qn)^2 - 1}{4n(n+1)} \quad , \tag{B.64}$$

where $1 \leq q \leq p \leq n-1$. The primary field having dimension $\Delta_{p,q}$ is denoted by $\psi_{(p,q)}$.

The central charge: general case

In the general case one obtains, setting the null fields equal to zero, the corresponding (infinite dimensional) degenerate representation. Inspection of $(B.61)$ shows that for a suitable choice of $\alpha_\pm$ one arrives at a finite dimensional degenerate representation of the conformal group. This is achieved for the choice

$$\frac{\alpha_-}{\alpha_+} = -\frac{m}{n} \quad , \tag{B.65a}$$

with the corresponding solutions $\frac{m-1}{n+1} = -\frac{p}{q}$ or $\frac{m+1}{n-1} = -\frac{p}{q}$. For this case, we solve $(B.62)$ and $(B.65a)$, obtaining for c

$$c = 1 - 6\frac{(p-q)^2}{pq} \quad ; \tag{B.65b}$$

this result reproduces $(B.63)$ if $q = p+1$, which is the unitary case. The non-unitary case is however also very important in statistical mechanics[215], and in two-dimensional gravity[130], which need not be unitary.

The primary fields $\psi_{p,q}$ are finitely many in this case, providing a realization of the conformal algebra in terms of a finite number of fields. This is the key ingredient to identify, in this case, statistical models of physical interest at criticality, using these ideas.

Constraints on correlators

When computing correlators containing null-fields, such as defined in $(B.57)$, one is led to differential equations, which must be obeyed by the correlators. Consider the expression

$$\langle L_{-2}\phi(z)\phi_1(z_1)\cdots\phi_n(z_n)\rangle = \frac{1}{2\pi i}\oint\frac{\mathrm{d}w}{w-z}\langle T(w)\phi(z)\phi_1(z_1)\cdots\phi_n(z_n)\rangle \quad . \tag{B.66}$$

where the contour encircles only the point z, and where ϕ_i are primary fields. We can deform it in such a way as to encircle the points $z_1\cdots z_n$ instead. Using the operator product expansion for the product of ϕ with the energy momentum tensor we obtain, after using $(B.7)$

$$\langle L_{-2}\phi(z)\phi_1(z_1)\ldots\phi_n(z_n)\rangle = \sum_a\{\frac{\Delta_a}{(z_a-z)^2} + \frac{1}{z-z_a}\partial_a\}\langle\phi(z)\phi_1(z_1)\ldots\phi_n(z_n)\rangle \quad . \tag{B.67}$$

For the level 2 constraint $(B.57)$ we thus find

$$\frac{3}{2(2\Delta+1)}\frac{\partial^2}{\partial z^2}\langle\phi(z)\phi_1(z_1)\ldots\phi_n(z_n)\rangle = \sum_a\{\frac{\Delta_a}{(z_a-z)^2} + \frac{1}{z-z_a}\partial_a\}\langle\phi(z)\phi_1(z_1)\ldots\phi_n(z_n)\rangle \quad , \tag{B.68}$$

where we specialized to $\phi(z) = \psi_{(1,2)}(z)$ (see definition following $(B.72)$); it has dimension $\Delta_{(1,2)}$ (see $(B.61)$), and obeys $(B.57)$.

The above second order differential equations can be used to determine completely the 4-point functions of the theory, in terms of the scale parameters Δ_a. Consider the most general conformally invariant four-point function, which is given by

$$\langle\phi_0(z_0)\phi_1(z_1)\phi_2(z_2)\phi_3(z_3)\rangle =$$
$$\prod(z_0 - z_i)^{-(\Delta+\Delta_i-\frac{1}{3}\sum_{j=0}^{3}\Delta_j)}(z_i - z_j)^{-(\Delta_i+\Delta_j-\frac{1}{3}\sum_{k=0}^{3}\Delta_k)}\mathcal{F}(z) \quad , \qquad (B.69)$$

where $z = \frac{(z_0-z_1)(z_2-z_3)}{(z_0-z_3)(z_2-z_1)}$. The function $\mathcal{F}(z)$ obeys a differential equation of the hypergeometric type.

From the completeness of the OPE[44] $(B.40)$ we obtain for $z_1 \approx z$,

$$\phi(z)\phi_1(z_1) = C(z - z_1)^{\lambda}\phi_{\Delta'}(z)\cdots \quad , \qquad (B.70)$$

where $\lambda = \Delta' - \Delta_1 - \Delta$ and Δ' is the scale dimension of the primary field $\phi_{\Delta'}(z)$. Consider now the most divergent term as $z_1 \to z$, in $(B.76)$; requiring its coefficient to vanish, we obtain

$$\frac{3}{2(2\Delta+1)}\lambda(\lambda-1) - \Delta + \lambda = 0 \quad . \qquad (B.71)$$

The above equation determines the critical behavior of the new field obtained by fusing ϕ and ϕ_1 . The so called fusion rules are fundamental in the study of conformal field theory. Their knowledge permits to obtain all relevant information of the theory. Ref. [44] contains tables for the fusion of some simple representations. The procedure permits the identification of the critical behavior of several statistical mechanics models. Procedures to obtain fusion rules are discussed in the literature[216].

Operator (Coulomb gas) realization of the conformal algebra

Let us consider now the issue of the realization of a conformal algebra in terms of field operators[29,30]. We discuss the role of exponential fields in conformal field theories, and show that they are natural objects to be considered in this case.

We take the exponential fields

$$V_{k\overline{k}}(z,\overline{z}) = C : e^{\left(ik\varphi(z)+i\overline{k}\overline{\varphi}(\overline{z})\right)} : \quad , \qquad (B.72)$$

which can be separated in right and left movers as

$$V_k(z,\overline{z}) = V_k(z)V_{\overline{k}}(\overline{z}) \quad . \qquad (B.73)$$

The fields φ and $\overline{\varphi}$ are the (Euclidean) right and left moving components of a free massless bosonic field

$$\phi(z,\overline{z}) = \varphi(z) + \overline{\varphi}(\overline{z}) \qquad (B.74)$$

with an action

$$S = \frac{1}{16\pi}\int \mathrm{d}z\mathrm{d}\overline{z}\partial_z\phi\partial_{\overline{z}}\phi = \frac{1}{8\pi}\int \mathrm{d}^2x(\partial_\mu\phi)^2 \quad . \tag{B.75}$$

which leads to the propagators

$$\begin{aligned} \underbrace{\phi(z,\overline{z})\phi(w,\overline{w})} &= -2\ln|z-w| \quad , \\ \underbrace{\varphi(z)\varphi(w)} &= -\ln(z-w) \quad , \end{aligned} \tag{B.76}$$

Because of the well-known selection rules[11], we have

$$\langle V_k(z)V_p(z')\rangle = (z-z')^{-k^2}\delta_{k,-p} \quad . \tag{B.77}$$

This is a conformally invariant two-point correlator. Further correlators can be computed analogously, and we obtain, using ($B.76$), and ($B.30d$)

$$\langle V_{k_1}(z_1)V_{k_2}(z_2)V_{k_3}(z_3)V_{k_4}(z_4)\rangle = \prod_{i>j}(z_i-z_j)^{k_ik_j} \quad , \tag{B.78}$$

with $\sum k_i = 0$, which is nothing but the super-selection rule.

The conformal dimension of $V_\alpha(z)$ follows immediately from ($B.77$) above,

$$\Delta_\alpha = \frac{1}{2}k^2 \quad . \tag{B.79}$$

The operator product expansion of the (holomorphic part of the) energy momentum tensor $T(z) = \frac{-1}{2}(\partial\phi)^2$ with the exponential field is given by the expression

$$T(z)V_k(w) = \frac{\Delta_k}{(z-w)^2}V_k + \frac{1}{z-w}\partial_w V_k + \text{regular terms} \quad , \tag{B.80}$$

where the first term corresponds to the double contraction of $(\partial\phi)^2$ with $\mathrm{e}^{ik\phi}$, the coefficient being $\frac{1}{2}k^2$. The second term corresponds to a single contraction, the non-contracted field giving rise to the derivative. The central charge has the special value $c = 1$, as one sees by computing the two-point function of the energy-momentum tensor,

$$\langle T(z)T(w)\rangle = \frac{1}{2(z-w)^4} \quad , \tag{B.81}$$

and comparing with Eq. ($B.37$).

The above example is particularly simple. It is however possible to modify the usual prescription and define a theory with $c \neq 1$[29,30].

The idea is to introduce background charges at infinity[29,30]. In fact, charges at infinity imply that we are considering special boundary conditions, since a pair of charges at infinity is equivalent to a constant background electric field. Consider as an example the correlator defined by

$$\langle V_{k_1}(z_1)\dots V_{k_n}(z_n)\rangle_{\alpha_o} = \lim_{R\to\infty} R^{4\alpha_o{}^2}\langle V_{k_1}(z_1)\dots V_{k_n}(z_n)V_{-2\alpha_o}(R)\rangle_o \tag{B.82}$$

where the right hand side is computed in the usual way. The "zero charge" selection rule $\sum k_i = 0$ in $(B.78)$ is now replaced by

$$\sum k_i = 2\alpha_o \quad . \tag{B.83}$$

In particular, for $n = 2$ we have

$$\langle V_k(z)V_{2\alpha_o-k}(z')\rangle_{\alpha_o} = (z-z')^{-k(k-2\alpha_o)} \quad . \tag{B.84}$$

But we know that the 2-point function of a conformal theory is non-vanishing only if the conformal dimensions of the fields are equal[195−197]. Thus we should have

$$\Delta_k = \Delta_{2\alpha_o-k} = \frac{1}{2}k(k-2\alpha_o) \quad . \tag{B.85}$$

We see that expression $(B.84)$ indeed reflects this property, being symmetric under the exchange $k \leftrightarrow 2\alpha_o - k$.

In order to obtain such an anomalous dimension, we have to modify the energy momentum tensor in an appropriate way. The boundary conditions for the fields at infinity are different now. They may be incorporated in the theory, by allowing for total derivatives in the energy momentum tensor. We thus take the Ansatz

$$T(z) = \frac{-1}{2}(\partial_z\phi)^2 + A\partial_z{}^2\phi \quad , \tag{B.86}$$

where the constant A is determined below. Notice that the term we added is a total derivative. Therefore we are allowing for non-trivial boundary conditions, if the new term is to be a relevant one. Consider the operator product expansion for $T(z)V_k$,

$$\begin{aligned} T(z)V_k(w) =&: \left(\frac{-1}{2}(\partial_z\phi)^2 + A\partial_z{}^2\phi\right): :e^{ik\phi(w)}:= \\ &= \frac{\frac{1}{2}(k^2+2ikA)}{(z-w)^2}:e^{ik\phi(w)}: + \frac{ik}{z-w}:\partial_z\phi(z)e^{ik\phi(w)}: + \\ &:\left(-\frac{1}{2}(\partial_z\phi)^2 + A\partial_z{}^2\phi\right)e^{ik\phi(w)}: \quad ; \end{aligned} \tag{B.87}$$

rewritting the above expansion in the form of equation $(B.80)$ we find

$$T(z)V_k = \frac{k^2+2ikA}{2(z-w)^2}V_k + \frac{1}{z-w}\partial_w V_k \quad . \tag{B.88}$$

Comparing with $(B.45\text{-}47)$ one obtains from above the dimension of V_k,

$$\Delta_k = \frac{1}{2}k(k+2ikA) \quad . \tag{B.89}$$

Thus, to match $(B.89)$ with $(B.85)$ we need

$$A = i\alpha_o \quad . \tag{B.90}$$

Correspondingly the properties of $\phi(z)$ under a conformal transformation

$$z \to f(z) = z + \epsilon(z) \quad . \tag{B.91}$$

also change. The field $V_k(z)$ has conformal dimension given in $(B.85)$, that is, under the transformation $(B.91)$, V_k transforms as

$$V_k(z) \to (f'(z))^{\frac{1}{2}k(k-2\alpha_o)} V_k(f(z)) = (f')^{\frac{1}{2}k^2} : \mathrm{e}^{\{ik(\phi(f(z))+i\alpha_o \ln f'(z))\}}: \quad . \tag{B.92}$$

The quantity between the parentheses in Eq. $(B.92)$ gives the transformation law of $\phi(z)$:

$$\phi(z) \to \phi(f(z)) + i\alpha_o \ln f'(z) \quad . \tag{B.93}$$

We can also calculate the central charge of the Virasoro algebra by calculating the energy-momentum two-point function

$$\langle T(z)T(w)\rangle = \frac{\frac{c}{2}}{(z-w)^2} \quad . \tag{B.94}$$

Using $(B.86)$, with $A = i\alpha_o$ we find

$$\left\langle \left(\frac{-1}{2} :(\partial_z\phi)^2: + i\alpha_o \partial_z{}^2 \phi(z)\right) \left(\frac{-1}{2} :(\partial_w\phi)^2: + i\alpha_o \partial_w{}^2 \phi(w)\right) \right\rangle =$$

$$= \frac{1}{8}\left(\frac{2}{(z-w)^2}\right)^2 + \alpha_o{}^2 \partial_z^2 \partial_w^2 \ln(z-w) = \frac{\frac{1}{2}}{(z-w)^4}\{1 - 12\alpha_o{}^2\} \quad , \tag{B.95}$$

where the first term corresponds to the double contraction of the terms $(\partial\phi)^2$, and the second, to the contraction of the terms $\partial^2\phi$.

Thus we get, for the central charge, the value

$$c = 1 - 12\alpha_o{}^2 \quad . \tag{B.96}$$

Four-point correlators for c≠1

With the above results, we are able to obtain integral representation for the various correlators. We analyze now the 4-point correlators $\langle V_{k_1}(z_1)V_{k_2}(z_2)V_{k_3}(z_3)V_{k_4}(z_4)\rangle$, subject to the condition

$$\sum_{i=1}^{4} k_i = 2\alpha_o \quad . \tag{B.97}$$

The conformal dimension of a multivalued function in a conformal field theory cannot be fixed at will, for all fields. In the case of the four-point correlator $(B.78)$ we must have equality (at least two by two) of the conformal dimensions. In the case where the conformal dimensions are all equal, we have $k_i = k$, or $k_i = 2\alpha_o - k$. Thus we have the following possibilities

$$i)\ k_1 = k_2 = k_3 = k_4 = \alpha \tag{B.98a}$$
$$ii)\ k_1 = k_2 = k_3 = k, \quad k_4 = 2\alpha_o - k \tag{B.98b}$$
$$iii)\ k_1 = k_2 = k, \quad k_3 = k_4 = 2\alpha_o - k \quad . \tag{B.98c}$$

None of them satisfies $(B.97)$. The alternative found in [29,30] is to use conformally invariant operators not depending on space-time, but with non-zero charge, in order that $(B.97)$ be met. To introduce such operators we make some digression.

In two-dimensional quantum field theory, charge can be defined with purely bosonic real fields by means of the construction of exponentials fields. Fixing the value of the background charge at infinity, and computing correlators of operators with restricted values for the conformal dimensions, the possibility of choosing the values of the charges of the fields is also rather restricted and sometimes there are no possible value, as we showed before.

However, one may always compute the correlators which do not change the fundamental conformal properties, but changes the background configuration. Thus they are integrated (vertex) operators, with zero conformal dimension, and the corresponding vertex operator has dimension one, that is such an operator is provided by

$$Q = \oint \mathrm{d}z J(z) \quad , \tag{B.99}$$

Since these operators are integrated, they can be interpreted as new vertices in the Lagrangian, that is

$$\delta\mathcal{L}_z = \int \mathrm{d}z J(z) \quad . \tag{B.100}$$

An arbitrary number of screening charges permits to extract a non-zero result for several four-point correlators.

The screening charges acquire also importance in the gravitationally interacting theory, since there is no dressing of such operators, as one readily see from a quick computation of the Liouville charge β, which turns out to vanish.

Therefore, external charges may be introduced by means of a term $\int \mathrm{d}z J(z)$ in the Lagrangian, with no gravitational dressing.

In order that Q to be conformally invariant $(\Delta_Q = 0)$, $J(z)$ must have conformal dimension one. We take

$$J_\pm(z) = V_{d_\pm}(z) \quad , \tag{B.101a}$$

where

$$\Delta_{d_\pm} = \frac{1}{2} d_\pm(d_\pm - 2\alpha_0) = 1 \quad . \tag{B.101b}$$

Thus

$$d_\pm = \alpha_0 \pm \sqrt{2 + \alpha_0{}^2} \quad . \tag{B.101c}$$

If we express α_0^2 in terms of c using equation $(B.96)$, we immediately recognize in $(B.101c)$ Kac's formula. The introduction of these charges can resolve the above inconsistency with $(B.98)$ only if $\sum_{i=1}^{4} k_i - 2\alpha_0 > 0$. This eliminates possibility $(B.98c)$. For $(B.98a)$ we need to cancel

$$\sum k_i - 2\alpha_0 = 4k - 2\alpha_0 \quad . \tag{B.102}$$

Adding $(n-1)\, Q_+{}'s$ and $(m-1)\, Q_-{}'s$ we have a quantization of $k(n,m \geq 1)$:

$$k_{mn} = \frac{1}{4}(1-n)d_+ + \frac{1}{4}(1-m)d_- \quad . \tag{B.103}$$

For $(B.98b)$ we have

$$\sum_{i=1}^{4} k_i - 2\alpha_0 = 2k \quad , \tag{B.104}$$

and the quantization of k reads instead

$$k_{mn} = \frac{1}{2}(1-n)d_+ + \frac{1}{2}(1-m)d_- \quad . \tag{B.105}$$

Now we can see this result used in (2.63). Let us deal with this last case, following [99]. The conformal dimension obeys the Kac's formula $(B.61)$, that is

$$\Delta_{mn} = \frac{1}{2}k_{mn}(k_{mn} - 2\alpha_o) = \frac{1}{8}[(nd_+ - md_-)^2 - (d_+ + d_-)^2] \quad . \tag{B.106}$$

We finally obtain an integral representation for the four-point function, given by

$$\begin{aligned}\langle V_{m_1n_1}(z_1)V_{m_2n_2}(z_2)V_{m_3n_3}(z_3)V_{2\alpha_o-k_{m_4n_4}}(z_4)\rangle &= \oint \mathrm{d}w_1\cdots \mathrm{d}w_M \mathrm{d}v_1\cdots \mathrm{d}v_N \times \\ \langle J_+(w_1)\cdots J_+(w_M)J_-(v_1)\cdots J_-(v_N)V_{m_1n_1}(z_1)&\cdots V_{m_4n_4}(z_4)\rangle_{2\alpha_o} \quad ,\end{aligned} \tag{B.107a}$$

where

$$M = \frac{m_1+m_2+m_3-m_4}{2} - 1 \tag{B.107b}$$

$$N = \frac{n_1+n_2+n_3-n_4}{2} - 1 \quad . \tag{B.107c}$$

The correlators inside the integral are immediately computed, taking into account the factorization of the integrand in terms of two-point functions using $(B.30\ c,d)$ as well as $(B.77)$ and $(B.84)$. We make a specific choice for three of the arguments in the correlator, using the projective invariance. We know that the four-point function is given by the scaling factors (powers of $z_i - z_j$) times a function of $z = \frac{z_{12}z_{43}}{z_{13}z_{42}}$.

We can extract these scaling factors, and choose $z_1 = 0, z_2 = z, z_3 = 1$ and $z_4 = \infty$. We do not perform the general correlator but we consider a simple example,

$$G^{mn}_{12}\,{}^{\tilde{m}\tilde{n}}_{12}(z) = \langle V_{12}(z)V_{12}(1)V_{mn}(0)V_{2\alpha_o - k_{mn}}(\infty)\rangle^{n.s.} =$$
$$\oint \mathrm{d}w \left\langle V_{12}(z)V_{12}(1)V_{mn}(0)V_{2\alpha_o - k_{mn}}(\infty)V_{\alpha_+}(w)\right\rangle^{n.s.} \quad , \qquad (B.108)$$

where the superscript $n.s.$ means omission of the scaling factors. In this case we obtain

$$G(z) = \oint \mathrm{d}w z^{k_{12}k_{mn}}(1-z)^{k_{12}}w^{d+k_{mn}}(w-1)^{d+k_{12}}(w-z)^{d+k_{12}} \quad . \qquad (B.109)$$

The integral

$$I_C(a,b,c;z) = \oint_C \mathrm{d}w w^a (w-1)^b (w-z)^c \qquad (B.110)$$

is a hypergeometric function[198]. There exists an explicit dependence on the path of integration, and I_C vanishes if no singularity is included. The various possible integration paths correspond to different solutions of the hypergeometric equation. Two independent solutions are

$$I_1(a,b,c;z) = \int_1 \mathrm{d}w w^a (w-1)^b (w-z)^c =$$
$$= \frac{\Gamma(-a-b-c-1)\Gamma(b+1)}{\Gamma(-a-c)}\mathcal{F}(-c,-a-b-c-1,-a-c;z) \quad (B.111a)$$

and

$$I_2(a,b,c;z) = \int_0^z \mathrm{d}w w^a(-w+1)^b(-w+z)^c = z^{1+a+c}\int_0^1 \mathrm{d}w w^a(1-w)^c(1-zw)^b$$
$$= z^{1+a+c}\frac{\Gamma(a+1)\Gamma(c+1)}{\Gamma(a+c+2)}\mathcal{F}(-b,a+1,a+c+2;z) \quad , \qquad (B.111b)$$

where $\mathcal{F}(\alpha,\beta,\gamma;z)$ is the usual hypergeometric function, defined by

$$\mathcal{F}(\alpha,\beta,\gamma;z) = 1 + \frac{\alpha\beta}{1!\gamma}z + \frac{\alpha(\alpha+1)\beta(\beta+1)}{2!\gamma(\gamma+1)}z^2 + \cdots \quad . \qquad (B.111c)$$

Our problem has thus been reduced to the solution of the hypergeometric equation. Along this monograph, we actually consider integrated functions, and the results are even simpler.

The above procedure is equivalent to the introduction of a gravitational interaction of the bosonic field ϕ, described by

$$S = \int \mathrm{d}^2x \left\{ \frac{1}{8\pi}(\partial_\mu\phi)^2 + \frac{i\alpha_0}{4\pi}R\phi \right\} \quad . \qquad (B.112)$$

Indeed, the energy-momentum tensor obtained from the above action in the conformal gauge is simply given by (B.86), up to an overall constant.

Veneziano formula can be obtained using the exponential fields, in the simplified case where background charges are absent, considering the correlators of four exponentials fields.

Non-Abelian Conformally Invariant Theory

The non-abelian generalization of a conformally invariant field theory in two dimensions is realized by the principal σ model with a Wess-Zumino term[217], the so called Wess-Zumino-Novikov-Witten action[218–220]

$$S = \frac{1}{2\lambda^2}\mathrm{tr}\int \mathrm{d}^2\xi\, \partial^\mu g^{-1}\partial_\mu g + kS_{WZ}(g) \quad , \tag{B.113}$$

where

$$S_{WZ}(g) = \frac{1}{4\pi}\epsilon^{\mu\nu}\int_0^1 \mathrm{d}r \int \mathrm{d}^2\xi\, g^{-1}\dot{g}g^{-1}\partial_\mu g g^{-1}\partial_\nu g \quad . \tag{B.114}$$

It is also known[219] that the corresponding β-function has a zero at the point $\lambda^2 = \frac{4\pi}{k}$. We suppose that λ assumes this critical value, from now on.
The action (B.113) at the critical point is invariant under the transformation[148,221]

$$g(\xi) \to \Omega(z)g(\xi)\overline{\Omega}^{-1}(\overline{z}) \quad , \tag{B.115}$$

where $z = \xi_1 + i\xi_2$ and $\overline{z} = \xi_1 - i\xi_2$, and Ω is independent of $\overline{\Omega}$.

The Noether currents associated with these symmetries are

$$J = J^a\tau^a = -\frac{1}{2}k\partial_z g g^{-1} \tag{B.116a}$$

and

$$\overline{J} = \overline{J}^a\tau^a = -\frac{1}{2}kg^{-1}\partial_{\overline{z}}g \quad , \tag{B.116b}$$

where τ^a are the generators of the group. These currents satisfy

$$\partial_{\overline{z}}J = 0 = \partial_z\overline{J} \quad , \tag{B.117}$$

as can be deduced from the equations of motion obtained from (B.113) with $\lambda^2 = \frac{4\pi}{k}$. Thus their dependence on the space-time variables is that of a holomorphic, or antiholomorphic function, respectively. We can generate a loop algebra for the currents by developping $J(z)$ and $\overline{J}(\overline{z})$ in a Laurent series

$$J^a(z) = \sum_{n=-\infty}^{\infty} J_n^a z^{-n-1} \tag{B.118a}$$

$$\overline{J}^a(\overline{z}) = \sum_{n=-\infty}^{\infty} \overline{J}_n^a \overline{z}^{-n-1} \quad . \tag{B.118b}$$

The quantum theory associated with (B.113) at the critical point, may be analyzed by various methods. Witten has discussed its equivalence to a free theory, as well as the algebra obeyed by the currents (B.118).

From the computation of the β function we expect the theory to have a conformally invariant fixed point. However, it is clear that the computation of the β function does not prove that the fixed point is exactly at the value $\lambda^2 = 4\pi/k$.

In order to further investigate the properties of the WZNW action (at the critical point), we proceed with a canonical quantization of the theory.

The WZ action ($B.114$) only depends linearly on the time derivative of g. Hence it proves useful to write it in the form[222]

$$S_{WZ}[g] = \frac{n}{4\pi}\int \mathrm{d}^2x\, tr A(g)\partial_0 g \quad , \tag{B.119}$$

where we have formally integrated over r, and $A(g)$ is a matrix valued function of g and $\partial_1 g$. We thus obtain for the momentum conjugate to g_{ij}

$$\Pi_{ij} = \frac{n}{4\pi}\partial_0 g^{-1}_{ji} + \frac{n}{4\pi}A_{ji} \equiv \hat{\Pi}_{ij} + \frac{n}{4\pi}A_{ji} \quad . \tag{B.120}$$

Taking into account that $A(g)$ is a function of g and its space derivatives (but not of the time derivative of g), we obtain, from the usual Poisson bracket formulation[10)]

$$\left\{\hat{\Pi}_{ij}(x), \hat{\Pi}_{kl}(y)\right\} = -\frac{n}{4\pi}F_{ji;lk}(x)\delta(x^1 - y^1) \tag{B.121a}$$

$$\left\{g_{ij}(x), \hat{\Pi}_{kl}(x)\right\} = \delta_{ik}\delta_{jl}\delta(x^1 - y^1) \quad . \tag{B.121b}$$

where[222]

$$F_{ij;kl} = \frac{\partial A_{ij}}{\partial g_{lk}} - \frac{\partial A_{kl}}{\partial g_{ji}} = \partial_1 g^{-1}_{il} g^{-1}_{kj} - g^{-1}_{il}\partial_1 g^{-1}_{kj} \quad , \tag{B.121c}$$

where the derivatives of $A(g)$ are obtained comparing the variation of ($B.119$) with that of ($B.114$).

At the critical point, the two conserved currents are given by

$$j_+(x) = -\frac{in}{2\pi}g^{-1}\partial_+ g = i2\hat{\Pi}^T g - \frac{in}{2\pi}g^{-1}\partial_1 g \quad , \tag{B.122a}$$

$$j_-(x) = -\frac{in}{2\pi}g\partial_- g^{-1} = -i2g\hat{\Pi}^T + \frac{in}{2\pi}g\partial_1 g^{-1} \quad , \tag{B.122b}$$

where $j_+ \equiv (j^0_L - j^1_L)$, $j_- \equiv (j^0_R + j^1_R)$. Using ($B.121a,b$), we find the Poisson bracket relations,

$$\{j^a_+(x), j^b_+(y)\} = 2f^{abc}j^c_+(x)\delta(x^+ - y^+) + \frac{2n}{\pi}\delta^{ab}\delta'(x^+ - y^+) \tag{B.123a}$$

$$\{j^a_-(x), j^b_-(y)\} = 2f^{abc}j^c_-(x)\delta(x^- - y^-) + \frac{2n}{\pi}\delta^{ab}\delta'(x^- - y^-) \tag{B.123b}$$

$$\{j^a_+(x), j^b_-(y)\} = 0 \quad . \tag{B.123c}$$

with $j^a = \mathrm{tr}\, j\tau^a$, and τ^a the $SU(N)$ generators. Here "prime" on δ represents the derivative with respect to the argument of the delta function. Thus, we have two

10) Poisson brackets are always understood to be computed at equal times.

Kac-Moody algebras with central extension $k = n$. The $O(N)$ case is little clumsy concerning technical details, but the conclusions are the same[219].

We now show which is the algebra one obtains in a field theory of N free Majorana fermions, with the $O(N)$ symmetric action $S = \frac{i}{2}\int d^2x\,\overline{\psi}^i\,\partial\!\!\!/\psi^i$. The (Hermitian) $O(N)$ currents $j^{ij}_\mu = \overline{\psi}^j\gamma_\mu\psi^i$, $(i,j = 1,\cdots,N)$, may be separated into right and left movers as

$$j^{ij}_+ = -2\psi^j_2\psi^i_2 \qquad (B.124a)$$
$$j^{ij}_- = 2\psi^j_1\psi^i_1 \quad . \qquad (B.124b)$$

The current $j^{ij}_+(j^{ij}_-)$ is a function of $x^+(x^-)$ only, as we see from the equation of motion for ψ.

The above objects satisfy a current algebra relation, which can be computed using the anti-commutation relations for the fermionic fields; indeed, we have for an $O(N)$ invariant Majorana field, the Dirac brackets (see [11])

$$\left\{\psi^i_2(x),\psi^j_2(y)\right\}_D = -i\delta^{ij}\delta(x^1-y^1) \Rightarrow \left\{\psi^i_2(x),\psi^j_2(y)\right\}_{AC} = \delta^{ij}\delta(x^1-y^1) \quad . \qquad (B.125)$$

From the above we compute[11)]

$$\begin{aligned}\left[j^{ij}_+(x), j^{kl}_+(y)\right] &= 4\left[\psi^j_2(x)\psi^i_2(x), \psi^l_2(y)\psi^k_2(y)\right]\\ &= 2(\delta^{il}j^{kj}_+(x) + \delta^{jk}j^{li}_+(x) - \delta^{jl}j^{ki}_+(x) - \delta^{ik}j^{lj}_+(x))\delta(x^+-y^+)\\ &\quad + \frac{2i}{\pi}(\delta^{jk}\delta^{il} - \delta^{ik}\delta^{jl})\delta'(x^+-y^+) \,. \qquad (B.126)\end{aligned}$$

One checks that the above algebra corresponds exactly to the Poisson algebras (B.123) (the current (BB.124b) obeys the analogous algebra, and (B.124a) commutes with (B.124b)).

---

11) The central term can be obtained as follows. Consider the most divergent term in the Wick expansion of

$$\{\psi^j_2(x)\psi^i_2(x)\psi^l_2(y)\psi^k_2(y)\}_{div} = -\frac{1}{(2\pi)^2}\frac{1}{x^+-y^+-i\epsilon}\frac{1}{x^+-y^+-i\epsilon}(\delta^{jk}\delta^{il} - \delta^{ik}\delta^{jl})$$

and

$$\{\psi^l_2(y)\psi^k_2(y)\psi^j_2(x)\psi^i_2(x)\}_{div} = -\frac{1}{(2\pi)^2}\frac{1}{x^+-y^++i\epsilon}\frac{1}{x^+-y^++i\epsilon}(\delta^{jk}\delta^{il} - \delta^{ik}\delta^{jl}) \quad .$$

The difference reads

$$\frac{1}{(x^+-y^++i\epsilon)^2} - \frac{1}{(x^+-y^+-i\epsilon)^2} = 2i\pi\delta'(x^+-y^+) \quad .$$

The algebra $(B.126)$ above is an infinite extension of the $O(N)$ Lie algebra with a central extension given by the Schwinger term, and is called a Kac-Moody algebra, which in general, is defined by the following commutation relation

$$[j^a(x), j^b(y)] = 4if^{abc}j^c(x)\delta(x-y) + \frac{2ik}{\pi}\delta^{ab}\delta'(x-y) \quad . \tag{B.127}$$

The *level* k of the algebra may differ from one if a higher representation is used for the fermions.

Since a Kac-Moody algebra defines a conformally invariant theory uniquely for $k = \pm 1$ (unitary irreducible representations are unique), we are thus led to identify the currents $(B.122)$ with $\mathrm{tr}\,(\tau^a j_\pm)$. In order to complete this identification, we look for the fermionic representation of the bosonic field g_{ij}, as well as the identification of the energy-momentum tensor in both theories.

Notice that under an infinitesimal transformation of the form $\Omega(z) = 1 + w^a(z)\tau^a$, the current $J(z)$ transforms as

$$\delta_\omega J(z) = \left[\omega(z), J(z)\right] + \frac{1}{2}k\omega'(z) \quad . \tag{B.128}$$

which in turn implies the commutation relations $(B.127)$, for consistency. The energy-momentum tensor is found to be of the Sugawara form.

$$T(z) = \frac{1}{2\chi} : J^a(z)J^a(z) := \frac{1}{2\chi} \sum_{n,m=-\infty}^{\infty} J^a_{n-m}J^a_m z^{-n-2} = \sum_{n=-\infty}^{\infty} L_n z^{-n-2} \quad , \tag{B.129}$$

with $\chi = c_V + k$, and c_V the Casimir, defined by $f^{abc}f^{dbc} = c_V\delta^{ad}$.

The normalization of the energy-momentum tensor above, has been chosen such that the L_n satisfy the usual Kac-Moody-Virasoro algebra. Indeed, proceeding as before, one finds

$$\left[L_n, J^a_m\right] = -mJ^a_{n+m} \quad . \tag{B.130}$$

as well as

$$\left[L_n, L_m\right] = (n-m)L_{n+m} + \frac{c}{12}n(n^2-1)\delta_{n+m,0} \quad . \tag{B.131}$$

The first term in $(B.131)$ follows from the classical algebra. For the central term we have

$$c = \frac{k\,\mathrm{dim}G}{c_V + k} \tag{B.132}$$

We now construct a representation of the (non-abelian) conformal algebra, by seeking primary fields ϕ_j transforming as[148,221]

$$\delta_\epsilon\phi_j(z) = \frac{1}{2\pi i}\oint T(\xi)\epsilon(\xi)\phi_j(z)\mathrm{d}\xi \tag{B.133}$$

$$\delta_\omega\phi_j(z) = \frac{1}{2\pi i}\oint J^a(\xi)\omega^a(\xi)\phi_j(z)\mathrm{d}\xi \quad , \tag{B.134}$$

under reparametrizations and local group transformation, respectively, with $\delta_\epsilon \phi_j$ given by $(B.45)$, and $\delta_\omega \phi_j$ by

$$\delta_\omega \phi_j(z) = \tau^a \omega^a \phi_j(z) \tag{B.135a}$$

This implies the following operator product expansion[148,221]

$$J^a(\omega)\phi_j(z) = \frac{\tau^a}{\omega - z}\phi_j(z) + \text{regular terms} \tag{B.135b}$$

which one obtains by comparing $(B.134)$ and $(B.135a)$. Since ϕ_j is a primary field, we have also

$$T(\omega)\phi_j(z) = \frac{\Delta_j}{(\omega - z)^2}\phi_j(z) + \frac{1}{\omega - z}\frac{\partial}{\partial z}\phi_j(z) + \text{regular terms} \quad , \tag{B.136}$$

where Δ_j is the scale dimension of ϕ_j.

Notice the similarity of the results of the last section with the steps followed in the discussion of $SL(2,R)$ symmetry in Chapter 1.

As last remarks we can say that conformally invariant field theory in two dimensions has turned into an extremely powerful tool to handle with a rich and important class of problems. It has encountered applications to fundamental interactions (string theory), critical theory in two-dimensional statistical models, random surfaces, polymer physics and provides a method for obtaining solutions to important models, such as the WZNW theory, two dimensional gravity and supergravity. In the framework of mathematics, several new and elegant developments have been achieved.

To quote same of them, general conformally invariant field theory in two dimensions could solve, and classify an extended class of statistical models, starting with the Ising model. They serve to classify general models showing critical behavior, supersymmetric models, generalizations to models invariant under non-abelian symmetry groups, as well as new identifications. A new mathematical branch of physics emerged from the fusion rules dictated by associativity in conformally invariant models.

As it turns out, the quantum group is a symmetry group of rational conformal field theories; the conformal blocks are the equivalent of invariant tensors of the quantum group, and the fusion algebra is given by the decomposition rules of tensors product of irreducible representations. A common language between conformally invariant systems and integrable systems emerges quite naturally.

Furthermore, the other face of the problem, from the side of string theory, provides representations of conformal algebra with central charge $c \geq 1$, but in a different context. Thus we have a mathematical framework for the study of extended objects represented by a quantum field .

The fact that both structures are underlined by the same theory has only very recently been better understood. As we have seen in the study of $2D$ gravity, the value $c = 1$ is a "borderline case for the applicability of the matrix model approach". In fact, the models with $c < 1$ have a finite number of degrees of freedom, a number which diverges as c approaches 1. Therefore, the theory with $c > 1$ must be a very difficult problem. The bordercase, $c = 1$ has been still treated in the literature, as we have seen . In fact, there are singularities when trying to study the string problem ($c > 1$) using matrix models techniques. There is a possibility that beyond the value $c = 1$, a strong coupling emerges, where the string is in another phase. Usual string (and superstring) theory in the critical dimension (26,10,etc) has been studied extensively in the course of the years, and represents the other side of the problems presented here. Further insight can also be gained introducing alternative formulations where the string action also contains terms depending on the intrinsic curvature. From the side of string theories however there are still many problems to be solved; especially the fact that the string perturbation theory is not Borel summable, requires non-perturbative treatments.

In this Appendix we reviewed the construction of a general conformally invariant theory, realizing it in terms of the exponential of free fields. Afterwards, we generalized the theory to include the non-abelian case, realizing the above construction both for the bosonic WZNW theory and for a non-abelian fermionic theory, thereby obtaining as a bonus, a generalization of the non-abelian bosonization prescription to the interacting case.

The quite extensive use of the methods presented here, in the course of the book, supports the presentation of the material appearing in this Appendix. Moreover, the discussion at the end of this Appendix provides also a crossed support for the inter-relation of both sides of this scenario.

Appendix C: Conformally Invariant Integrals

Whenever we calculate scattering amplitudes in string theory at tree level (sphere), we are led to very involved integrals over the complex plane (stereographic projection of the sphere). The general form of such integrals is given for N particles scattering, by the expression

$$\mathcal{A}_N = \prod_{i=1}^{N} \int_{\mathbb{C}} \mathrm{d}^2 z_i \prod_{i<j=1}^{N} |z_i - z_j|^{2\theta_{ij}} \quad , \tag{C.1}$$

where $\theta_{ij}(= \theta_{ji})$ are functions of the momenta and spin polarization of the particles. The aim of this Appendix is to present some techniques which can be used to compute (C.1) at least in some specific cases, since a general treatment is not available.

In fact, as it stands, the expression (C.1), is ill defined, due to the volume of the $SL(2,C)$ group. This problem should have been fixed earlier, at the level of gauge fixing. Indeed, we can fix three of the N variables z_i at $0, 1$ and ∞. We choose $z_1 = 0$, $z_2 = 1$ and $z_3 = \infty$, and (C.1) turns into

$$\mathcal{A}_N = \prod_{i=4}^{N} \int_{\mathbb{C}} \mathrm{d}^2 z_i \prod_{i=4}^{N} |z_i|^{2\theta_{1i}} |1 - z_i|^{2\theta_{2i}} \prod_{i<j=4}^{N} |z_i - z_j|^{2\theta_{ij}} \quad . \tag{C.2}$$

Another general feature of the N-coupled integral (C.2) is the existence of infrared (IR) and ultraviolet (UV) divergences. The sources of UV divergences are the coincident points where some of the z_i approach each other. The IR divergences come from the fact that we are integrating a conformally invariant function over the whole complex plane. Such divergencies can be avoided by restricting the values of θ_{ij} as we will see later on, but for arbitrary θ_{ij} the integrals (C.2) need a proper regularization. In what follows we discuss two apparently different ways to avoid such problems, and some other general features of the integrals (C.2) analysing the simplest case of the four-particle scattering ($N = 4$), i.e.,

$$\mathcal{A}_4 = \int \mathrm{d}^2 z |z|^{2\alpha} |1 - z|^{2\beta} \quad , \tag{C.3}$$

where we defined $\theta_{14} = \alpha$ and $\theta_{24} = \beta$.

As a first example we introduce a gaussian factor in (C.3) through the identity

$$|z|^{2\alpha} = \frac{1}{\Gamma(-\alpha)} \int_0^\infty \mathrm{d}t t^{-\alpha-1} \mathrm{e}^{-t|z|^2} \quad , \tag{C.4}$$

with $\Gamma(-\alpha)$ being the usual analytic continuation of the Euler's gamma function. Using an analogous identity for the factor $|1 - z|^{2\beta}$ and performing the gaussian integrals we end up with

$$\mathcal{A}_4 = \frac{\pi}{\Gamma(-\alpha)\Gamma(-\beta)} \int_0^\infty \int_0^\infty \frac{\mathrm{d}t\mathrm{d}u}{(t+u)} t^{-\alpha-1} u^{-\beta-1} \mathrm{e}^{-\frac{tu}{t+u}} \quad . \tag{C.5}$$

Changing variables from u to $v = u/(t+u)$ and integrating first over t we have finally:

$$\mathcal{A}_4 = \pi\Delta(1+\alpha)\Delta(1+\beta)\Delta(-1-\alpha-\beta) \quad , \tag{C.6}$$

with $\Delta(x) = \Gamma(x)/\Gamma(1-x)$. Such result has been first obtained by Veneziano[178] and Virasoro[223]. We give below a second derivation of $(C.6)$ due to Dotsenko[224]. First we rewrite $(C.3)$ in terms of real variables with $z = x + iy$.

$$\mathcal{A}_4 = \int_{-\infty}^{+\infty} dx \int_{-\infty}^{+\infty} dy (x^2+y^2)^{\alpha}((x-1)^2+y^2)^{\beta} \quad . \tag{C.7}$$

Since the terms in the brackets above are all positive the integral will diverge for most of the values of α and β. In order to define the integral in the complex plane, we transform y into an imaginary number with a small real part,

$$y \to iy + 2\epsilon y \quad , \tag{C.8}$$

where ϵ is a small and positive real number. Now $\mathcal{A}_4$ becomes:

$$\begin{aligned}\mathcal{A}_4 = &\frac{i}{2}\int_{-\infty}^{+\infty} du_+[u_+ - i\epsilon(u_+ - u_-)]^{\alpha}[u_+ - 1 - i\epsilon(u_+ - u_-)]^{\beta} \\ &\times \int_{-\infty}^{+\infty} du_-[u_- + i\epsilon(u_+ - u_-)]^{\alpha}[u_- - 1 + i\epsilon(u_+ - u_-)]^{\beta} \quad ,\end{aligned} \tag{C.9}$$

where $u_{\pm} = x \pm y$. Next we split the integral over u_+ into three regions:

$$\begin{aligned}\mathcal{A}_4 = &\frac{i}{2}\int_{-\infty}^{+\infty} du_-[u_- + i\epsilon(u_+ - u_-)]^{\alpha}[u_- - 1 + i\epsilon(u_+ - u_-)]^{\beta} \\ &\times \Big[\int_{-\infty}^{0} du_+[u_+ - i\epsilon(u_+ - u_-)]^{\alpha}[u_+ - 1 - i\epsilon(u_+ - u_-)]^{\beta} \\ &+ \int_{0}^{1} du_+[u_+ - i\epsilon(u_+ - u_-)]^{\alpha}[u_+ - 1 - i\epsilon(u_+ - u_-)]^{\beta} \\ &+ \int_{1}^{+\infty} du_+[u_+ - i\epsilon(u_+ - u_-)]^{\alpha}[u_+ - 1 - i\epsilon(u_+ - u_-)]^{\beta}\Big] \quad .\end{aligned} \tag{C.10}$$

In the limit where $\epsilon \to 0$ we have branch points at $u_- = 0, 1$ so we must know how to surround such points in the u_- integral. For the first term above $\epsilon(u_+ - u_-) < 0$ for $u_- \sim 0$ and $u_- \sim 1$, thus the contour of the u_- integral must pass below such points as indicated in the Fig. $C.1a$. When we close the contour $(C.1a)$ at infinity the integral over u_- will vanish since we do not have singularities in the lower half-plane. For analogous reasons the third term above (see Fig. $C.1c$) also vanishes and we are left with the contour $C.1b$ of Fig. $C.1$ which corresponds to the second term of $(C.10)$:

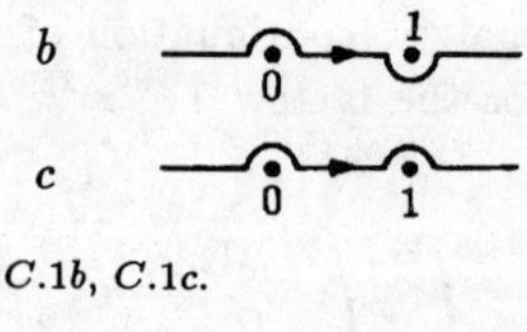

Figure $C.1$: Contours $C.1a$, $C.1b$, $C.1c$.

$$\mathcal{A}_4 = \frac{i}{2}\int_0^1 du_+ u_+^\alpha (u_+ - 1)^\beta \int_{C.1b} du_- u_-^\alpha (u_- - 1)^\beta \quad . \tag{C.11}$$

The contour $C.1b$ may be deformed into the contour C shown in Fig. $C.2$.

Figure $C.2$: Change of contour.

The left and right moving paths of Fig. $C.2$ belong to different Riemann sheets so we pick up a factor $e^{i2\pi\beta}$ for the left moving path due to the branch point $u_- = 1$, thus we have:

$$\mathcal{A}_4 = \frac{i}{2}(1 - e^{2\pi i\beta})\int_0^1 du_+ u_+^\alpha (u_+ - 1)^\beta \int_1^\infty du_- (u_- - 1)^\beta \quad . \tag{C.12}$$

Taking $u_- \to 1/u_-$ we get, up to an overall phase factor

$$\mathcal{A}_4 = \sin \pi b \int_0^1 du_+ u_+^\alpha (1 - u_+)^\beta \int_0^1 du_- u_-^{-2-\alpha-\beta}(1 - u_-)^\beta \quad . \tag{C.13}$$

Using $\sin \pi\beta = \pi/\Gamma(-\beta)\Gamma(\beta + 1)$ we recover formula $(C.6)$:

$$\begin{aligned}\mathcal{A}_4 &= \int d^2z |z|^{2\alpha} |1 - z|^{2\beta} \\ &= \pi\Delta(1+\alpha)\Delta(1+\beta)\Delta(-1-\alpha-\beta) \quad .\end{aligned} \tag{C.14}$$

It may seem even surprising that by a different procedure one arrives at exactly the same result for $\mathcal{A}_4$, but this is actually a consequence of the various constraints on the result. One of those constraints comes from the symmetries of $\mathcal{A}_4$ under the changes $z \to 1 - z$ and $z \to 1/z$ which leads respectively to:

$$\mathcal{A}_4(\alpha, \beta) = \mathcal{A}_4(\beta, \alpha) \quad , \tag{C.15}$$

$$\mathcal{A}_4(\alpha, \beta) = \mathcal{A}_4(-2 - \alpha - \beta, \beta) \quad . \tag{C.16}$$

Another powerful constraint over $\mathcal{A}_4$ is its asymptotic behaviour for large $\alpha\,(\alpha \to \infty)$ and large $\beta\,(\beta \to \infty)$. To determine the large-α behaviour of the integral $\mathcal{A}_4$ we make a change of variables $z = e^{\eta/\alpha}$ after which we have

$$\mathcal{A}_4(\alpha \to \infty) \approx \alpha^{-2-2\beta} \int d^2\eta\, e^{(\eta+\eta^*)} |\eta|^{2\beta} \quad . \tag{C.17}$$

Since the integral above is α-independent we have a powerlike asymptotic behaviour:

$$\mathcal{A}_4(\alpha \to \infty) \sim \alpha^{-2-2\beta} \quad . \tag{C.18}$$

Due to $(C.15)$ we have:

$$\mathcal{A}_4(\beta \to \infty) \sim \beta^{-2-2\alpha} \quad . \tag{C.19}$$

In order to check that the result $(C.6)$ satisfies the asymptotic behaviour given by $(C.18)$ and $(C.19)$ it is useful to use $\lim_{n\to\infty} \Gamma(n+B) = n^B\Gamma(n)$ which can be deduced from Stirling formula, $\Gamma(n \to \infty) \sim \mathrm{e}^{-n}n^{n-\frac{1}{2}}$. The large-$\alpha$ and -β behaviours, and the knowledge of the position of the poles at $z = 0, 1$ for $\alpha = -1, -2, -3, \cdots$, and $\beta = -1, -2, \cdots$ respectively, fix the result $(C.6)$ completely, if we assume that the amplitude $\mathcal{A}_4$ is an analytical function of α and β, by means of Carlson's theorem. This is the reason why we have obtained the same result using different regularizations.

The next simplest amplitude to be computed is the 5-particle scattering,

$$\mathcal{A}_5 = \int \mathrm{d}^2 w \int \mathrm{d}^2 z |z|^{2\theta_{14}} |w|^{2\theta_{15}} |1-z|^{2\theta_{24}} |1-w|^{2\theta_{25}} |z-w|^{2\theta_{45}} \quad . \tag{C.20}$$

We could try to apply the same techniques that we have used to compute $\mathcal{A}_4$ on $\mathcal{A}_5$, but it turns out that for arbitrary values of θ_{ij} we are not able to get a meaningful result for $\mathcal{A}_5$. Actually, for arbitrary θ_{ij} a closed expression for $\mathcal{A}_5$ is not known. However, for the special case $\theta_{14} = \theta_{15} = \alpha$ and $\theta_{24} = \theta_{25} = \beta$ with $\theta_{45} = 2\rho$ we have

$$\mathcal{A}_5 = \int \mathrm{d}^2 w \int \mathrm{d}^2 z |z|^{2\alpha} |w|^{2\alpha} |1-z|^{2\beta} |1-w|^{2\beta} |z-w|^{4\rho} \quad . \tag{C.21}$$

This integral is an example of a more general integral whose calculation has been performed in [30] and that we reproduce below, namely:

$$I_n = \prod_{i=1}^{n} \int \mathrm{d}^2 z_i \prod_{i=1}^{n} |z_i|^{2\alpha} |1-z_i|^{2\beta} \prod_{i<j=1}^{n} |z_i - z_j|^{4\rho} \quad . \tag{C.22}$$

It turns out that the first regularization procedure that we have presented to calculate $\mathcal{A}_4$ does not work for $\mathcal{A}_n$ with $n \geq 5$, thus following [30] we have to generalize Dotsenko's technique to perform the integral using contour integrals to compute I_n. The first step is to implement the rotation $(C.8)$ in the complex plane for each variable z_i and rewrite I_n as a product of n integrals over u_+^i $(i = 1, 2, \cdots)$ times n integrals over u_-^i $(i = 1, 2, \cdots n)$. The integration range of the u_+^i integrals can be split and for symmetries reasons we can reorder the variables. We end up with:

$$\begin{aligned} I_n = & \int_0^1 \mathrm{d}u_+^1 \int_0^{u_+^1} \mathrm{d}u_+^2 \cdots \int_0^{u_+^{i-1}} \mathrm{d}u_+^i \prod_{i=1}^{n} (u_+^i)^\alpha (u_+^i - 1)^\beta \prod_{i<j=1}^{n} (u_+^i - u_+^j)^{2\rho} \\ & \times \prod_{i=1}^{n} \int_{C_i} \mathrm{d}u_-^i \prod_{i=1}^{n} (u_-^i)^\alpha (u_-^i - 1)^\beta \prod_{i<j=1}^{n} (u_-^i - u_-^j)^{2\rho} \quad , \end{aligned} \tag{C.23}$$

where the contours C_i are shown in Fig. $C.3$.

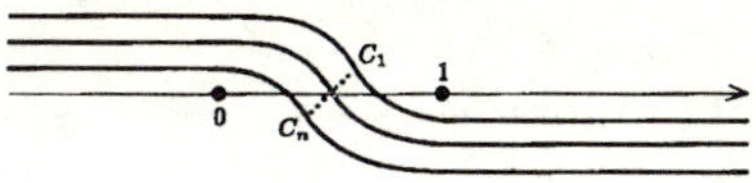

Figure $C.3$: Contours of I_n.

The contours C_i can be deformed into integrals from 1 to ∞ as well as the ordered integrals over u_+^i can be written in terms of integrals from 0 to 1, up to a numerical factor ($n!$) such that we have:

$$
\begin{aligned}
I_n = & \prod_{i=1}^{n} \int_0^1 du_+^i \prod_{i=1}^{n} (u_+^i)^\alpha (1-u_+^i)^\beta \prod_{i<j} (u_+^i - u_+^j)^{2\rho} \\
& \times \prod_{i=1}^{n} (\sin\beta\pi + \sin(\beta+2\rho)\pi + \cdots + \sin(\beta+2(i-1)\rho)\pi) \qquad (C.24a) \\
& \times \prod_{i=1}^{n} \int_1^\infty du_-^i \prod_{i=1}^{n} (u_-^i)^\alpha (u_-^i - 1)^\beta \prod_{i<j} (u_-^i - u_-^j)^{2\rho} \quad ,
\end{aligned}
$$

where $\sin\beta\pi$, $\sin(\beta+2\rho)\pi\ldots$, come respectively from surrounding the cuts $(1\,,\,u_+^1)$, (u_+^1, u_+^2), etc., in such a way that the full integral can be related to the ordered integral $u_+^i < u_+^{i+1}$, by means of a "cut-crossing" procedure. Therefore we have

$$
\begin{aligned}
I_n = & \prod_{i=1}^{n} \left(\frac{\sin\pi(\beta+(i-1)\rho)\sin i\pi\rho}{\sin\pi\rho} \right) \\
& \times \widetilde{I}_n(\alpha,\beta,\rho) \times \widetilde{I}_n(-2-\alpha-\beta-2(n-1)\rho,\beta,\rho) \qquad (C.24b)
\end{aligned}
$$

with

$$
\widetilde{I}_n(\alpha,\beta,\rho) = \prod_{i=1}^{n} \int_0^1 du_i \prod_{i=1}^{n} u_i^\alpha (1-u_i)^\beta \prod_{1<j=1}^{n} (u_i - u_j)^{2\rho} \quad , \qquad (C.25)
$$

where we have used

$$
\begin{aligned}
&(\sin\beta\pi + \sin(\beta+2\rho)\pi + \cdots + \sin(\beta+2(i-1)\rho)\pi) = \\
&\qquad = \left(\frac{\sin\pi(\beta+(i-1)\rho)\sin i\pi\rho}{\sin\pi\rho} \right) \quad . \qquad (C.26)
\end{aligned}
$$

Thus, we are left with the integrals $\widetilde{I}_n$ which have been calculated by Selberg[225] as follows. Using

$$
(\Delta(u))^{2\rho} = \left(\prod_{i<j} (u_i - u_j) \right)^{2\rho} = \sum_{\{a\}} C_{\{a\}} u_1^{a_1} \cdots u_p^{a_p} \quad , \qquad (C.27)
$$

we have

$$\widetilde{I}_n(\alpha,\beta,\rho) = \sum_{\{a\}} C_{\{a\}} \widetilde{I}_{\{a\}} \quad , \tag{C.28}$$

with each term given by the expression

$$\widetilde{I}_{\{a\}} = \prod_{i=1}^{n} \int_0^1 du_i \, u_i^{\alpha+a_i}(1-u_i)^{\beta} = \prod_{i=1}^{n} \frac{\Gamma(1+\alpha+a_i)\Gamma(1+\beta)}{\Gamma(2+\alpha+\beta+a_i)} \quad . \tag{C.29}$$

If we order a_i such that $0 \le a_1 \le a_2 \le \cdots \le a_n$ and use that (see $(C.27)$) $\sum_{i=1}^{\nu} a_i = 2\rho \times \frac{1}{2}\nu(\nu-1) = \rho\nu(\nu-1)$ we deduce

$$\begin{aligned} \nu a_\nu &\ge \rho\nu(\nu-1) \quad , \\ a_\nu &\ge \rho(\nu-1) \quad . \end{aligned} \tag{C.30}$$

Next we can use the property

$$(\Delta(u))^{2\rho} = \Delta\left(\frac{1}{u}\right)^{2\rho} \prod_{i=1}^{n} (u_i)^{2(n-1)\rho} \tag{C.31}$$

to derive

$$a'_\nu = 2(n-1)\rho - a_{n-\nu+1} \ge \rho(\nu-1) \quad . \tag{C.32}$$

Therefore we arrive at the bound

$$(\nu-1)\rho \le a_\nu \le (n+\nu-2)\rho \quad . \tag{C.33}$$

This permits us to rewrite $(C.29)$ as

$$\widetilde{I}_\alpha = \prod_{\nu=1}^{n} \frac{\Gamma(1+\alpha+(\nu-1)\rho)}{\Gamma(2+\alpha+\beta+(n+\nu-2)\rho)} q_{a_\nu}(\alpha,\beta) \quad , \tag{C.34}$$

where $q_{a_\nu}(\alpha,\beta)$ is a polynomial in β of degree $(n+\nu-2)\rho - a_\nu$, that means that we can write the result in two alternative ways,

$$\begin{aligned} \widetilde{I}_n(\alpha,\beta,\rho) &= Q(\alpha,\beta) \prod_{\nu=1}^{n} \frac{\Gamma(1+\alpha+(\nu-1)\rho)\Gamma(1+\beta)}{\Gamma(2+\alpha+\beta+(n+\nu-2)\rho)} \\ &= \left[\frac{Q(\alpha,\beta)}{\prod_{\nu=1}^{n} (1+\beta)(1+\beta+1)\cdots(1+\beta+(\nu-1)\rho-1)} \right] \times \\ &\qquad \times \prod_{\nu=1}^{n} \frac{\Gamma(1+\alpha+(\nu-1)\rho)\Gamma(1+\beta+(\nu-1)\rho)}{\Gamma(2+\alpha+\beta+(n+\nu-2)\rho)} \end{aligned} \tag{C.35}$$

with $Q(\alpha,\beta)$ a polynomial of degree $\sum_{i=1}^{n}((n+i-2)\rho - a_i) = \frac{n\rho}{2}(n-1)$ in β (we have used $\sum_{i=1}^{n} a_i = \rho n(n-1)$, see $(C.27)$). Therefore the term in brackets in

$(C.35)$ is β-independent for $\beta \to \infty$. If we assume that it is also analytic, which is certainly true for $\beta > -1$, we conclude that it must be β-independent as well as α-independent by symmetry. Thus we write

$$\widetilde{I}_n(\alpha,\beta,\rho) = C_n(\rho) \prod_{\nu=1}^{n} \frac{\Gamma(1+\alpha+(\nu-1)\rho)\Gamma(1+\beta+(\nu-1)\rho)}{\Gamma(2+\alpha+\beta+(n+\nu-2)\rho)} \quad . \tag{C.36}$$

Taking $\alpha = 0 = \beta$ and making a change of variables $u_i \to u_n u_i \, (i = 1, 2, \cdots, n-1)$ we get the recurrence relation

$$C_n(\rho) = C_{n-1}(\rho) \frac{\Gamma(1+n\rho)}{\Gamma(1+\rho)} \quad . \tag{C.37}$$

Since $C_1(\rho) = 1$ we have finally:

$$\widetilde{I}_n(\alpha,\beta,\rho) = \prod_{i=1}^{n} \frac{\Gamma(1+i\rho)}{\Gamma(1+\rho)} \prod_{i=0}^{n-1} \frac{\Gamma(1+\alpha+i\rho)\Gamma(1+\beta+i\rho)}{\Gamma(2+\alpha+\beta+(n+i-1)\rho)} \quad . \tag{C.38}$$

Plugging back in $(C.24b)$ using $\sin \pi z = \pi / \Gamma(-z)\Gamma(z+1)$ we reproduce formula $(B.9)$ of Dotsenko and Fatteev[30]:

$$\begin{aligned} I_n &= \prod_{i=1}^{n} \int d^2 z_i \prod_{i=1}^{n} |z_i|^{2\alpha} |1-z_i|^{2\beta} \prod_{i<j} |z_i - z_j|^{4\rho} \\ &= n! \, [\Delta(1-\rho)]^n \prod_{i=1}^{n} \Delta(i\rho) \prod_{i=0}^{n-1} \Delta(1+\alpha+i\rho)\Delta(1+\beta+i\rho)\Delta(-1-\alpha-\beta-(n-1+i)\rho) \end{aligned} \tag{C.39}$$

Notice that the result $(C.39)$ satisfies the following constraints

$$\lim_{\rho \to 0} I_n(\alpha,\beta,\rho) = (\mathcal{A}_4(\alpha,\beta))^n \tag{C.40}$$

$$I_n(\alpha,\beta,\rho) = I_n(\beta,\alpha,\rho) \tag{C.41}$$

$$I_n(\alpha,\beta,\rho) = I_n(-\alpha-\beta-2-2\rho(n-1),\beta,\rho) \tag{C.42}$$

where $(C.41)$ and $(C.42)$ are consequence of the symmetries $z_i \to 1 - z_i$; $z_i \to 1/z_i$.

The result $(C.39)$ does also have the correct asymptotic behaviour for large α and β as it can be easily verified verified by a change of variables $z_i \to e^{\eta_i/\alpha}$.

Let us now compute the integral I_G^s which appeared in the 3-point function of super Liouville theory, Eq. (5.23), which is the supersymmetric counterpart of $(C.39)$. Making the substitutions

$$z_s \equiv w \quad , \quad z_{2i-1} \equiv \zeta_i \quad , \quad z_{2i} \equiv \eta_i \quad , \tag{C.43}$$

we arrive at the expression

$$\begin{aligned} J^n(\alpha,\beta;\rho) &= \frac{\alpha^2}{2^n n!} I_G^{s=2n+1} = \frac{\alpha^2}{2^n n!} \int d^2 w \prod_{i=1}^{n} d^2\zeta_i d^2\eta_i |w|^{2\alpha-2} |1-w|^{2\beta} \\ &\prod_{i=1}^{n} |w-\zeta_i|^{4\rho} |w-\eta_i|^{4\rho} |\zeta_i - \eta_i|^{-2} \prod_{i=1}^{n} |\zeta_i|^{2\alpha} |\eta_i|^{2\alpha} |1-\zeta_i|^{2\beta} |1-\eta_i|^{2\beta} \\ &\times \prod_{i,j} |\zeta_i - \eta_j|^{4\rho} \prod_{i<j} |\zeta_i - \zeta_j|^{4\rho} |\eta_i - \eta_j|^{4\rho} \quad . \end{aligned} \tag{C.44}$$

We might try to compute J_n by using Dotsenko and Fatteev's technique once again, actually this has been done in ref. [81], but since there are so many constraints on such integrals we could try to calculate it by some guess work as presented in ref. [79] where the integral I_n was first calculated. The first constraint is based on the residual $SL(2,\mathbb{C})$ symmetries of I_n. Although not explicitly, J_n satisfies:

$$J_n(\alpha,\beta,\rho)=J_n(\beta,\alpha,\rho) \quad . \tag{C.45}$$

To convince oneself, the reader may look at an analogous four-point function:

$$J_0=\alpha^2\int \mathrm{d}^2w|w|^{2\alpha-2}|1-w|^{2\beta} \tag{C.46}$$

$$\begin{aligned}&=\pi\alpha^2\Delta(\alpha)\Delta(1+\beta)\Delta(-\alpha-\beta)\\ &=-\pi\Delta(1+\alpha)\Delta(1+\beta)\Delta(-\alpha-\beta) \quad ,\end{aligned} \tag{C.47}$$

where we have used the result $(C.6)$ for the integral. Thus, we end up with an $\alpha\leftrightarrow\beta$ symmetric integral. Moreover, by taking successively $w\to 1-w$, $w\to 1/w$ and $w\to 1-w$ again we get from $(C.46)$:

$$J_0=\alpha^2\int \mathrm{d}^2w|w|^{2\alpha-2}|1-w|^{-2\beta-2\alpha-2} \quad . \tag{C.48}$$

Comparing with $(C.46)$ and considering the $\alpha\longleftrightarrow\beta$ symmetry of J_0 we have another symmetry, which is clear from $(C.47)$, namely:

$$J_0(\alpha,\beta)=J_0(-\alpha-\beta-1,\beta) \quad . \tag{C.49}$$

Repeating this procedure for J_n we have[79]

$$J_n(\alpha,\beta,\rho)=J_n(-\alpha-\beta-1-4n\rho,\beta,\rho) \quad . \tag{C.50}$$

Another important information about J_n is its asymptotic behaviour:

$$J_n(\alpha\to\infty,\beta,\rho)\sim\alpha^{-2n-2(2n+1)\beta-4\rho n(2n+1)} \quad . \tag{C.51}$$

Collecting all these informations together we have an Ansatz for J_n

$$\begin{aligned}J_n(\alpha,\beta;\rho)=C_n(\alpha,\beta;\rho)&\prod_{i=0}^{n}\Delta(1+\alpha+2i\rho)\Delta(1+\beta+2i\rho)\Delta(-\alpha-\beta+(2i-4n)\rho)\\ &\times\prod_{i=1}^{n}\Delta(\frac{1}{2}+\alpha+(2i-1)\rho)\Delta(\frac{1}{2}+\beta+(2i-1)\rho)\\ &\times\Delta(-\frac{1}{2}-\alpha-\beta+(2i-4n-1)\rho) \quad ,\end{aligned} \tag{C.52}$$

where $C_n(\alpha,\beta;\rho)$ has the same symmetries as $J_n(\alpha,\beta;\rho)$. By looking at the large-α behavior as given in $(C.51)$, one can confirm that $C_n(\alpha,\beta;\rho)$ is, as a function of α, bounded as $|\alpha| \to \infty$ and if we assume it to be analytic, we conclude that $C_n(\alpha, \beta; \rho)$ is independent of α, and by symmetry, of β as well,

$$C_n(\alpha, \beta, \rho) = C_n(\rho) \quad . \tag{C.53}$$

In order to compute $C_n(\rho)$ we have to learn more about the pole structure of these conformally invariant integrals. Again we will make use of simpler amplitudes like $\mathcal{A}_4$, Eq. $(C.3)$ to draw some important and general conclusions about further integrals. We start taking $\alpha = -1 + \epsilon\,(\epsilon \to 0)$ in $(C.6)$ and using $\Gamma(\epsilon) \sim 1/\epsilon$ we have

$$\begin{aligned} \mathcal{A}_4 &= \int \mathrm{d}^2 z |z|^{-2+2\epsilon} |1-z|^{2\beta} \\ &= \pi/\epsilon \quad . \end{aligned} \tag{C.54}$$

This is consistent with the identification of distributions

$$|z|^{-2+\epsilon} \sim \frac{\pi}{\epsilon} \delta^{(2)}(z) \quad , \tag{C.55}$$

where $\delta^{(2)}(z)$ is the two-dimensional Dirac delta function. The same identification holds for the pole $z \sim 1$ when $\beta = -1 + \epsilon$. The next simplest amplitude ($\mathcal{A}_5$), in $(C.2)$, corresponds to I_2 of $(C.39)$, i.e.,

$$\begin{aligned} I_2 &= \int \mathrm{d}^2 z \int \mathrm{d}^2 w |z|^{2\alpha} |w|^{2\alpha} |1-z|^{2\beta} |1-w|^{2\beta} |w-z|^{4\rho} \\ &= \pi^2 \Delta(2\rho) \Delta(1+\alpha) \Delta(1+\beta) \Delta(-1-\alpha-\beta-\rho) \\ &\times \Delta(1+\alpha+\rho) \Delta(1+\beta+\rho) \Delta(-1-\alpha-\beta-2\rho) \quad . \end{aligned} \tag{C.56}$$

If we take $\rho = -1/2 + \epsilon$ we have from $(C.56)$

$$\begin{aligned} I_2(\alpha, \beta, \rho = -\frac{1}{2} + \epsilon) &\approx \frac{\pi^2}{\epsilon} \Delta(1+2\alpha) \Delta(1+2\beta) \Delta(-1-2\alpha-2\beta) \\ &= \frac{\pi}{\epsilon} \mathcal{A}_4(2\alpha, 2\beta) \quad , \end{aligned} \tag{C.57}$$

where we have used $\Delta(2n) = \Delta(n)\Delta(n+1/2)2^{4n-1}$ and the result $(C.6)$ for $\mathcal{A}_4$. The above result is consistent with the identification of distributions:

$$|w-z|^{-2+\epsilon} \sim \frac{\pi}{\epsilon} \delta^{(2)}(w-z) \quad . \tag{C.58}$$

It turns out that the identifications $(C.55)$ and $(C.58)$ hold true not only for the first poles of the gamma functions but also for higher poles. In order to check the higher poles we just have to integrate by parts; i.e., up to numerical factors:

$$\begin{aligned} |w|^{-2k+\epsilon} &\sim (\partial_w \overline{\partial}_w)^{k-1} |w|^{-2+\epsilon} \\ &\sim \frac{\pi}{\epsilon} (\partial_w \overline{\partial}_w)^{k-1} \delta^{(2)}(w) \quad . \end{aligned} \tag{C.59}$$

Behind the identification (C.59) there is a well known property of string scattering amplitudes, namely the residue of a pole corresponding to an intermediate state of a n particle scattering can be calculated in terms of a $n-1$ scattering amplitude. For our purposes (C.59) will be of special relevance.

Coming back to the calculation of $C_n(\rho)$ in (C.52), if we take $\alpha = \epsilon (\epsilon \to 0)$ in (C.44) and use $|w|^{-2+\epsilon} = (\pi/\epsilon)\delta^{(2)}(w)$, after integrating over w, we have:

$$J_n(\alpha=\epsilon,\beta,\rho) = -\frac{\pi}{2^n n!}\Delta(1+\epsilon)\Delta(1+\beta)\Delta(-\epsilon-\beta)\widetilde{J}_n(2\rho,\beta,-\frac{1}{2},\rho) \qquad (C.60)$$

with

$$\tilde{J}_n(\alpha,\beta;\gamma;\rho) = \int \prod_{i=1}^{n} \mathrm{d}^2\zeta_i \mathrm{d}^2\eta_i \prod_{i=1}^{n} |\zeta_i|^{2\alpha}|\eta_i|^{2\alpha}|1-\zeta_i|^{2\beta}|1-\eta_i|^{2\beta} \prod_{i,j}^{n} |\zeta_i-\eta_j|^{4\rho}$$
$$\times \prod_{i<j}^{n} |\zeta_i-\zeta_j|^{4\rho}|\eta_i-\eta_j|^{4\rho} \prod_{i=1}^{n} |\zeta_i-\eta_i|^{4\gamma} \quad . \qquad (C.61)$$

It turns out that after the introduction of the extra parameter γ the integral $\widetilde{J}_n$ in (C.61) satisfies (up to symmetry factors), besides the usual constraints based on symmetries and asymptotic behaviours, the following extra constraints (see (C.38))

$$\begin{aligned} \widetilde{J}_1(\alpha,\beta,\gamma,\rho) &= I_2(\alpha,\beta,\rho+\gamma) \quad , \\ \widetilde{J}_n(\alpha,\beta,\gamma,\rho=0) &= (I_2(\alpha,\beta,\gamma))^n \quad , \\ \widetilde{J}_n(\alpha,\beta,\gamma=0,\rho) &= I_{2n}(\alpha,\beta,\rho) \quad . \end{aligned} \qquad (C.62)$$

Such extra constraints permit us, after some guess work, to fully determine $\widetilde{J}_n(\alpha,\beta,\gamma,\rho)$, including its ρ-dependent coefficient $\widetilde{C}_n(\rho)$:

$$\tilde{J}_n(\alpha,\beta;\gamma;\rho) =$$
$$= \frac{\pi^{2n}}{2^n} n! \left[\Delta\left(-(\gamma+\rho)\right)\right]^{2n} \prod_{i=1}^{n} \Delta\left(1+2(\gamma+i\rho)\right)\Delta\left(1+\gamma+(2i-1)\rho\right)$$
$$\times \prod_{i=0}^{n-1} \Delta(1+\alpha+2i\rho)\Delta(1+\beta+2i\rho)\Delta(-1-\alpha-\beta-2\gamma+(2i-4n+2)\rho)$$
$$\times \prod_{i=1}^{n} \Delta(1+\alpha+\gamma+(2i-1)\rho)\Delta(1+\beta+\gamma+(2i-1)\rho)\Delta(-1-\alpha-\beta-\gamma+(2i-4n+2)\rho). \qquad (C.63)$$

Plugging back the above result in (C.60) we can fix $C_n(\rho)$ and consequently J_n:

$$J_n(\alpha,\beta,\rho) = -\frac{\pi^{2n+1}}{2^{2n}} \left[\Delta\left(\frac{1}{2}-\rho\right)\right]^{2n+1} \prod_{i=1}^{n} \Delta\left(2i\rho\right) \prod_{i=0}^{n} \Delta\left(\frac{1}{2}+(2i+1)\rho\right)$$
$$\times \prod_{i=0}^{n} \Delta(1+\alpha+2i\rho)\Delta(1+\beta+2i\rho)\Delta(-\alpha-\beta+(2i-4n)\rho)$$
$$\times \prod_{i=1}^{n} \Delta(\frac{1}{2}+\alpha+(2i-1)\rho)\Delta(\frac{1}{2}+\beta+(2i-1)\rho)$$
$$\times \Delta(-\frac{1}{2}-\alpha-\beta+(2i-4n-1)\rho) \quad . \qquad (C.64)$$

Up to numerical factors the above result agrees with the one obtained by Alvarez-Gaumé and Zaugg in ref. [81] and therefore we have succeeded in obtaining the supersymmetric generalization of I_n without getting into details of any regularization scheme. The integrals I_n and J_n appear in conformal field theory when we calculate correlation functions of 3 primary fields with the inclusion of screening charges of just one kind, say $(u_-)^n$. Wherever the two different kinds of $SU(2)$ screening charges are needed, say $(u_-)^n(u_+)^m$, we have to generalize I_n, J_n to I_{mn}, J_{mn}. The integral I_{mn} has been calculated by Dotsenko and Fatteev[30], see formula $(B.10)$ of ref. [30], as well as formula $(C.75)$ in this Appendix, while its supersymmetric generalization has been calculated in ref. [80] (see also refs. [81] and [82]. Motivated by the success of the calculation of J_n we apply basically the same strategy to compute the integral J_{mn} which can be written as (notice that J_{mn} is nothing but I_M^{nm} of Chapter 5, as given in (5.42)),

$$
\begin{aligned}
J_{mn} = I_M^{nm}(\alpha,\beta;\rho) = & \prod_{i=1}^{n}\int d^2t_i|t_i|^{2\alpha}|1-t_i|^{2\beta}\prod_{i<j}^{n}|t_i-t_j|^{2\rho} \\
& \times \prod_{i=1}^{m}\int d^2r_i|r_i|^{2\alpha'}|1-r_i|^{2\beta'}\prod_{i<j}^{m}|r_i-r_j|^{2\rho'}\prod_{i=1}^{n}\prod_{j=1}^{m}|t_i-r_j|^{-2} \\
& \times \left\langle \prod_{i=1}^{n}\bar{\xi}\xi(t_i)\prod_{i=1}^{m}\bar{\xi}\xi(r_i)\right\rangle_0 \quad , \qquad (C.65)
\end{aligned}
$$

where, $\xi,\bar{\xi}$ are free fermions $\langle\bar{\xi}\xi(t)\bar{\xi}\xi(r)\rangle = |t-r|^{-2}$ and $\alpha' = -\rho'\alpha\,;\ \beta' = -\rho'\beta$ with $\rho' = \rho^{-1}$. The symmetries and asymptotic behaviour below

$$
\begin{aligned}
J_{mn}(\alpha,\beta,\rho) &= J_{mn}(\beta,\alpha;\rho) \quad , \\
J_{mn}(\alpha,\beta;\rho) &= J_{mn}(m-1-\alpha-\beta-\rho(n-1),\beta;\rho) \quad , \qquad (C.66) \\
J_{mn}(\alpha\to\infty,\beta;\rho) &= \alpha^{(2nm-n-m-2n\beta-2m\beta'-\rho(n-1)n-\rho'(m-1)m)} \quad ,
\end{aligned}
$$

can be used together with the constraint that for $\rho=\rho'=-1\,(\alpha'=\alpha\,;\ \beta'=\beta)$ the integral J_{mn} is a function of n and m through the combination $n+m$ to suggest us the following Ansatz: (for n and m even)

$$
\begin{aligned}
& J_{mn}(\alpha,\beta;\rho) = C_{mn}(\rho)\times \\
& \prod_{i=0}^{\frac{n}{2}-1}\Delta(1+\alpha+i\rho)\Delta(1+\beta+i\rho)\Delta(m-\alpha-\beta+\rho(i-n+1)) \\
& \prod_{i=1}^{\frac{n}{2}}\Delta(\frac{1}{2}+\alpha+(i-\frac{1}{2})\rho)\Delta(\frac{1}{2}+\beta+(i-\frac{1}{2})\rho)\Delta(-\frac{1}{2}-\alpha+m-\beta+\rho(i-n+\frac{1}{2})) \\
& \prod_{i=0}^{\frac{m}{2}-1}\Delta(1+\alpha'-\frac{n}{2}+i\rho')\Delta(1-\frac{n}{2}+\beta'+i\rho')\Delta(\frac{n}{2}-\alpha'-\beta'+\rho'(i-m+1)) \\
& \prod_{i=1}^{\frac{m}{2}}\Delta(\frac{1}{2}-\frac{n}{2}+\alpha'+(i-\frac{1}{2})\rho')\Delta(\frac{1}{2}-\frac{n}{2}+\beta'+(i-\frac{1}{2})\rho') \\
& \Delta(\frac{-1}{2}+\frac{n}{2}-\alpha'-\beta'+\rho'(i-m+\frac{1}{2})) \quad . \qquad (C.67)
\end{aligned}
$$

The coefficient $C_{mn}(\rho)$ can be determined by using the constraints:

$$J_{on} = (-)^{n/2}(n!!)\widetilde{J}_{n/2}(\alpha,\beta,\gamma=-1/2,\rho/2) \quad , \tag{C.68}$$

$$J_{mo} = (-)^{m/2}(m!!)\widetilde{J}_{m/2}(\alpha',\beta',\gamma=-1/2,\rho'/2) \quad , \tag{C.69}$$

where $\widetilde{J}_n$ is given in $(C.63)$. Finally, we get

$$\begin{aligned}
&J_{mn}(\alpha,\beta;\rho) = \\
&I_M^{mn}(\alpha,\beta,\rho) = (-)^{\frac{n+m}{2}}\frac{\pi^{n+m}}{2^{n+m}}n!m!\left(-\frac{\rho}{2}\right)^{-2nm}\left[\Delta\left(\frac{1}{2}-\frac{\rho}{2}\right)\right]^n\left[\Delta\left(\frac{1}{2}-\frac{\rho'}{2}\right)\right]^m \\
&\times\prod_1^{\frac{n}{2}}\Delta(i\rho)\Delta\left(\frac{1}{2}+\rho\left(i-\frac{1}{2}\right)\right)\prod_1^{\frac{m}{2}}\Delta(i\rho'-\frac{n}{2})\Delta\left(\frac{1}{2}-\frac{n}{2}-\rho'\left(i-\frac{1}{2}\right)\right) \\
&\times\prod_{i=0}^{\frac{n}{2}-1}\Delta(1+\alpha+i\rho)\Delta(1+\beta+i\rho)\Delta(m-\alpha-\beta+\rho(i-n+1)) \\
&\times\prod_{i=1}^{\frac{n}{2}}\Delta(\frac{1}{2}+\alpha+(i-\frac{1}{2})\rho)\Delta(\frac{1}{2}+\beta+(i-\frac{1}{2})\rho)\Delta(-\frac{1}{2}-\alpha+m-\beta+\rho(i-n+\frac{1}{2})) \\
&\times\prod_{i=0}^{\frac{m}{2}-1}\Delta(1+\alpha'-\frac{n}{2}+i\rho')\Delta(1-\frac{n}{2}+\beta'+i\rho')\Delta(\frac{n}{2}-\alpha'-\beta'+\rho'(i-m+1)) \\
&\times\prod_{i=1}^{\frac{m}{2}}\Delta(\frac{1}{2}-\frac{n}{2}+\alpha'+(i-\frac{1}{2})\rho')\Delta(\frac{1}{2}-\frac{n}{2}+\beta'+(i-\frac{1}{2})\rho')\Delta(\frac{-1}{2}+\frac{n}{2}-\alpha'-\beta'+\rho'(i-m+\frac{1}{2})).
\end{aligned} \tag{C.70}$$

Our result $(C.70)$ matches with other results in the literature up to numerical overall factors. Therefore we have succeeded again in calculating a complicated conformally invariant integral without referring explicitly to any regularization scheme.

As the last example of such calculation we have computed also the integral $\overline{J}_m$ below,

$$\begin{aligned}
&\overline{J}_m(\alpha,\beta,\epsilon,\gamma,\rho) = \prod_{i=1}^m\int d^2w_i d^2z_i|w_i|^{2\alpha}|z_i|^{2\alpha}|1-w_i|^{2\beta}|1-z_i|^{2\beta} \\
&\times\prod_{i<j=1}^m|w_i-w_j|^{4\epsilon}|z_i-z_j|^{4\epsilon}\prod_{i,j=1}^m|w_i-z_j|^{4\rho}\prod_{i=1}^m|w_i-z_i|^{4\gamma} \\
&=\frac{m!}{2^m}[\pi\Delta(-\gamma-\epsilon)]^{2m}\prod_{i=1}^m\Delta(\frac{1}{2}+\gamma+i\rho)\Delta(1+\gamma+i\epsilon+(i-1)\rho)\Delta(1+\gamma+i\epsilon) \\
&\times\prod_{i=0}^{m-1}\Delta(1+\alpha+i(\rho+\epsilon))\Delta(1+\beta+i(\rho+\epsilon))\Delta(-2\gamma-1-\alpha-\beta+\rho(i-2m)+\epsilon(i+2-2m)) \\
&\times\prod_{i=1}^m\Delta(1+\gamma+\alpha+i\rho+(i-1)\epsilon)\Delta(1+\gamma+\beta+i(\rho+\epsilon)-\epsilon) \\
&\times\Delta(-\gamma-1-\alpha-\beta+\rho(i-2m)+\epsilon(i+1-2m)) \quad .
\end{aligned} \tag{C.71}$$

We leave as an exercise for the reader to check that $(C.71)$ satisfies all required constraints, such as

$$\overline{J}_m(\alpha,\beta,\epsilon,\gamma,\rho) = \overline{J}_m(\beta,\alpha,\epsilon,\gamma,\rho) \tag{C.72}$$
$$\overline{J}_m(\alpha,\beta,\epsilon,\gamma,\rho) = \overline{J}_m(-\alpha-\beta-2\rho-2\epsilon(m-1)-2(\gamma+1),\beta,\epsilon,\gamma,\rho) \tag{C.73}$$
$$\overline{J}_m(\alpha,\beta,\epsilon,\gamma,\epsilon) = \widetilde{J}_m(\alpha,\beta,\gamma,\epsilon) \quad . \tag{C.74}$$

Formulae $(C.64)$, $(C.70)$, and $(C.71)$ are the main results of this Appendix.

For completeness we add here, without proof, the expression found by Dotsenko and Fatteev[30] (their formula $(B.10)$):

$$J^{2D}_{nm}(\alpha,\beta;\rho) = \frac{1}{n!m!}\int\prod_1^n\left(\frac{1}{i}\mathrm{d}t_i\mathrm{d}\bar{t}_i\right)\int\prod_1^m\left(\frac{1}{i}\mathrm{d}\tau_i\mathrm{d}\bar{\tau}_i\right)$$
$$\times\prod_1^n|t_i|^{2\alpha'}|1-t_i|^{2\beta'}\prod_1^m|\tau_i|^{2\alpha}|1-\tau_i|^{2\beta}\prod_{i<j}^n|t_i-t_j|^{4\rho'}\prod_{i<j}^m|\tau_i-\tau_j|^{4\rho}\prod_{i,j}|t_i-\tau_j|^{-4}$$
$$=\pi^{n+m}(\rho)^{-4nm}\left(\frac{\Gamma(1-\rho')}{\Gamma(\rho')}\right)^n\left(\frac{\Gamma(1-\rho)}{\Gamma(\rho)}\right)^m$$
$$\times\prod_1^n\frac{\Gamma(i\rho'-m)}{\Gamma(1-i\rho'+m)}\prod_1^n\frac{\Gamma(i\rho)}{\Gamma(1-i\rho)}$$
$$\times\prod_0^{n-1}\frac{\Gamma(1-m+\alpha'+i\rho')\Gamma(1-m+\beta'+i\rho')\Gamma(-1+m-\alpha'-\beta'-(n-1+i)\rho')}{\Gamma(m-\alpha'-i\rho')\Gamma(m-\beta'-i\rho')\Gamma(2-m+\alpha'+\beta'+(n-1+i)\rho')}$$
$$\times\prod_0^{m-1}\frac{\Gamma(1+\alpha+i\rho)\Gamma(1+\beta+i\rho)\Gamma(-1+2n-\alpha-\beta-(m-1+i)\rho)}{\Gamma(-\alpha-i\rho)\Gamma(-\beta'-i\rho)\Gamma(2-2n+\alpha+\beta+(m-1+i)\rho)} \quad . \tag{C.75}$$

Appendix D: Loop Equations

One-cut solution of the planar loop equations

In Chapter 3 we have shown how the Virasoro constraints lead, in the planar limit, to a set of non-linear equations for the microscopic loops $\{w^{(k)}\}$. If we consider a polynomial potential of degree a,

$$V(p) = \sum_{k=0}^{a} g_k p^k \quad , \quad g_a \neq 0 \quad , \tag{D.1}$$

the loop equations (3.36) are reduced to

$$\sum_{k=1}^{a} k g_k w^{(k-1)} = 0 \quad , \tag{D.2a}$$

$$\sum_{k=1}^{a} k g_k w^{(k-n)} = \sum_{k=0}^{n} w^{(k)} w^{(n-k)} \quad , \quad n \geq 0 \quad , \tag{D.2b}$$

with the initial condition

$$w^{(0)} = \Lambda \quad . \tag{D.3}$$

Given the a coupling constants, $\{g_1, \cdots, g_a\}$ (notice the absence of g_0) and the initial value Λ, one has to calculate the coefficients $w^{(k)}$. Since $w^{(0)}$ is known and $g_a \neq 0$, the first equation ($D.2a$) allows one to solve $w^{(a-1)}$ in terms of the loops $w^{(1)}, \cdots, w^{(a-2)}$, as follows

$$w^{(a-1)} = -\frac{1}{a g_a} \sum_{k=1}^{a-1} k g_k w^{(k-1)} \quad . \tag{D.4}$$

The other equation, ($D.2b$) with $n = 0$, fixes the loop $w^{(a)}$

$$w^{(a)} = -\frac{1}{a g_a} \sum_{k=1}^{a-1} k g_k w^{(k)} + (w^{(0)})^2 \tag{D.5}$$

and the following equations, ($D.2b$) for $n = 1, 2, \cdots$, can be used to determine $w^{(a+n)}$ recursively.

We conclude that there are $(a-2)$ unknown coefficients, namely,

$$w^{(1)} \ldots w^{(a-2)} \quad , \tag{D.6}$$

which are not determined by Eq. ($D.2$), The remaining loops, $w^{(0)}$ and $w^{(a+n)}$, $n \geq -1$, are fixed by the initial condition and the loop equations, respectively.

We have also mentioned in Chapter 3 that the loop generator

$$w(p) = \sum_{n\geq 0} \frac{w^{(n)}}{p^{n+1}} = \frac{\Lambda}{p} + \frac{w^{(1)}}{p^2} + \cdots \qquad (D.7)$$

satisfies the quadratic equation

$$w^2(p) = V'(p)w(p) - Q(p) \quad , \qquad (D.8)$$

as a consequence of Eq. $(D.2)$. Above we have defined

$$V'(p) = \sum_{k=1}^{a} k g_k p^{k-1} \qquad (D.9)$$

and $Q(p)$ is a polynomial of degree $(a-2)$ whose coefficients depend on the unknown loops $(D.6)$, as follows

$$\begin{aligned} Q(p) &= \sum_{k=2}^{a} k g_k \left(\sum_{n=1}^{k-1} w^{(k-n-1)} p^{n-1} \right) = \sum_{i=0}^{a-2} q_i p^i \quad , \\ q_i &= \sum_{j=2}^{a-i} (i+j) g_{i+j} w^{(j-2)} \quad . \end{aligned} \qquad (D.10)$$

Indeed we remark that

$$q_{a-2} = a g_a w^{(0)} = a g_a \Lambda \qquad (D.11)$$

is fixed by the initial condition $(D.3)$, but the remaining terms, $q_0 \cdots q_{a-3}$, are linear combinations of the unknown coefficients.

Equation $(D.8)$ implies

$$w(p) = \frac{1}{2} \left\{ V'(p) \pm \sqrt{(V'(p))^2 - 4Q(p)} \right\} \quad , \qquad (D.12)$$

and we choose the correct sign by studying the large p behavior of the above expression,

$$w(p) = \frac{2Q}{V' \mp \sqrt{V'^2 - 4Q}} \overset{p\to\infty}{\longrightarrow} \frac{2 a g_a \Lambda p^{a-2}}{a(g_a \mp |g_a|)p^{a-1}} = \frac{2}{\left(1 \mp \frac{|g_a|}{g_a}\right)} \frac{\Lambda}{p} \quad , \qquad (D.13)$$

By comparison with $(D.7)$ we conclude that the proper solution must be taken according to the sign of the coupling g_a.

Although the unknown coefficients in $(D.6)$ are present on both sides of Eq. $(D.12)$, on the left through the loop $w(p)$ and on the right through the polynomial $Q(p)$, the previous analysis has shown that in fact they are *not* determined by the loop equation.

Consider now the expression inside the square root in $(D.12)$: it is a polynomial in p of degree $2(a-1)$, whose coefficients in $p^{2(a-1)} \cdots p^{a-2}$ are known (they come from $V'(p)^2$ and q_{a-2}), while the terms in $p^{a-3} \cdots p^0$ depend on the unknown coefficients through $q_{a-3} \cdots q_0$. Suppose that one chooses the $(a-2)$ undetermined coefficients of Q in such a way that $2(a-2)$ among the $2(a-1)$ roots of the polynomial $V'^2 - 4Q$ become double roots, that is

$$V'^2 - 4Q = M^2(p-A)(p-B) \quad , \tag{D.14}$$

where $M(p)$ is a polynomial of degree $(a-2)$,

$$M(p) = \sum_{k=0}^{a-2} M_k p^k \quad . \tag{D.15}$$

This has effectively exhausted our freedom of choices, since the unknown coefficients have been fixed in the following way:

– comparing the known coefficients in $p^{2a-2} \cdots p^{a-2}$ on the l.h.s. of $(D.14)$ with corresponding coefficients on the r.h.s., we get a system of $(a+1)$ equations for the $(a+1)$ coefficients $\{M_k, A, B\}$, which would then be solved in terms of g_k and Λ.

– then compare the lower powers in p to determine the coefficients of Q, which corresponds to fixing the coefficients $w^{(1)} \cdots w^{(a-2)}$.

Actually there is an explicit solution[129] for the polynomial $M(p)$, namely

$$M(p) = \oint_C \frac{dp'}{2\pi i} \frac{V'(p) - V'(p')}{p - p'} \frac{1}{\sqrt{(p'-A)(p'-B)}} \quad , \tag{D.16}$$

where C goes anti-clockwise around the cut $[A, B]$ in the complex p-plane. With this choice the loop generator reads

$$\begin{aligned} w(p) &= \frac{1}{2}\left\{V'(p) - M(p)\sqrt{(p-A)(p-B)}\right\} \\ &= \frac{1}{2}\sqrt{(p-A)(p-B)} \oint_C \frac{dp'}{2\pi i} \frac{V'(p')}{(p-p')\sqrt{(p'-A)(p'-B)}} \quad . \end{aligned} \tag{D.17}$$

The condition $\lim_{p\to\infty} w(p) = 0$ implies

$$0 = \oint_C \frac{dp'}{2\pi i} \frac{V'(p')}{\sqrt{(p'-A)(p'-B)}} \quad , \tag{D.18a}$$

which relates the cuts A and B; moreover the condition $\lim_{p\to\infty} pw(p) = \Lambda$ gives

$$\Lambda = \frac{1}{2} \oint_C \frac{dp'}{2\pi i} \frac{p'V'(p')}{\sqrt{(p'-A)(p'-B)}} \quad , \tag{D.18b}$$

finally solving A and B in terms of Λ and the coupling constants $\{g_k\}$. Eq. $(D.18b)$ is the discrete version of the string equation (3.70) and is sometimes called "pre-string equation".

For even potentials, i.e. when $V(-p) = V(p)$, Eq. $(D.18a)$ implies that the cut is symmetric: $A = -B \equiv \sqrt{R}$. In this case Eq. $(D.16)$ indicates that M becomes a polynomial in p^2 and the solution $(D.17)$ reduces to

$$w(p) = \frac{1}{2}\left\{V'(p) - M(p^2)\sqrt{p^2 - R}\right\} = \frac{\Lambda}{p} + \frac{w^{(2)}}{p^3} + \cdots \quad . \tag{D.19}$$

If one considers an odd perturbation around an even potential,

$$V(p) = V_+(p) + V_-(p) \quad , \tag{D.20a}$$

$$V_+(p) = \sum_{k\geq 0} g_{2k} p^{2k} \quad , \tag{D.20b}$$

$$V_-(p) = \sum_{k\geq 0} g_{2k+1} p^{2k+1} \quad , \tag{D.20c}$$

where the odd couplings g_{2k+1} are supposed to be small, the one cut solution acquires a first-order correction, as follows

$$w(p) = \frac{1}{2}\left\{V'_+(p) + V'_-(p) - M_+(p^2)\sqrt{p^2 - R} - pM_-(p^2)\frac{1}{\sqrt{p^2 - R}}\right\} . \tag{D.21}$$

By splitting the loop into even and odd pieces,

$$w(p) = w_+(p) + w_-(p) \quad , \tag{D.22a}$$

$$w_+(p) = \sum_{k\geq 0} \frac{w^{(2k)}}{p^{2k+1}} = \frac{\Lambda}{p} + \frac{w^{(2)}}{p^3} + \cdots \quad , \tag{D.22b}$$

$$w_-(p) = \sum_{k\geq 0} \frac{w^{(2k+1)}}{p^{2k+2}} = \frac{w^{(1)}}{p^2} + \frac{w^{(3)}}{p^4} + \cdots \quad , \tag{D.22c}$$

Eq. $(D.21)$ can be separated in two independent equations too,

$$w_+(p) = \frac{1}{2}\left\{V'_+(p) - M_+(p^2)\sqrt{p^2 - R}\right\} \quad , \tag{D.23a}$$

$$w_-(p) = \frac{1}{2}\left\{V'_-(p) - pM_-(p^2)\frac{1}{\sqrt{p^2 - R}}\right\} \quad . \tag{D.23b}$$

The solution of the even equation, as discussed in detail in Chapter 3, is given by

$$\frac{\partial}{\partial \Lambda} w_+(p) = \frac{1}{\sqrt{p^2 - R}} \quad , \tag{D.24}$$

where Λ and R are related by the string equation $(D.18b)$. As for the odd equation $(D.23b)$, one finds[84] the solution

$$\frac{\partial}{\partial R} w_-(p) = \frac{p}{(p^2 - R)^{3/2}} \frac{\partial}{\partial R} W_-(R) \quad , \tag{D.25}$$

where

$$\frac{\partial}{\partial R} W_-(R) = -\frac{1}{4} \sum_{k\geq 0} \binom{2k}{k} \frac{R^k}{4^k} (2k+1) g_{2k+1} \quad . \tag{D.26}$$

contains all the dependence on the odd couplings. Notice that this dependence is linear, which holds only at the first-order perturbation.

One-cut solution of planar superloop equations

In the super-eigenvalue model one finds the loop equations (6.54) and (6.56). We assume that the bosonic part of the potential is a polynomial as in (D.1). In order to solve those equations we expand the loops in the fermionic couplings,

$$\begin{aligned} \omega^{(k)} &= \omega_0^{(k)} + \omega_2^{(k)} + \omega_4^{(k)} + \cdots \quad , \\ \nu^{(k)} &= \nu_1^{(k)} + \nu_3^{(k)} + \nu_5^{(k)} + \cdots \quad , \end{aligned} \tag{D.27}$$

where the lower indices show the order in powers of ξ's. In this case we can rewrite the superloop equations as follows

$$\begin{aligned} \sum_{k=0}^{a} k g_k \omega_{2\alpha}^{(k+n)} &= \frac{1}{2}\sum_{k=0}^{n}\left(\sum_{\beta=0}^{\alpha}\omega_{2\beta}^{(k)}\omega_{2\alpha-2\beta}^{(n-k)}\right) + \\ &+ \frac{1}{2}\sum_{k=0}^{n-1}\frac{(n-2k-1)}{2}\left(\sum_{\beta=0}^{\alpha-1}\nu_{2\beta+1}^{(k)}\nu_{2\alpha-2\beta-1}^{(n-k-1)}\right) + \\ &- \sum_{k\geq 0}\frac{(n+2k+1)}{2}\xi_{k+1/2}\nu_{2\alpha-1}^{(k+n)} \quad , \quad n\geq -1, \alpha \geq 0 \end{aligned} \tag{D.28}$$

$$\sum_{k=0}^{a} k g_k \nu_{2\alpha+1}^{(k+n)} = \sum_{k=0}^{n}\left(\sum_{\beta=0}^{\alpha}\omega_{2\beta}^{(n-k)}\nu_{2\alpha-2\beta+1}^{(k)}\right) - \sum_{k\geq 0}\xi_{k+1/2}\omega_{2\alpha}^{(k+n+1)} \,,\; n\geq -1, \alpha\geq 0. \tag{D.29}$$

In particular, the zeroth-order bosonic equations are (up to a factor 1/2) identical to the purely bosonic model:

$$\sum_{k=0}^{a} k g_k \omega_0^{(k+n)} = \frac{1}{2}\sum_{k=0}^{n}\omega_0^{(k)}\omega_0^{(n-k)} \quad , \quad n \geq -1 \quad , \tag{D.30}$$

which allow us to solve $\{\omega_0^{(a+n)}, n\geq -1\}$ in terms of $(a-2)$ unknown coefficients, namely

$$\omega_0^{(1)}, \cdots, \omega_0^{(a-2)} \quad . \tag{D.31}$$

Next we substitute this solution into the first-order fermionic equations:

$$\sum_{k=0}^{a} k g_k \nu_1^{(k+n)} = \sum_{k=0}^{n}\omega_0^{(n-k)}\nu_1^{(k)} - \sum_{k\geq 0}\xi_{k+1/2}\omega_0^{(k+n+1)} \quad , \quad n\geq -1 \tag{D.32}$$

from where we can solve $\{\nu_1^{(a+n)}, n \geq -1\}$, leaving undetermined another set of $(a-1)$ fermionic coefficients:

$$\nu_1^{(0)}, \cdots, \nu_1^{(a-2)} \quad . \tag{D.33}$$

For the following second-order bosonic equations, the above pattern will be repeated:

$$\sum_{k=0}^{a} k g_k \omega_2^{(k+n)} = \frac{1}{2}\sum_{k=0}^{n}(\omega_0^{(k)}\omega_2^{(n-k)} + \omega_2^{(k)}\omega_0^{(n-k)}) +$$
$$+\frac{1}{2}\sum_{k=0}^{n-1}\frac{(n-2k-1)}{2}\nu_1^{(k)}\nu_1^{(n-k-1)} + \tag{D.34}$$
$$-\sum_{k\geq 0}\frac{(n+2k+1)}{2}\xi_{k+1/2}\nu_1^{(k+n)} \quad , \quad n \geq -1 \quad ,$$

solving all $\omega_2^{(k)}$ except $\omega_2^{(1)}, \cdots, \omega_2^{(a-2)}$, and so on. The conclusion is:

– there are $(a-2)$ unknown bosonic coefficients, $\omega^{(1)}, \cdots, \omega^{(a-2)}$; and $(a-1)$ unknown fermionic ones, $\nu^{(0)}, \cdots, \nu^{(a-2)}$;

– $\omega^{(0)} = \Lambda$ is an initial condition;

– the remaining 1-point functions are fixed by the super-loop equations.

Now we fix the unknown coefficients and solve completely the superloop equations, using the same strategy of the purely bosonic model. In terms of the superloop generators defined in (6.58) and (6.59), the equations $(D.28)$ and $(D.29)$ imply

$$(\omega(p) - V'(p))^2 + (\nu(p) - \xi(p))'(\nu(p) - \xi(p)) = \Delta_0(p) \quad , \tag{D.35}$$
$$(\omega(p) - V'(p))(\nu(p) - \xi(p)) = \Delta_1(p) \quad , \tag{D.36}$$

where

$$V'(p) = \sum_{k=0}^{a} k g_k p^{k-1} \quad , \tag{D.37}$$
$$\xi(p) = \sum_{k\geq 0}\xi_{k+1/2}p^k \quad , \tag{D.38}$$

and

$$\Delta_0(p) = V'(p)^2 + \xi(p)'\xi(p) - Q_0(p) \quad , \tag{D.39a}$$
$$Q_0(p) = 2\sum_{k\geq 2} k g_k \sum_{n=2}^{k}\omega^{(k-n)}p^{n-2} - \sum_{k\geq 2}\xi_{k+1/2}\sum_{n=2}^{k}(2k+1-n)\nu^{(k-n)}p^{n-2} \tag{D.39b}$$
$$\Delta_1(p) = V'(p)\xi(p) - Q_1(p) \quad , \tag{D.39c}$$
$$Q_1(p) = -\sum_{k\geq 2} k g_k \sum_{m=2}^{k}\nu^{(k-m)}p^{m-2} - \sum_{k\geq 1}\xi_{k+1/2}\sum_{m=2}^{k+1}\omega^{(k-m+1)}p^{m-2} \tag{D.39d}$$

We start from the zeroth-order bosonic equation, taken from $(D.35)$

$$\omega_0^2(p) - 2V'(p)\omega_0(p) = -Q_0^0(p) \quad , \tag{D.40}$$

with

$$Q_0^0(p) = 2\sum_{k=2}^{a} k g_k \sum_{n=2}^{k}\omega_0^{(k-n)}p^{n-2} \quad . \tag{D.41}$$

Above we are stressing that an ath-order polynomial form of the potential provides an $(a-2)$th-order polynomial $Q_0^0(p)$, whose coefficients depend on the unknown $\omega_0^{(1)} \cdots \omega_0^{(a-2)}$. Notice the similarity with the purely bosonic equations: with the same initial condition $\omega_0^{(0)} = \Lambda$, they are related by a simple rescaling of couplings, $g_k \to 2g_k$. Therefore, instead of $(D.17)$, we have

$$\omega_0(p) = \sqrt{(p-A)(p-B)} \oint \frac{dp'}{2\pi i} \frac{V'(p')}{(p-p')\sqrt{(p'-A)(p'-B)}} \quad , \qquad (D.42)$$

where the cut points A and B are related to the coupling constants and initial condition by the following pair of equations:

$$0 = \oint \frac{dp'}{2\pi i} \frac{V'(p')}{\sqrt{(p'-A)(p'-B)}} \quad , \qquad (D.43)$$

$$\Lambda = \oint \frac{dp'}{2\pi i} \frac{p'V'(p')}{\sqrt{(p'-A)(p'-B)}} \quad . \qquad (D.44)$$

Now we use the above results to calculate $\nu(p)$ at first-order in the fermionic couplings. From $(D.36)$ we obtain

$$\nu_1(p) = \xi(p) + \frac{\Delta_1^1(p)}{\omega_0(p) - V'(p)} \quad , \qquad (D.45)$$

where

$$\begin{aligned} \Delta_1^1(p) &= V'(p)\xi(p) - Q_1^1(p) \quad , \\ Q_1^1(p) &= -\sum_{k\geq 2} k g_k \sum_{m=2}^{k} \nu_1^{(k-m)} p^{m-2} - \sum_{k\geq 1} \xi_{k+1/2} \sum_{m=2}^{k+1} \omega_0^{(k-m+1)} p^{m-2} . \end{aligned} \qquad (D.46)$$

Notice that Δ_1^1, as a series in p, includes the $(a-1)$ undetermined coefficients $\nu_1^{(0)} \cdots \nu_1^{(a-2)}$. As we chose the coefficients $\omega_0^{(1)} \cdots \omega_0^{(a-2)}$ to get a 1-cut solution at zeroth-order, i.e.

$$(\omega_0 - V')(p) = -M(p)\sqrt{(p-A)(p-B)} \quad , \qquad (D.47)$$

we can also simplify the solution $(D.45)$ by a suitable tuning of the coefficients $\nu_1^{(0)} \cdots \nu_1^{(a-2)}$. We begin with splitting Δ_1^1 in two pieces:

$$\Delta_1^1 = \Gamma_1 + \Gamma_2 \quad , \qquad (D.48)$$

where the first one,

$$\Gamma_1(p) = V'(p)\xi(p) + \sum_{k\geq 1} \xi_{k+1/2} \sum_{m=2}^{k+1} \omega_0^{(k-m+1)} p^{m-2}$$
$$= \sum_{n\geq 0} \Gamma_1^n p^n \tag{D.49a}$$

is a series in p with known and fixed coefficients, while the second piece,

$$\Gamma_2(p) = \sum_{k=2}^{a} k g_k \sum_{m=2}^{k} \nu_1^{(k-m)} p^{m-2} = \sum_{n=0}^{a-2} \left[\sum_{i=n}^{a-2} (i+2) g_{i+2} \nu_1^{(i-n)} \right] p^n$$
$$= \sum_{n=0}^{a-2} \Gamma_2^n p^n \tag{D.49b}$$

is an $(a-2)$th-order polynomial in p, whose coefficients are unknown. Since $M(p)$ is also an $(a-2)$th-order polynomial itself, we can write

$$\Gamma_1(p) = M(p)N(p) + r(p) \quad , \tag{D.50}$$

$$r(p) = \sum_{n=0}^{a-3} r_n p^n \quad , \tag{D.51}$$

i.e., $r(p)$ is the $(a-3)$th-order polynomial representing the rest of the division $\Gamma_1(p)/M(p)$; $N(p)$ is the quotient (a determined series in p). Now we can choose the coefficients $\nu_1^{(0)} \cdots \nu_1^{(a-2)}$ so that

$$r(p) + \Gamma_2(p) = \alpha M(p) \quad , \tag{D.52}$$

where α is any anti-commuting coefficient. Altogether the above choice implies that $\Delta_1^1(p)$ is exactly divisible by $M(p)$:

$$\Delta_1^1(p) = M(p)(N(p) + \alpha) \quad , \tag{D.53}$$

α remaining undetermined. Substituting $(D.47)$ and $(D.53)$ in $(D.45)$ we find

$$\nu_1(p) = \xi(p) - \frac{N(p)}{\sqrt{(p-A)(p-B)}} \quad , \tag{D.54}$$

a solution whose non-analytical terms come from the same cut $[A, B]$ found in the bosonic solution. For simplicity, we have replaced $N(p) + \alpha$ by $N(p) = \sum_i n_i p^i$, letting n_0 undetermined.

Now we read from $(D.35)$ the equation for the 2nd-order bosonic loop $\omega_2(p)$,

$$2(\omega_0 - V')(p)\omega_2(p) + (\nu_1 - \xi)'(p)(\nu_1 - \xi)(p) = \Delta_0^2(p) \quad , \tag{D.55}$$

with

$$\Delta_0^2(p) = \xi'(p)\xi(p) + \sum_{k\geq 2} \xi_{k+1/2} \sum_{n=2}^{k} (2k+1-n)\nu_1^{(k-n)} p^{n-2} - 2\sum_{k=1}^{a} k g_k \sum_{n=2}^{k} \omega_2^{(k-n)} p^{n-2} . \tag{D.56}$$

The first term on the l.h.s. of $(D.55)$ reads

$$2(\omega_0 - V')(p)\omega_2(p) = -2M(p)\sqrt{(p-A)(p-B)} \sum_{n\geq 0} \frac{\omega_2^{(n)}}{p^{n+1}} \quad , \tag{D.57}$$

and contains a non-analytical piece in p; the same holds for the second term,

$$(\nu_1 - \xi)'(\nu_1 - \xi) = \frac{N'(p)N(p)}{(p-A)(p-B)} \quad . \tag{D.58}$$

These non-analytical contributions must cancel each other, as the r.h.s. of $(D.55)$ is analytical in p. This requirement is enough to determine all loops $\omega_2^{(k)}$ in terms of the coefficients of $M(p), N(p)$ and the unknown loops $\omega_2^{(1)} \cdots \omega_2^{(a-2)}$. At this point we could fix the latter (which are present in $\Delta_0^2(p)$) by requiring that

$$N'(p)N(p) - (p-A)(p-B)\Delta_0^2(p) = 2M(p)C(p) \quad , \tag{D.59}$$

for some analytical function $C(p) = \sum_{n\geq 0} C_n p^n$. Hence the polynomial $M(p)$ can be factorized from equation $(D.55)$, which we write as

$$\begin{aligned} 2M(p)\sqrt{(p-A)(p-B)}\omega_2(p) &= \frac{N'(p)N(p) - (p-A)(p-B)\Delta_0^2(p)}{(p-A)(p-B)} \\ &= \frac{2M(p)C(p)}{(p-A)(p-B)} \quad , \end{aligned} \tag{D.60}$$

implying that

$$\omega_2(p) = \frac{C(p)}{[(p-A)(p-B)]^{3/2}} \quad . \tag{D.61}$$

As before the cut $[A, B]$ undertakes the non-analytic properties of the loop $\omega_2(p)$. Since $\omega^{(0)} = \Lambda$ at all orders, so that $\omega_2^{(0)} = 0$ and $\omega_2(p) = \frac{\omega_2^{(1)}}{p^2} + \frac{\omega_2^{(2)}}{p^3} + \cdots$, and using the expansion $[(p-A)(p-B)]^{3/2} = \frac{a_1}{p^3} + \frac{a_2}{p^4} + \cdots$, we conclude that $C(p)$ must be (at most) linear in p:

$$C(p) = C_0 + C_1 p \quad , \tag{D.62}$$

where C_0 and C_1 can be calculated from $(D.59)$, by solving the system

$$\begin{aligned} 2M(A)C(A) &= 2M(A)(C_0 + AC_1) = N'(A)N(A) \quad , \\ 2M(B)C(B) &= 2M(B)(C_0 + BC_1) = N'(B)N(B) \quad . \end{aligned} \tag{D.63}$$

Therefore the function $C(p)$ reads

$$C(p) = \frac{1}{2(A-B)} \left[\frac{N'(A)N(A)}{M(A)}(p-B) - \frac{N'(B)N(B)}{M(B)}(p-A) \right] \quad . \tag{D.64}$$

Now we take the 3rd-order fermionic equation:

$$(\omega_0 - V')\nu_3 + \omega_2(\nu_1 - \xi) = \Delta_1^3(p) \quad , \tag{D.65}$$

where

$$\Delta_1^3(p) = \sum_{k=1}^{a} k g_k \sum_{n=2}^{k} \nu_3^{(k-n)} p^{n-2} + \sum_{k\geq 1} \xi_{k+1/2} \sum_{n=2}^{k+1} \omega_2^{(k-n+1)} p^{n-2} \quad . \tag{D.66}$$

Let us analyze the second term on the l.h.s of (D.65): from (D.54) and (D.61) we have

$$\omega_2(\nu_1 - \xi) = -\frac{C(p)N(p)}{[(p-A)(p-B)]^2} \quad . \tag{D.67}$$

Taking (D.64) and the following alternative expansions of the analytical function $N(p)$,

$$N(p) = N(A) + (p-A)N'(A) + (p-A)^2 \left(\frac{N(B)-N(A)}{(B-A)^2} - \frac{N'(A)}{(B-A)} \right) + \mathcal{O}\left((p-A)^2(p-B)\right) \quad , \tag{D.68a}$$

$$N(p) = N(B) + (p-B)N'(B) + (p-B)^2 \left(\frac{N(A)-N(B)}{(B-A)^2} + \frac{N'(B)}{(B-A)} \right) + \mathcal{O}\left((p-B)^2(p-A)\right) \quad , \tag{D.68b}$$

we find

$$C(p)N(p) = \frac{N(A)N(B)}{2(A-B)^3} \left[(p-A)^2(p-B)\frac{N'(A)}{M(A)} + (p-A)(p-B)^2 \frac{N'(B)}{M(B)} \right] + \mathcal{O}\left([(p-A)(p-B)]^2\right) \quad . \tag{D.69}$$

The coefficient n_0, undetermined up to now, can be chosen in order to cancel the first term in (D.69): it is enough to impose

$$N(A) = \text{const.} N(B) \Longrightarrow N(A)N(B) = 0 \quad , \tag{D.70}$$

implying that (D.67) is analytical in p.

Equation $(D.65)$ is then reduced to an homogeneous equation for ν_3:

$$(\omega_0 - V')\nu_3(p) = \sum_{k=1}^{a} k g_k \sum_{n=2}^{k} \nu_3^{(k-n)} p^{n-2} \quad , \tag{D.71}$$

admitting the solution

$$\nu_3 = 0 \quad . \tag{D.72}$$

As a consequence, the next order equation is homogeneous too:

$$2(\omega_0 - V')\omega_4(p) = -2 \sum_{k=1}^{a} k g_k \sum_{n=2}^{k} \omega_4^{(k-n)} p^{n-2} \tag{D.73}$$

with possible solution

$$\omega_4(p) = 0 \quad , \tag{D.74}$$

and this pattern repeats itself at the subsequent orders: $\nu_{2i+1} = 0$ provides an homogeneous equation for ω_{2i+2} with possible solution $\omega_{2i+2} = 0$, which gives an homogeneous equation for ν_{2i+3} with solution $\nu_{2i+3} = 0$, and so on.

Gathering the results we conclude that the superloop equations admit a 1-cut solution which is of 2nd-order in the fermionic couplings:

$$\begin{aligned} \omega(p) &= V'(p) - M(p)\sqrt{(p-A)(p-B)} + \frac{C(p)}{[(p-A)(p-B)]^{3/2}} \quad , \\ \nu(p) &= \xi(p) - \frac{N(p)}{\sqrt{(p-A)(p-B)}} \quad , \end{aligned} \tag{D.75}$$

thus justifying the Ansatz proposed in refs. [84,152].

Appendix E: Properties of Special Functions

Legendre polynomials

The Legendre Polynomials can be defined as the set of polynomials $P_n(x)$, of degree n, orthogonal with respect to the internal product

$$\langle P_i, P_j \rangle = \int_{-1}^{+1} \mathrm{d}x P_i(x) P_j(x) \quad . \tag{E.1}$$

With the normalization $P_n(1) = 1$, they are given by

$$\begin{aligned} P_i(x) &= \sum_{s=0}^{[i/2]} a_s^{(i)} x^{i-2s} \quad , \\ a_s^{(i)} &= (-1)^s \frac{2^{i-2s}}{\Gamma(1/2)} \frac{\Gamma(i-s+1/2)}{\Gamma(s+1)\Gamma(i-2s+1)} = \frac{(-1)^s}{2^i} \frac{(2i-2s)!}{s!(i-s)!(i-2s)!} \quad . \end{aligned} \tag{E.2}$$

They satisfy the Legendre equation

$$\frac{\mathrm{d}}{\mathrm{d}x}\left[(1-x^2)\frac{\mathrm{d}}{\mathrm{d}x}P_n\right] = -n(n+1)P_n \quad . \tag{E.3}$$

The integrals in $(E.1)$ read

$$\int_{-1}^{+1} \mathrm{d}x P_i(x) P_j(x) = \left(\frac{2}{2i+1}\right)\delta_{ij} \quad , \tag{E.4}$$

which implies

$$\begin{aligned} &\sum_{s=0}^{[i/2]} \sum_{r=0}^{[j/2]} a_s^{(i)} a_r^{(j)} \int_{-1}^{+1} \mathrm{d}x x^{i+j-2s-2r} = \\ &= \sum_{s=0}^{[i/2]} \sum_{r=0}^{[j/2]} \frac{(1+(-1)^{i+j})}{(i+j+1-2s-2r)} = \frac{2}{(2i+1)}\delta_{ij} \quad . \end{aligned} \tag{E.5}$$

The definition $(E.2)$ can be inverted: using the results

$$\int_{-1}^{+1} \mathrm{d}x x^i P_j(x) = \begin{cases} 0 & , \quad i<j \\ 0 & , \quad i \geq j \quad , \quad i-j \ \text{odd} \\ \frac{i!}{2^{j-1}(i-j)!(i+j+1)} \frac{\Gamma(\frac{i-j+1}{2})}{\Gamma(\frac{i+j+1}{2})} & , \quad i \geq j \quad , \quad i-j \ \text{even} \end{cases} \tag{E.6}$$

in particular,

$$\int_{-1}^{+1} \mathrm{d}x x^i P_{i-2s}(x) = 2^{1-i} \frac{i!}{s!} \frac{\Gamma(1/2)}{\Gamma(i-s+3/2)} \quad , \tag{E.7}$$

one finds the expansion

$$x^i = \sum_{s=0}^{[i/2]} \frac{\Gamma(1/2)i!}{2^i} \frac{(2i-4s+1)}{s!\Gamma(i-s+3/2)} P_{i-2s}(x) \quad . \tag{E.8}$$

Formulae (E.2) and (E.8) tell us how to change the basis $\{x^i\}$ and $\{P_i\}$ into each other.

π-polynomials

They are defined as the polynomials orthogonal with respect to the internal product

$$\langle \pi_i, \pi_j \rangle = \int_0^1 \mathrm{d}x \ \pi_i(x)\pi_j(x) = \frac{1}{2}\int_{-1}^{+1} \mathrm{d}y \ \pi_i\left(\frac{y+1}{2}\right)\pi_j\left(\frac{y+1}{2}\right) . \tag{E.9}$$

Comparing (E.1) with (E.9) we can relate π_i to the Legendre polynomials as follows

$$\pi_i\left(\frac{y+1}{2}\right) = P_i(y) \quad \Longleftrightarrow \quad \pi_i(x) = P_i(2x-1) \ , \tag{E.10}$$

and with (E.4) we find

$$\int_0^1 \mathrm{d}x \ \pi_i(x)\pi_j(x) = \frac{1}{2}\int_{-1}^{+1} \mathrm{d}y \ P_i(y)P_j(y) = \frac{1}{2i+1}\delta_{ij} \quad . \tag{E.11}$$

Using the expressions

$$\pi_i(x) = \sum_{n=0}^{i} \frac{x^n}{n!}\left(\frac{\mathrm{d}^n}{\mathrm{d}x^n}\pi_i\right)_{x=0} \quad , \tag{E.12a}$$

$$\left(\frac{\mathrm{d}^n}{\mathrm{d}x^n}\pi_i\right)_{x=0} = \left(2^n \frac{\mathrm{d}^n}{\mathrm{d}y^n}P_i\right)_{y=-1} \quad , \tag{E.12b}$$

$$\left(\frac{\mathrm{d}^n}{\mathrm{d}y^n}P_i(y)\right)_{y=\pm 1} = \begin{cases} 0 & n > i \\ \frac{(\pm 1)^{i-n}}{2^n}\frac{(i+n)!}{n!(i-n)!} & n \leq i \end{cases} \quad , \tag{E.12c}$$

We find the expressions

$$\begin{aligned} \pi_i(x) &= \sum_{s=0}^{i} b_s^{(i)} x^{i-s} \quad , \\ b_s^{(i)} &= \frac{(-1)^s}{s!}\frac{(2i-s)!}{[(i-s)!]^2} \quad . \end{aligned} \tag{E.13}$$

In order to invert $(E.13)$, we take the recursion relation

$$P'_{n+1}(y) = yP'_n(y) + (n+1)P_n(y) \quad , \tag{E.14}$$

obeyed by the Legendre polynomials, and translate it to the π-polynomials as

$$\pi'_{n+1}(x) = (2x-1)\pi'_n(x) + 2(n+1)\pi_n(x) \quad . \tag{E.15}$$

We need the integrals

$$I_{m,n} \equiv \int_0^1 \mathrm{d}x\; x^m \pi_n(x) \quad ; \tag{E.16}$$

multiplying $(E.15)$ by x^m and integrating over x we find the following recurrence formula

$$-mI_{m-1,n+1} + \pi_{n+1}(1) = -2(m-n)I_{m,n} + mI_{m-1,n} + \pi_n(1) \quad . \tag{E.17}$$

Using $\pi_n(1) = P_n(1) = 1$ we conclude that

$$I_{m,n+1} = 2\frac{(m-n+1)}{(m+1)}I_{m+1,n} - I_{m,n} \quad . \tag{E.18}$$

Explicit integration gives

$$I_{m,0} = \frac{1}{(m+1)} \tag{E.19a}$$

and using $(E.18)$ we find

$$I_{m,1} = \frac{2}{(m+2)} - \frac{1}{(m+1)} = \frac{m}{(m+2)(m+1)} \quad , \tag{E.19b}$$

$$I_{m,2} = \frac{2m}{(m+3)(m+2)} - \frac{m}{(m+2)(m+1)} = \frac{m(m+1)}{(m+3)(m+2)(m+1)} \tag{E.19c}$$

and by induction

$$I_{m,n} = \frac{m(m-1)\cdots(m-n+1)}{(m+n+1)(m+n)\cdots(m+1)} = \frac{[m!]^2}{(m-n)!(m+n+1)!} \quad , \quad n < m \quad . \tag{E.20}$$

Therefore

$$\int_0^1 \mathrm{d}x\; x^m \pi_n(x) = \begin{cases} 0 & m < n \quad , \\ \frac{[m!]^2}{(m-n)!(m+n+1)!} & m \geq n \quad , \end{cases} \tag{E.21}$$

and in particular

$$\int_0^1 \mathrm{d}x\; x^m \pi_{m-s}(x) = \frac{[m!]^2}{s!(2m+1-s)!} \quad . \tag{E.22}$$

Finally, using $(E.22)$ one obtains the expansion

$$x^i = \sum_{s=0}^{i}(2i+1-2s)\frac{[i!]^2}{s!(2i+1-s)!}\pi_{i-s}(x) \quad . \tag{E.23}$$

Modified Bessel functions

The modified Bessel functions of first kind

$$I_\alpha(x) = \sum_{s=0}^{\infty} \frac{1}{s!\Gamma(s+\alpha+1)} \left(\frac{x}{2}\right)^{\alpha+2s} \quad , \tag{E.24}$$

and second kind

$$K_\alpha(x) = \frac{\pi}{2\sin\alpha\pi}\left[I_{-\alpha}(x) - I_\alpha(x)\right] \quad , \tag{E.25}$$

are linearly independent solutions of the Bessel equation

$$\left[-\left(x\frac{\partial}{\partial x}\right)^2 + x^2 + \alpha^2\right] Z_\alpha(x) = 0 \quad . \tag{E.26}$$

When the index α is a half-integer the function $K_\alpha(x)$ can be written as

$$\begin{aligned} K_{n+1/2}(x) &= \frac{\pi}{2}(-1)^n\left[I_{-n-1/2}(x) - I_{n+1/2}(x)\right] \\ &= \sqrt{\pi}\mathrm{e}^{-x}\sum_{k=0}^{n}\frac{(n+k)!}{k!(n-k)!}(2x)^{-k-1/2} \quad , \end{aligned} \tag{E.27}$$

and admits the expansion

$$K_{n+1/2}(x) = \sqrt{\frac{\pi}{2}}\sum_{s=0}^{[n/2]} a_s^{(n)} \frac{x}{(n-2s+1)}\psi_{n-2s+1}(x) \quad , \tag{E.28}$$

where $a_s^{(n)}$ are the coefficients in the expansion $(E.2)$ of the Legendre polynomials and

$$\frac{x}{(n+1)}\psi_{n+1}(x) = n!x^{-n-1/2}\mathrm{e}^{-x}\sum_{s=0}^{n}\frac{x^s}{s!} \tag{E.29}$$

are the special functions found in ref. [31]. They relate to the functions $\overline{\psi}_n(x)$, defined in (3.88), in a simple way: $\psi_n(x) = \frac{x}{(n+1)}\overline{\psi}_{n+1}(x)$. We also mention the identities

$$x\frac{\mathrm{d}}{\mathrm{d}x}I_\alpha(x) = xI_{\alpha\pm1}(x) \pm \alpha I_\alpha(x) \quad , \tag{E.30}$$

$$x\frac{\mathrm{d}}{\mathrm{d}x}\left[I_{\alpha-2}(x) - I_\alpha(x)\right] = (\alpha-2)I_{\alpha-2}(x) - \alpha I_\alpha(x) \quad , \tag{E.31}$$

useful in the calculation of wave functions.

Degenerate hypergeometric $\varphi(\alpha, \gamma; x)$ functions

They are defined by the expansion

$$\varphi(\alpha,\gamma;x) = \sum_{s\geq 0} (-1)^s \frac{\Gamma(s+\alpha)}{\Gamma(\alpha)} \frac{\Gamma(\gamma)}{\Gamma(s+\gamma)} \frac{x^{s+\gamma/2}}{s!} \quad , \tag{E.32}$$

and satisfy the equation

$$\left[\left(x\frac{\partial}{\partial x} \right)^2 + (x-1)\frac{\partial}{\partial x} + \left(\alpha - \frac{\gamma}{2} \right) x + \frac{\gamma}{2}\left(1 - \frac{\gamma}{2}\right) \right] \varphi = 0 \quad . \tag{E.33}$$

When $\gamma = 2\alpha$, they can be related to the modified Bessel functions:

$$\varphi(\alpha, 2\alpha; x) = 4^{\alpha-1/2}\Gamma\left(\alpha + \frac{1}{2}\right) \mathrm{e}^{-x/2} x^{1/2} I_{\alpha-1/2}\left(\frac{x}{2}\right) \tag{E.34}$$

and using the identity

$$x^{-1/2}\varphi(\alpha, \gamma+1; x) = \varphi(\alpha,\gamma;x) + \frac{\alpha}{\gamma(\gamma+1)}\varphi(\alpha+1, \gamma+2; x) \quad , \tag{E.35}$$

we have, for n integer,

$$\varphi(n, 2n+1; x) = 4^n\Gamma\left(n+\frac{1}{2}\right)\mathrm{e}^{-x/2}\left(\frac{x}{2}\right)\left[I_{n-1/2}\left(\frac{x}{2}\right) - I_{n+1/2}\left(\frac{x}{2}\right)\right] \quad , \tag{E.36}$$

a useful formula in the study of superloop 1-point functions.

Appendix F: Contact Terms and Redefinition of Couplings

The comparison between continuum and discrete approaches to (super)conformal theories coupled to (super)gravity is based on the identification of the proper set of operators and their correlation functions. In the continuum, one considers operators like (2.20), (5.11) or (5.88), which involve integrals over world-sheet coordinates. Let us represent such an operator briefly by $\int \Phi_i(z)$ and add source terms $t_i \int \Phi_i(z)$ to the action ($\{t_i\}$ is the set of coupling constants), so that correlation functions are calculated taking derivatives as follows

$$\prod_i \frac{\partial}{\partial t_i} \left\langle e^{\sum_k t_k \int \Phi_k} \right\rangle = \left\langle \left[\prod_i \int \Phi_i \right] e^{\sum_k t_k \int \Phi_k} \right\rangle \quad . \tag{F.1}$$

Due to the integrals, the r.h.s of the above equation includes local correlation functions at coincident points, $\langle \cdots \Phi_i(z) \Phi_j(z) \cdots \rangle$, which are typically singular. One needs a prescription to deal with those singularities: in general, two prescriptions will differ by δ-functions supported on the coincidence point z. These ambiguous δ-function differences are referred to as *contact terms.*

In the operator product expansions[139]

$$\Phi_i(z)\Phi_j(0) = \frac{g_{ij}(t)}{|z|^4} + \Gamma^k_{ij}(t)\Phi_k(0)\delta^2(z) + \cdots \quad , \tag{F.2}$$

the contact terms are represented by the coefficients Γ^k_{ij}: different definitions of correlators at coincident points correspond to different values of Γ.

On the other hand, g_{ij} and Γ^k_{ij} can be interpreted[139] as the metric and respective connection in the space of couplings $\{t_i\}$: indeed, let us consider an analytic transformation of coupling constants, $\{t_i\} \to \{\hat{t}_i\}$, given by

$$t_i = \hat{t}_i + A^i_{kl}\hat{t}_k\hat{t}_l + \cdots \tag{F.3}$$

and let $\int \hat{\Phi}_i$ represent the operator coupled to $\hat{t}_i$. From Eq. (F.1) one finds

$$\begin{aligned} \left\langle \int \hat{\Phi}_i \int \hat{\Phi}_j \right\rangle &= \frac{\partial t_k}{\partial \hat{t}_i} \left\langle \int \Phi_k \int \Phi_l \right\rangle \frac{\partial t_l}{\partial \hat{t}_j} + \frac{\partial^2 t_k}{\partial \hat{t}_i \partial \hat{t}_j} \left\langle \int \Phi_k \right\rangle \\ &= \left\langle \int \Phi_i \int \Phi_j \right\rangle + A^k_{ij} \left\langle \int \Phi_k \right\rangle + \mathcal{O}(t_k) \end{aligned} \tag{F.4}$$

Above we have emphasized the terms at lowest order in the t_k-couplings: comparing (F.2) with (F.4) we see that the transformation (F.3) implies a correction $\Gamma^k_{ij} \to \Gamma^k_{ij} + A^k_{ij}$. Therefore we conclude that a change of contact terms is equivalent to an analytic redefinition of coupling constants.

Evidently the metric $g_{ij}(t)$ also changes under a transformation as $(F.3)$,

$$\hat{g}_{ij}(\hat{t}) = \frac{\partial t_k}{\partial \hat{t}_i} g_{kl}(t) \frac{\partial t_l}{\partial \hat{t}_j} \tag{F.5}$$

and one can look for a basis of couplings/operators where the metric g_{ij} is diagonal, as proposed in Chapters 4 and 6. In particular, if one finds two basis whose metrics are non-degenerate and differ only by analytic terms in the couplings, they can be considered equivalent[31]: the analytic changes are thus seen as non-universal pieces.

We also mention that some extra properties, required by the model under investigation, may eventually constrain the transformations $(F.3)$. In Chapters 4 and 6, for instance, we have imposed compatibility of the scaling dimensions of operators. Moreover, in supersymmetric theories some contact terms turn out to be necessary[143] to ensure world-sheet supersymmetry, tying up the transformations of bosonic and fermionic couplings.

References

[1] J.H. Schwarz, ed. Superstrings, *The First Fifteen Years of Superstring Theory*, World Scientific, 1985.

[2] M.B. Green, J.H. Schwarz and E. Witten, *Superstring Theory*, Cambridge University Press, 1987.

[3] D.J. Gross, J.A. Harvey, E. Martinec and R. Rohm, Phys. Rev. Lett. **54** (1985) 502.

[4] M. Dine, V. Kaplunovsky, M. Mangano, C.R. Nappi and N. Seiberg, Nucl. Phys. **B259** (1985) 549.

[5] K.S. Narain, Phys. Lett. **169B** (1986) 41.

[6] J. H. Schwarz, Phys. Rep. **C8** (1973) 269.

[7] M. Lüscher, K. Symanzik and P. Weisz, Nucl. Phys. **B173** (1980) 365.

[8] A. M. Polyakov, Phys. Lett. **103B** (1981) 207.

[9] J. Scherk, Rev. Mod. Phys. **47** (1975) 123.

[10] O. Alvarez, Nucl. Phys. **B216** (1983) 125.

[11] E. Abdalla, M.C.B. Abdalla and K. Rothe, *Non Perturbative Methods in Two Dimensional Quantum Field Theory*, World Scientific, 1991.

[12] P. B. Gilkey, Proc. Symp. Pure Math. **27** (1973); J. Diff. Geom. **10** (1975) 601.

[13] J. Distler and H. Kawai, Nucl. Phys. **B321** (1988) 171;
F. David, Mod. Phys. Lett. **A3** (1988) 1651.

[14] J. Liouville, J. Math. Pure Appl. **18** (1853) 71;
H. Poincaré, Journal de Mathematique **V-4** (1898) 137;
R. Courant, *Dirichlet's Principle, Conformal Mapping, and Minimal Surfaces*, Springer Verlag, 1950.

[15] F. Klein, *Gesammelte Abhandlungen*, Bd. III. Berlin: Springer Verlag, 1923.

[16] P.G. Zograf and L.A. Takhtajan, Russian Math. Surveys **42** (1987) 169.

[17] P. Koebe, Mathematische Annalen **69** (1910) 1; Mathematische Zeitschrift **7** (1920) 235.

[18] J. Hempel, Bulletin of London Math. Society **20** (1988) 97.

[19] L. Alvarez-Gaumé, C. Gomez, Lectures at Trieste Spring School, April, 1991.

[20] E. Braaten, T. Curtright and C.B. Thorn, Phys. Lett. **118B** (1982) 115.

[21] N. Seiberg, Lecture at 1990 Yukawa Int. Sem. Common Trends in Math. and Quantum Field Theory, and Cargèse Meeting on Random Surfaces, Quantum Gravity and Strings, May 27, June 2, 1990.

[22] J.L. Gervais and A. Neveu, Nucl. Phys. **B199** (1982) 59; Phys. Lett. **123B** (1983) 86.

[23] J.L. Gervais and A. Neveu, Nucl. Phys. **B209** (1982) 125; *ibid* **B224** (1983) 329.

[24] E. Braaten, T. Curtright, G. Ghandour and C.B. Thorn, Annals of Phys. **147** (1983) 365.

[25] B. Durhuus, H.B. Nielsen, P. Olesen and J.L. Petersen, Nucl. Phys. **B196** (1982) 498.

B. Durhuus, P. Olesen and J.L. Petersen, Nucl. Phys. **B198** (1982) 157; *ibid* **B201** (1982) 176.

[26] H. O. Girotti and K. D. Rothe, J. Mod. Phys. **A4** (1989) 3041.

[27] T. Curtright and C.B. Thorn, Phys. Rev. Lett. **48** (1982) 1309;
J.L. Gervais and A. Neveu, Commun. Math. Phys. **100** (1985) 15.

[28] J.L. Gervais and A. Neveu, Nucl. Phys. **B238** (1984) 125,396; Phys. Lett. **151B** (1985) 27.

[29] Vl. Dotsenko and V. Fatteev, Nucl. Phys. **B240** (1984) 312.

[30] Vl. Dotsenko and V. Fatteev, Nucl. Phys. **B251** (1985) 691.

[31] G. Moore, N. Seiberg and M. Staudacher, Nucl. Phys. **B362** (1991) 665.

[32] J.B. Hartle and S.W. Hawking, Phys. Rev. **D28** (1983) 2960.

[33] B.S. DeWitt, Phys. Rev. **160** (1967) 1113;
J.A. Wheeler, in *Battelle Rencontres* ed. by C. DeWitt and J.A. Wheeler, Benjamin, New York, 1968.

[34] C. Teitelboim, Phys. Rev. Lett. **50** (1983) 705.

[35] E. d'Hoker and R. Jackiw, Phys. Rev. Lett. **50** (1983) 1719.

[36] S. Das, S. Naik and S. Wadia, Mod. Phys. Lett. **A4** (1989) 1033;
T. Banks and J. Lykken, Nucl. Phys. **B331** (1990) 173.

[37] A.M. Polyakov, Mod. Phys. Lett. **A2** (1987) 893.

[38] A. M. Polyakov, *Les Houches*, Field Theory and Critical Phenomena, 1988, ed. E. Brézin and J. Zinn Justin, Elsevier 1989;
E. Abdalla, M.C.B. Abdalla and A. Zadra, Trieste/IC/89/56 (unpublished).

[39] E. Abdalla, M.C.B. Abdalla, J. Gamboa and A. Zadra, Phys. Lett. **273B** (1991) 222.

[40] J. Balog, L. Feher, L. O'Raifeartaigh, P. Forgacs and A. Wipf, Ann. Phys. (NY) **203** (1990) 76; Phys. Lett. **251B** (1990) 435.

[41] M. Bershadski and H. Ooguri, Commun. Math. Phys. 126 (1989)49.

[42] V.G. Knizhnik, A.M. Polyakov and A.B. Zamolodchikov, Mod. Phys. Lett. **A3** (1988) 819.

[43] A.M. Polyakov, *Gauge Fields and Strings, Contemporary Concepts in Physics*, vol. 3, Harwood Scientific Publishing.

[44] A. A. Belavin, A. M. Polyakov and A. B. Zamolodchikov, Nucl. Phys. **B241** (1984) 333.

[45] V.G. Kac, *Infinite Dimensional Lie Algebras*, Prog. in Math. **44** Birkhäuser, 1984.

[46] V.A. Fatteev, A.B. Zamolodchikov, Sov. J. Nucl. Phys. **43** (1986) 657; Sov. Phys. JETP **62** (1985) 215.

[47] A.M. Polyakov and A.B. Zamolodchikov, Mod. Phys. Lett. **A3** (1988) 1213;
E. Abdalla, M.C.B. Abdalla and A. Zadra, Mod. Phys. Lett. **A4** (1989) 849.

[48] I. Fukai, Nucl. Phys. **B299** (1988) 346.

[49] E. Abdalla and R.S. Jasinschi, Nucl. Phys. **B232** (1984) 426.

[50] J. Distler, Z. Hlousek and H. Kawai, Int. J. Mod. Phys. **A5** (1990) 391.

[51] D.J. Gross and N. Miljković, Phys. Lett. **238B** (1990) 217.

[52] F. David, Nucl. Phys. **B257** (1985) 45.

[53] G.'t Hooft, Nucl. Phys. **B72** (1974) 461.

[54] H.B. Nielsen (unpublished).

[55] V.A. Kazakov, Phys. Lett. **119A** (1986) 140;
D. Boulatov and V.A. Kazakov, Phys. Lett. **186B** (1987) 379.
[56] D. Gross and A.A. Migdal, Phys. Rev. Lett. **64** (1990) 717.
[57] V.A. Kazakov, Nucl. Phys. **B4** (1988) 93.
[58] B. Duplantier and I.K. Kostov, Phys. Rev. Lett. **61** (1988) 1433.
[59] I.K. Kostov, Mod. Phys. Lett. **A4** (1989) 217.
[60] T.D. Lee and C.N. Yang, Phys. Rev. **87** (1952) 404,410.
[61] M. Staudacher, Nucl. Phys. **B336** (1990) 349.
[62] D.J. Gross and I. Klebanov, Nucl. Phys. **B344** (1990) 475.
[63] M.R. Douglas, Phys. Lett. **238B** (1990) 176.
[64] T. Banks, M. Douglas, N. Seiberg, S. Shenker, Phys. Lett. **238B** (1990)279;
C. Crnković, P. Ginsparg and G. Moore, Phys. Lett. **237B** (1990) 196;
E. Brézin, M. Douglas, V. Kasakov, S. Shenker, Phys. Lett. **237B** (1990) 43.
[65] M.A. Awada and Z. Qiu, Phys. Lett. **245B** (1990) 359;
A. Bilal and J.L. Gervais, Nucl. Phys. **B295** (1988) 277;
J.L. Gervais, Commun. Math. Phys. **130** (1990) 257; Int. J. Mod. Phys. **B6** (1992) 1917.
[66] V.A. Kazakov and A.A. Migdal, Nucl. Phys. **B311** (1988) 171.
[67] E. Witten, Nucl. Phys. **B340** (1990) 281.
[68] D. Weingarten, Phys. Lett. **B90** (1980) 285;
T. Eguchi and H. Kawai, Phys. Lett. **114B** (1982) 247.
[69] S. Chaudhuri, H. Kawai and S.-H. Henry Tye, Phys. Rev. **D36** (1986) 1148.
[70] D.J. Gross and A.A. Migdal, Nucl. Phys. **B340** (1990) 333;
M.R. Douglas and S.H. Shenker, Nucl. Phys. **B335** (1990) 635;
E. Brézin and V.A. Kazakov, Phys. Lett. **236B** (1990) 144.
[71] F. David, Mod. Phys. Lett. **A5** (1990) 1019.
[72] A. Gupta, S.P. Trivedi and M. Wise, Nucl. Phys. **B340** (1990) 475.
[73] M. Goulian and M. Li, Phys. Rev. Lett. **66** (1991) 2051.
[74] P. Di Francesco and D. Kutasov, Phys. Lett. **B261** (1991) 385.
[75] P. Di Francesco and D. Kutasov, Nucl. Phys. **B375** (1992) 119.
[76] K. Aoki and E. D'Hoker, Mod. Phys. Lett. **A7** (1992) 235.
[77] Vl. Dotsenko, Mod. Phys. Lett. **A6** (1991) 3601.
[78] Y. Kitazawa, Phys. Lett. **B265** (1991) 262.
[79] E. Abdalla, M.C.B. Abdalla, D. Dalmazi and K. Harada, Phys. Rev. Lett. **68** (1992) 1641; Int. J. Mod. Phys. **A7** (1992)2437.
[80] E. Abdalla, M.C.B. Abdalla, D. Dalmazi and K. Harada, IFT-Prep. 042/91.
[81] L. Alvarez-Gaumé and Ph. Zaugg, Phys. Lett. **B273** (1991) 81.
[82] K. Aoki and E. D'Hoker, Mod. Phys. Lett. **A7** (1992) 333.
[83] P. Di Francesco and D. Kutasov, Nucl. Phys. **B342** (1990) 589.
[84] L. Alvarez-Gaumé, H. Itoyama, J.L. Mañez and A. Zadra, Int. J. Mod. Phys. **A7** (1992) 5337.
[85] I.R. Klebanov, A.M. Polyakov, Mod. Phys. Lett. **A6** (1991) 3273.
[86] E. Witten, Nucl. Phys. **B373** (1992) 187.
[87] E. Abdalla, M.C.B. Abdalla and D. Dalmazi, Phys. Lett. **291B** (1992) 32.
D. Dalmazi, Workshop on Superstrings and Related Topics, ICTP-Trieste, 1992.

[88] A. D'Adda and F. Lizzi, Phys. Lett. **B191** (1987) 85.
[89] H. Ooguri and C. Vafa, Nucl. Phys. **B361** (1991) 469; *ibid* **B367** (1991) 83.
[90] G. Moore, Nucl. Phys. **B368** (1992) 557.
[91] D.J. Gross and I.R. Klebanov, Nucl. Phys. **B354** (1991) 459.
[92] M. Bershadsky and I.R. Klebanov, Phys. Rev. Lett. **65** (1990) 3088.
[93] D.J. Gross, I.R. Klebanov and M.J. Newman, Nucl. Phys. **B350** (1991) 621.
[94] D.J. Gross and I.R. Klebanov, Nucl. Phys. **B344** (1990) 475.
[95] P. Di Francesco, H. Saleur and J. Zuber, J. Stat. Phys. **49** (1987) 57.
[96] A.M. Polyakov, Mod. Phys. Lett. **A6** (1991) 635.
[97] A.M. Polyakov, *Two Dimensional Quantum Gravity*, Jerusalem Winter School for Theoretical Physics, ed. D.J. Gross, T.Piran and S. Weinberg, World Scientific, 1991.
[98] J. Polchinski, *Strings '90 Conference*, College Station, Texas, 1990; Nucl. Phys. **B357** (1991) 241;
E. D'Hoker, *VI Swieca Summer School*, Brazil, World Scientific, 1991.
[99] J. Goldstone, unpublished;
V.G. Kac, *Contravariant form for infinite-dimensional Lie algebras and superalgebras*, in *Group Theoretical Methods in Physics*, ed. W. Beiglböck et al., Lecture Notes in Physics, Vol. 94, Springer, Berlin, 1979.
[100] B. Lian and G. Zuckerman, Phys. Lett. **B254** (1991) 417; *ibid* **B266** (1991) 21.
[101] B. Lian and G. Zuckerman, Commun. Math. Phys. **145** (1992) 561.
[102] A. Rocha, *Vertex Operators in Mathematics and Physics*, Springer, Berlin, 1983; *10th Johns Hopkins Workshop on Current Problems in Particle Theory*, ed. V. Rittenberg, World scientific, Singapore, 1986.
[103] H. Kanno, M.H. Sarmadi, ICPT-preprint IC-92-150, hepth/9207078.
[104] P. Bouwknegt, J.Mc Carthy and K. Pilch, Nucl. Phys. **B377** (1992) 541; Lett. Math. Phys. **23** (1991) 193; Commun. Math. Phys. **145** (1992) 541.
[105] G. Felder, J. Fröhlich and G. Keller, Commun. Math. Phys. **124** (1989) 417; *ibid* **130** (1990) 1.
[106] S. Govindarajan, T. Jayaraman, V. John and P. Majumdar, Mod. Phys. Lett. **A7** (1992) 1063.
[107] J. González and A.H. Vozmediano, Phys. Lett. **278B** (1992) 428.
[108] Vl. Dotsenko, Int. J. Mod. Phys. **B7** (1993) 3531; Mod. Phys. Lett. **A7** (1992) 2505.
[109] I.R. Klebanov and A. Pasquinucci, Nucl. Phys. **B393** (1993) 261.
[110] I.R. Klebanov, Mod. Phys. Lett. **A7** (1992) 723.
[111] T. Regge, Nuovo Cim. **19** (1961) 558.
[112] V.A. Kazakov, Phys. Lett. **150B** (1985) 282.
[113] J. Ambjorn, B. Durhuus and J. Frölich, Nucl. Phys. **B259** (1985) 433.
[114] J. Frölich, Lecture Notes in Physics, vol. 216, ed. L. Garrido (1985) Springer.
[115] F. David, Nucl. Phys. **B257** (1985) 543.
[116] V.A. Kazakov, I.K. Kostov and A.A. Migdal, Phys. Lett. **157B** (1985) 295.
[117] B. Durhuus, J. Frölich and J. Jonsson, Nucl. Phys. **B240** (1984) 453.
[118] D. Boulatov, V.A. Kazakov, I.K. Kostov and A.A. Migdal, Nucl. Phys. **B275** (1986) 641.

[119] A. Ambjorn, B. Durhuus, J. Frölich and P. Orland, Nucl. Phys. **B270** (1986) 457.

[120] I. Kostov and A. Krzywicki, Phys. Lett. **187B** (1987) 149.

[121] L. Alvarez-Gaumé, Helv. Phys. Acta **64** (1991) 359.

[122] F. David, *Introduction to the Statistical Mechanics of Random Surfaces and Membranes*, 8th Jerusalem Winter School for Theoretical Physics, *Two-Dimensional Gravity and Random Surfaces*, Jerusalem, Israel, 1990/1991.

[123] F. David, *Les Houches*, Session LVII, 1992.

[124] E. Brézin, C. Itzykson, G. Parisi and J.B. Zuber, Commun. Math. Phys. **59** (1979) 35.

[125] A. Bilal, *John Hopkins Worshop*, Hungary, 1990.

[126] M.L. Mehta, *Random Matrices*, Academic Press, 1967; Commun. Math. Phys. **79** (1981) 327;
S. Chadha, G. Mahoux and M.L. Mehta, J. Phys. **A14** (1981) 587.

[127] A. Mironov and A. Morosov, Phys. Lett. **255B** (1990) 47;
H. Itoyama and Y. Matsuo, Phys. Lett. **255B** (1991) 202.

[128] Y-X. Cheng, Phys. Lett. **284B** (1992) 253.

[129] A.A. Migdal, Phys. Reports **102** (1983) 199.

[130] V.A. Kazakov, Mod. Phys. Lett. **A3** (1989) 2125.

[131] M. Fukuma, H. Kawai and R. Nakayama, Int. J. Mod. Phys. **A6** (1991) 1385.

[132] R. Dijkgraaf, E. Verlinde and H. Verlinde, Nucl. Phys. **B348** (1991) 435.

[133] Yu. M. Makeenko, A. Marshakov, A. Mironov and A. Morosov, Nucl. Phys. **B356** (1991) 574.

[134] L. Alvarez-Gaumé, C. Gomez and J. Lacki, Phys. Lett. **B253** (1991) 56;
L. Alvarez-Gaumé, Talk at Nato Advanced Research Workshop on Quantum Field Theory, Statistical Mechanics, Quantum Groups and Topology, Coral Gables, FL, Jan. 7-12, 1991;
E. Martinec, Commun. Math. Phys. **138** (1991) 56.

[135] J. Ambjorn, J. Jurkiewicz and Yu. M. Makeenko, Phys. Lett. **B251** (1991) 517.

[136] D.J. Gross, *Two Dimensional Quantum Gravity*, Jerusalem Winter School for Theoretical Physics, ed. D.J. Gross, T.Piran and S. Weinberg, World Scientific, 1991.

[137] I. Kostov, Phys. Lett. **B215** (1988) 499.

[138] L.D. Landau and E.M. Lifshitz, *Quantum Mechanics: Nonrelativistic Theory*, Oxford, Pergamon.

[139] D. Kutasov, Phys. Lett. **220B** (1989) 153.

[140] A. Polyakov, Phys. Lett. **103B** (1981) 211.

[141] E. Martinec, Phys. Rev. **D28** (1983) 2604.

[142] P.S. Howe, J. Phys. **A12** (1979) 393.

[143] M. Green and N. Seiberg, Nucl. Phys. **B299** (1988) 559.

[144] E. D'Hoker and D.H. Phong, Rev. of Mod. Phys. **60** (1988) 917.

[145] P. Di Francesco, J. Distler and D. Kutasov, Mod. Phys. Lett. **A5** (1990) 2135.

[146] D. Friedan, E. Martinec and S. Shenker, Nucl. Phys. **B241** (1986) 93.

[147] M.E. Peskin. *Proceedings of the Theoretical Advanced Study Institute in Elementary Particle*, Univ. of California, Santa Cruz vol. 2 (1986) 277, ed. by H.E. Hober, World Scientific.

[148] V. Knizhnik and A.B. Zamolodchikov, Nucl. Phys. **B247** (1984) 83.

[149] P. Goddard and D. Olive, *Vertex Operators in Mathematics and Physics* , ed. J. Lepowsky et al., Springer, 1984.

[150] E. Abdalla and D. Dalmazi, Phys. Lett. **B312** (1993) 398.

[151] L. Alvarez-Gaumé and J.L. Mañes, Mod. Phys. Lett. **A6** (1991) 2039.

[152] L. Alvarez-Gaumé, K. Becker, M. Becker, R. Emparán and J.L. Mañes, Int. J. Mod. Phys. **A8** (1993) 2297.

[153] H. Itoyama, Phys. Lett. **B299** (1993) 64.

[154] K. Becker and M. Becker, Mod. Phys. Lett. **A8** (1993) 1205.

[155] J.M. Figueroa - O'Farril and S. Stanciu, Mod. Phys. Lett. **A8** (1993) 2125; Phys. Lett. **B316** (1993) 282.

[156] A. Zadra and E. Abdalla, Trieste Summer School and Workshop on High Energy Physics and Cosmology, 1993.

[157] C. Vafa, Simposium on Fields, Strings and Quantum Gravity, Beijing, 1989, and Summer School on High Energy Physics and Cosmology, ICTP, Trieste, 1989, 1990;
P. Fendley, S.D. Mathur, C. Vafa and N.P. Warner, Phys. Lett. **B243** (1990) 257.

[158] A. Giveon and A. Shapere, Nucl. Phys. **B386** (1992) 43.
M. Li, Nucl. Phys. **B395** (1993) 129.

[159] M.K. Prasad, A. Sinha, L. L. Chau Wang, Phys. Rev. Lett. **43** (1979) 750.

[160] E. Witten, Nucl. Phys. **B266** (1986) 255;
J. Avan, H.J. de Vega and J.M. Maillet, Phys. Lett. **171B** (1986) 255;
E. Abdalla, M. Forger and M. Jacques, Nucl. Phys. **B307** (1988) 198.

[161] V.A. Fatteev, Int. J. Mod. Phys. **A12** (1991) 2109;
P. Fendley and K. Intriligator, Nucl. Phys. **B380** (1992) 265.

[162] S.J. Gates, M. Grisaru, M. Roček and W. Siegel, *Superspace or One Thousand and One Lessons in Supersymmetry*, Benjamin, Cummings, 1983.

[163] N. Seiberg and D. Kutasov, Phys. Lett. **B251** (1990) 67.

[164] S. Mukhi and C. Vafa, Nucl. Phys. **B407** (1993) 667.

[165] O. Aharony, O. Ganor, J. Sonnenschein and S. Yankielowicz, Phys. Lett. **B305** (1993) 35.

[166] R. Penner, J. Diff. Geom. **27** (1988) 35.

[167] M.R. Douglas, Proceedings of Cargése Workshop, 1990.

[168] J.-M. Daul, V.A. Kazakov and I.K. Kostov, Nucl. Phys. **B409** (1993) 311.

[169] M. Kontsevich, Commun. Math. Phys. **147** (1992) 1.

[170] E. Witten, On the Kontsevich model, IANSS-HE-91/24.

[171] D.J. Gross and M.J. Newman, Nucl. Phys. **B380** (1992) 168.

[172] A. Marshakov, A. Mironov and A. Morosov, Phys. Lett. **B274** (1992) 280.

[173] T. Hollowood, L. Miramontes, A. Pasquinucci and C. Nappi, Nucl. Phys. **B373** (1992) 247.

[174] H. Lehmann, K. Symanzik and W. Zimmermann, Nuovo Cimento **1** (1955) 205.

[175] E.E. Salpeter and H. Bethe, Phys. Rev. **84** (1951) 1232.

[176] M. Jacob, ed. *Dual theory, Physics Reports Reprint Volume I*, Noth Holland, Amsterdam, 1974.

[177] *A Passion for Physics*, Essays in Honor of Geoffrey Chew, Ed. C. Detar, J. Finkelstein and C.I.Tan, World Scientific, 1985.

[178] G. Veneziano, Nuovo Cimento **57A** (1968) 190.

[179] J. Schwinger, Phys. Rev. **128** (1962) 2425.

[180] J. Schwinger, Theor. Phys. (IAEA) Vienna (1963) 88.

[181] J. Lowenstein and J.A. Swieca, Annals of Phys. **68** (1971) 172.

[182] G. 't Hooft, Nucl. Phys. **B72** (1974) 461.

[183] S. Coleman, *Aspects of symmetry,* Cambridge University Press, 1985.

[184] T.T. Wu, Phys. Rev. Lett. **71B** (1977) 142.

[185] Y. Frishman, C. T. Sachrajda, H. Abarbanel and R. Blankenbecler, Phys. Rev. **D15** (1977) 2275.

[186] A. Patrascioiu, Phys. Rev. **D15** (1977) 3592.

[187] P. Mitra and P. Roy, Phys. Lett. **79B** (1978) 469.

[188] V. Baluni, Phys. Lett. **90B** (1980) 407.

[189] P. J. Steinhardt, Nucl. Phys. **B176** (1980) 100.

[190] A.A. Migdal, Zh. Eksp. Teor. Fiz. **69** (1975) 810.

[191] B.Y. Rusakov, Mod. Phys. Lett. **A5** (1990) 693.

[192] D.J. Gross, Nucl. Phys. **B400** (1993) 161;
D.J. Gross and W. Taylor, Nucl. Phys. **B403** (1993) 395; *ibid* **B400** (1993) 181.

[193] S. Das and A. Jevicki, Mod. Phys. Lett. **A5** (1990) 1639.

[194] J.A. Minahan, Phys. Rev. **D47** (1993) 3430;
J.A. Minahan and A.P. Polychronakos, Phys. Lett. **B3121** (1993) 155.

[195] A.A. Migdal, Phys. Lett. **37B** (1971) 98; *ibid* **37B** (1971) 386;
G. Mack and K. Symanzik, Commun. Math. Phys. **27** (1972) 247.
S. Ferrara, R. Gatto and A. Grillo, *Springer Tracts in Modern Physics* 67, *Conformal Algebra in Space Time*, Springer 1973;
G. Mack and A. Salam, Ann. Phys. (NY)**53** (1969) 174.

[196] M. Hortaçsu, R. Seiler and B. Schroer, Phys. Rev. **D5** (1972) 2519;
J.A. Swieca and A.H. Völkel, Commun. Math. Phys. **29** (1973) 319;
S. Cicciarello and G. Sartori, Nuovo Cimento **19A** (1974) 470.

[197] I.T. Todorov, M.C. Mintcheva and V.B. Petkova, *Conformal Invariance in Quantum Field Theory*, Pisa, Classe di Scienze 1978;
T. Eguchi, *Lectures Academia Sinica*, 1986.

[198] P.M. Morse and H. Feshbach, *Methods of Theoretical Physics*, Mac Graw Hill Book Comp. Inc., 1953.

[199] J. E. Wess, Nuovo Cimento, **18** (1960) 1086.

[200] K. Wilson, Phys. Rev. **179** (1969) 1499.

[201] B. Schroer, Nuovo Cimento Letters **2** (1971) 867.

[202] W. Marciano and H. Pagels, Phys. Rep. **C36** (1978) 137.

[203] J. L. Cardy, in *Phase Transitions and Critical Phenomena vol.11*, ed. by C. Domb, J.L. Lebowitz, Acad. Press, 1987.

[204] A. Polyakov, Jetp. Lett. **12** (1970) 381; Sov. Phys. Jetp. **39** (1974) 10.

[205] K. Wilson, Phys. Rev. **D2** (1970) 1473; Rev Mod. Phys. **47** (1975) 773.

[206] B. Zumino, *Brandeis Lectures on Theoretical Physics*, 1970.

[207] P. Furlan, G. M. Sotkov anf I. T. Todorov, Rivista del Nuovo Cimento **12** (1989) 1.
[208] F. Jegerlehner and B. Schroer, Nucl. Phys. **B68** (1974) 461.
[209] J. Schwinger, Phys. Rev. Lett. **3** (1959) 296.
[210] P. A. M. Dirac, Rev. Mod. Phys. **21** (1949) 392;
S. Fubini, A. J. Hansen and R. Jackiw, Phys. Rev. **D7** (1973) 1732.
[211] P. Ginsparg, *Applied Conformal Field Theory*, Les Houches, 1988.
[212] A. N. Turin, Usp. Math. Nauk. **33** (1978) 149.
[213] B. L. Feigin and D. B. Fucks, Funct. Anal. and Appl. **16** (1982) 114.
[214] D. Friedan, Z. Qiu and S. Shenker, Phys. Rev. Lett. **52** (1984) 1575.
[215] A. Capelli, C. Itzykson, J. B. Zuber, Commun. Math. Phys. **113** (1987) 1.
[216] E. Verlinde, Nucl. Phys. **B300** (1988) 360.
[217] J. Wess and B. Zumino, Phys. Lett. **37B** (1971) 95.
[218] A. M. Polyakov and P. B. Wiegman, Phys. Lett. **131B** (1983) 121; *ibid* **141B** (1984) 223.
[219] E. Witten, Commun. Math. Phys. **92** (1984) 455.
[220] S. P. Novikov, Usp. Mat. Nauk. **37** (1982) 3.
[221] D. Gepner and E. Witten, Nucl. Phys. **B278** (1986) 493.
[222] E. Abdalla, K. D. Rothe, Phys. Rev. **D36** (1987) 3190.
[223] M.A. Virasoro, Phys. Review **117** (1969) 2309.
[224] Vl. Dotsenko, Advanced Studies in Pure Math. **16** (1988) 123.
[225] A. Selberg, Norsk Mat. Tidsskrift **26** (1944) 71.
[226] Y. Lozano, J. L. Mañes, Fortsch. Phys. **41** (1993) 45.
[227] P. Ginsparg, G. Moore, *Lectures given at TASI Summer School*, Boulder CO, June 1992, YCTP-P23-92.
[228] P. Di Francesco, P. Ginsparg and J. Zinn-Justin, LA-UR-93-1722.

Index

Springer-Verlag and the Environment

We at Springer-Verlag firmly believe that an international science publisher has a special obligation to the environment, and our corporate policies consistently reflect this conviction.

We also expect our business partners – paper mills, printers, packaging manufacturers, etc. – to commit themselves to using environmentally friendly materials and production processes.

The paper in this book is made from low- or no-chlorine pulp and is acid free, in conformance with international standards for paper permanency.